FOOD SCIENCE AND TECHNOLOGY

A Series of Monographs

Series Editor

Steve L. Taylor
University of Nebraska

Advisory Board

John E. Kinsella
University of California, Davis

Douglas Archer
FDA, Washington, D.C.

Jesse F. Gregory III
University of Florida

Susan K. Harlander
University of Minnesota

Daryl B. Lund
Rutgers, The State University of New Jersey

Barbara O. Schneeman
University of California, Davis

Robert Macrae
University of Hull, United Kingdom

A complete list of the books in this series appears at the end of this volume.

POSTHARVEST HANDLING

POSTHARVEST HANDLING

A Systems Approach

Edited by

Robert L. Shewfelt

Department of Food Science and Technology
The University of Georgia
Griffin, Georgia

Stanley E. Prussia

Department of Biological and Agricultural Engineering
The University of Georgia
Griffin, Georgia

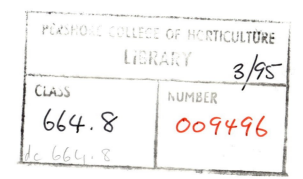

ACADEMIC PRESS, INC.
Harcourt Brace Jovanovich, Publishers
San Diego   New York   Boston   London   Sydney   Tokyo   Toronto

Front cover photograph: Pimiento pepper handling at a receiving station close to harvest. (Courtesy of Virgil Esensee.)

This book is printed on acid-free paper. ∞

Academic Press, Inc.
1250 Sixth Avenue, San Diego, California 92101-4311

United Kingdom Edition published by
Academic Press Limited
24–28 Oval Road, London NW1 7DX

Library of Congress Cataloging-in-Publication Data

Postharvest handling : a systems approach / edited by Robert L.
    Shewfelt and Stanley E. Prussia.
        p.   cm. – (Food science and technology)
    Includes bibliographical references and index.
    ISBN 0-12-639990-5
    1. Fruit–Handling.   2. Vegetables–Handling.   3. Fruit–
Postharvest technology.   4. Vegetables–Postharvest technology.
5. Fruit-United States–Marketing.   6. Vegetables–United States–
Marketing.   I. Shewfelt, Robert L.   II. Prussia, Stanley E.
III. Series.
SB360.P665   1992
631.5'6–dc20   92-23495
                                                                      CIP

PRINTED IN THE UNITED STATES OF AMERICA
92 93 94 95 96 97   QW   9 8 7 6 5 4 3 2 1

CONTENTS

CHAPTER 5
MEASURING QUALITY AND MATURITY   *Robert L. Shewfelt*

CHAPTER 6
MICROBIAL QUALITY   *Robert E. Brackett*

CHAPTER 7
MEASURING AND MODELING CONSUMER ACCEPTANCE
Stanley M. Fletcher, Anna V. A. Resurreccion, and Sukant K. Misra

CHAPTER 8
MODELING QUALITY CHARACTERISTICS   *Chi N. Thai*

CONTRIBUTORS

Numbers in parentheses indicate the pages on which the authors' contributions begin.

R. B. Beverly (73), Department of Horticulture, The University of Georgia Agricultural Experiment Station, Griffin, Georgia 30223

Frank Bollen (187), Research Engineering, New Zealand Agricultural Engineering Institute, Hamilton 3062, New Zealand

Robert E. Brackett (125, 301), Department of Food and Science Technology, The University of Georgia Agricultural Experiment Station, Griffin, Georgia 30223

J. K. Brecht (225), Vegetable Crops Department, University of Florida, Gainesville, Florida 32611

Stanley M. Fletcher (149, 293, 301), Department of Agricultural and Applied Economics, The University of Georgia Agricultural Experiment Station, Griffin, Georgia 30223-1797

Dan L. Horton (301), Extension Entomology Department, The University of Georgia, Athens, Georgia 30602

R. Brian How (1), Cornell University, Ithaca, New York 14853

Yen-Con Hung (211), Department of Food and Science Technology, Food Safety and Quality Enhancement Laboratory, The University of Georgia Agricultural Experiment Station, Griffin, Georgia 30223

J. G. Latimer (73), Department of Horticulture, The University of Georgia Agricultural Experiment Station, Griffin, Georgia 30223

Amos Lidror (187, 277), Agricultural Engineering Institute, Agricultural Research Organization, The Volcani Center, Bet Dagan 50-250, Israel

Sukant K. Misra (149), Department of Agricultural and Applied Economics, The University of Georgia Agricultural Experiment Station, Griffin, Georgia 30223-1797

Stanley E. Prussia (27, 43, 187, 277, 327), Department of Biological and Agricultural Engineering, The University of Georgia Agricultural Experiment Station, Griffin, Georgia 30223

Anna V. A. Resurreccion (149), Department of Food Science and Technology, The University of Georgia Agricultural Experiment Station, Griffin, Georgia 30223

Robert L. Shewfelt (27, 43, 99, 257, 327), Food Safety and Quality Enhancement Laboratory, Department of Food Science and Technology, The University of Georgia Agricultural Experiment Station, Griffin, Georgia 30223

David M. Smallwood (301), Economic Research Service, United States Department of Agriculture, Washington, D.C. 20005-4788

D. A. Smittle (73), Department of Horticulture, Costal Plains Experiment Station, Jiffen, Georgia 31793

Chi N. Thai (167), Department of Biological and Agricultural Engineering, The University of Georgia Argicultural Experiment Station, Griffin, Georgia 30223

E. W. Tollner (225), Department of Biological and Agricultural Engineering, The University of Georgia Agricultural Experiment Station, Griffin, Georgia 30223

B. L. Upchurch (225), USDA-ARS, Appalachian Fruit Research Station, Kearneysville, West Virginia 25430

FOREWORD

The world is shrinking. Once it was fragmented into countries and regions and divided rigidly by political, cultural, and social differences. Now we are learning that, despite these diversities, essentially similar basic motivations drive the enterprises of human values and economic infrastructures. Not too many decades ago, countries and regions were largely self-centered, with little interaction with the world as a whole. Modern rapid travel and communication technologies have changed our worldview. Events in a "remote" corner affect the entire world. This impact is particularly evident when we consider natural resources such as food, environment, energy, and climate.

The world of science is also shrinking—not in the volume of new knowledge and discovery—but in the sense of divisions and rigid boundaries among various branches of science. Like the countries of the world with rigid boundaries, the study of nature once was divided neatly and rigidly into different disciplines of science (e.g., physics, chemistry, and biology). The diversity of scientific disciplines shares a common basic motivation. Rapid ability to travel into (access) and communicate with (understand) other disciplines of science has changed the worldview of science also. Lowering the boundaries among countries and, similarly, among scientific disciplines permits the vision of a continuum. In such a vision, necessary and rigorous reductionist scientific inquires are more focused and are made in the context of the continuum to which they belong. Interestingly, changes in both the political and the scientific worlds are nearly parallel in space and time.

The concept of a "systems approach" has developed a way of thinking that addresses these changes, providing an approach for inquiries into the new order of the world of science, with few boundaries among the scientific disciplines. A systems approach permits an organized and focused view of the continuum while enabling study of one portion of the continuum for changes and improvements. It permits a process for selecting a focus with the greatest potential impact and evaluating the impact of change on the continuum. Also, this approach is being developed to include "soft" information, such as behavior, value, and judgment, which is usually beyond the scope of a scientific inquiry. A more detailed discussion of this aspect of a systems approach is presented in Chapter 3.

This book is a result of early applications of this integrative approach to the problems of postharvest handling. This work began at the University of Georgia Experimental Station in Griffin in 1980, when Tommy Nakayama, then head of the Food Science Department, and I, the newly appointed head of the Agricultural Engineering Department, entered into a brainstorming dialogue to project the future research needs of the agricultural and food-processing systems. We determined that, although more than 70 cents of each consumer dollar were spent on postproduction activities, research expenditures and priorities were disproportionate for production-related systems. Further, the world of food distribution was narrowing; we projected that imports and exports were going to be dependent on postharvest systems that were effective in maintaining and distributing commodities on an economically competitive basis.

Nakayama shared his experience of the 1950s at the Max Planck Institute in Germany, where he found that the most successful biological engineer was actually a team composed of a biologist and an electrical engineer, both studying a common problem. We chose to follow this model for the interdisciplinary postharvest research and hired a new faculty member in each of our departments to initiate the team research. A similar new faculty appointment by the Agricultural Economics Department completed the nucleus of the original postharvest team. These three new faculty members—Stanley E. Prussia, agricultural engineer; Robert L. Shewfelt, food scientist; and Jeffrey L. Jordan, agricultural economist—at the beginning of their professional careers, accomplished what is often talked about but seldom achieved—true interdisciplinary research. Merely soliciting the expertise of an individual in another discipline to review the experiments after they have been planned, or to evaluate the collected data, can never result in integration of the worldviews of these diverse disciplines. Prussia, Shewfelt, and Jordan are complimented for learning early that successful integrative research is accomplished by beginning at the beginning, that is, at the problem definition stage.

Intense discussions revealing an engineering, economic, or food science perspective of the postharvest continuum were successful in defining the problem and defining approaches for solutions. Without this integration, interdisciplinary research is not possible; the developing and defining of such concepts as, for example, latent damage, extension of time for acceptable quality, application of hedonic price relationships, and modifications of the weakest link in the chain would be unlikely.

Numerous other researchers since then have joined this original team to address postharvest concerns in quality, maturity, sorting, inspection, sensing, physiology, microbiology, safety, handling, management, and effects of preharvest conditions on postharvest systems. These researchers at the University of Georgia (including visiting scientists from Israel and New Zealand) have reported work in their areas of expertise in various chapters of this book. The original goal of the research team was the development of a comprehensive model of the postharvest system. This book reports progress toward this goal.

I recommend that this book be read with two objectives: (1) understanding the scientific concept in each subject areas and, (2) studying the application of a systems approach to a continuum and interdisciplinary research to identifying and defining problems and finding solutions in an integrative manner. I hope readers will find this integrative approach applicable to their own research and valuable for educating future scientists.

The political and scientific worlds are indeed shrinking. Integrative research using a systems approach to understanding the postharvest continuum, with the goal of a healthy and safe food supply, is timely since the boundaries among disciplines and among countries are simultaneously disappearing.

Griffin, Georgia
January 1992

Brahm P. Verma
Professor and Head
Department of Biological and
    Agricultural Engineering
The University of Georgia

PREFACE

This book represents a decade of experience moving away from traditional perspectives for postharvest research and toward the development of future research approaches using a systems approach to focus on problems uncovered by systems thinking. As we embarked on our research careers in the early 1980s, we saw a need for greater integration across disciplines in postharvest studies. In our early discussions, it became apparent that many postharvest problems were not being studied because the standard tools available to the technologist or physiologist were not suitable. We chose a systems approach as the tool to study the problems that were being ignored. This book reports on the insights gained, the knowledge needed to adapt a systems approach in research and commercial settings, and potential benefits to be derived.

We wrote *Postharvest Handling: A Systems Approach* for commercial handlers of produce and scientists conducting research with the objective of improving handling of fresh produce. Practical problems facing the produce industry frequently are too complex for simple solutions by a single investigator. Teams of scientists and engineers who can integrate knowledge at the stages of problem definition and experimental design stages are more likely to provide practical solutions to real problems. Systems thinking provides the means to integrate disciplinary perspectives.

We present here an integrated overview of the entire handling and distribution system from the field to the consumer. The book begins with an introduction to fresh fruit and vegetable distribution and to the many challenges facing the industry as viewed from marketing (Chapter 1) and material handling (Chapter 2) perspectives. Subsequent chapters discuss the historical development of a systems approach as applied to postharvest handling (Chapter 3), implications for production phase operations (Chapter 4), specific aspects of research methodology (Chapters 5 and 6), the use of mathematical modeling techniques (Chapters 7 and 8), a current perspective of critical issues in postharvest research (Chapters 9–15), and new approaches for solving postharvest problems (Chapter 16).

The decade of the 1990s presents a golden opportunity for postharvest research. Health-conscious consumers seek more fresh or minimally processed fruits and vegetables, while demanding fresher, fuller flavor. Systems thinking suggests that advances will occur as the market shifts from a production orientation to a consumer orientation. Postharvest research efforts must achieve greater integration among physiology, technology, and economics, adopt team efforts to solve postharvest problems, and provide a scientific base for quality management programs.

Postharvest Handling: A Systems Approach is unique in its integration of economic, engineering, food science, and plant science perspectives on fresh fruits and vegetables. It provides tools to answer questions that cannot be answered by traditional research. Other books are available that provide more in-depth presentations of specific aspects of fruit and vegetable quality and storage, harvesting and handling operations, marketing and distribution, plant production, postharvest physiology, postharvest technology, or systems theory. We hope that this book will help link these diverse topics into an united approach for future studies.

This book would not have been possible without the help and encouragement of numerous people. Michael O'Brien and Herb Hultin taught us to question assumptions and think independently. Brahm Verma and Tom Nakayama identified postharvest handling as an interdisciplinary research priority in the early 1980s, hired us to initiate new programs, and encouraged us to pursue a systems approach when many others were skeptical. Bill Hurst, Jeff Jordan, and Stephen Myers were active collaborators on research projects. Bill Bramlage, Tim Holt, Chris Hubbert, Adel Kader, Stan Kays, Barry McGlasson, Justin Morris, Dick Schoorl, Chein-Yi Wang, and Bruce Wasserman offered constructive criticism of our ideas as we formulated our approach. Tim Campbell and Joe Garner coordinated postharvest team research efforts. Marlene Brooks, Mike Dosier, Bob Flewellen, Larry Hitchcock, Sue Ellen McCullough, Haley Manley, Durward Smith, and Eddie Stone provided the technical assistance needed to collect experimental data. Ann Autry, Kim Santerre, and Doris Walton typed the bulk of the manuscripts. Our families endured the ordeal of lectures and haranguing on the benefits and drawbacks of interdisciplinary research and obstinate collaborators as well as deadlines associated with this book.

Griffin, Georgia   Robert L. Shewfelt
May 1992   Stanley E. Prussia

CHAPTER 1

MARKETING SYSTEM FOR FRESH PRODUCE IN THE UNITED STATES

R. Brian How

A major transformation has occurred in the marketing system for fresh produce in the United States over the past 25 years. The industry was once a relatively minor segment of the total food sector and consisted of many small family firms, each specializing in a few products or services and using traditional methods passed down from generation to generation. Now food processing has attained considerable economic importance and is dominated by large diversified firms that employ highly sophisticated technology to distribute a much larger volume and variety of fresh fruits and vegetables, obtained from many sources. This development was fueled largely by changes in consumer demand resulting from increased public interest in nutrition and health, but was facilitated by the application of new technology and management skills to the production and distribution of these highly perishable products. The industry response has resulted in major changes in the organization and structure of marketing firms—larger size, greater diversification, and increased vertical as well as horizontal integration. The characteristics of fresh produce—variety of items, weather-induced fluctuations in supply, highly changeable prices, great variability in quality attributes, extreme perishability in most cases, varying requirements for storage and handling— present real challenges to marketers. Seemingly profitable opportunities often have failed to materialize, and fierce competition has caused the demise of many established firms.

Recent changes in fresh produce markets and marketing have not taken place in isolation. They occurred while the food industry itself was in a period of great transition. The forces that affected the fresh produce sector also transformed the other sectors in this industry. Major changes also took place in the market for other products, as well as in the firms and facilities providing marketing services and in the channels of distribution.

The current marketing system for fresh produce is larger and more complex than generally is realized. The many products involved, the very many small firms operating in competition with multinational corporations, and the great variation in processing and packaging requirements are some of the characteristics that deter the collection of adequate data to describe this business fully. Existing statistics and educated guesses allow a tentative characterization of this sector.

Higher consumer expectations still motivate individuals involved in fresh produce marketing in the United States. Consumers in this country anticipate, when visiting their local supermarket or eating out, the availability of a wide range of produce items, all of impeccable quality, no matter what the season of the year. Great strides have been made toward achieving this goal, but it has not been attained. Efforts to insure that all produce items will be of ideal size, shape, and color and free of blemishes year round is thought by some to incur excessive costs and to place too great a burden on our environment. A small but significant minority of consumers has indicated a willingness to accept minor imperfections in return for a reduction in the use of chemical pesticides. Considerations of food safety as well as health may play a greater part in future fresh produce marketing.

I.   Changing Patterns in Food Consumption, Markets, and Supply Sources

A.   Food Consumption Patterns

To understand current developments in fresh produce markets and marketing fully, it is necessary to see them in the context of recent changes in the entire food industry. The capacity of the human stomach is somewhat limited and the number of consumers in the market changes only by a small percentage from one year to the next; thus, any significant increase in the consumption of some foods must be accompanied by a decrease in the consumption of others. Many researchers believe that changes have occurred in consumer demand for food as a result of greater knowledge of and concern for nutrition and health, and of changes in lifestyle, tastes, and preferences (How, 1990). Authors of a recent econometric analysis concluded, however, that changes in fruit consumption have resulted primarily from changes in prices, incomes, and demographics rather than from changes in consumer preferences (Thompson *et al.*, 1990). Whatever the reason, the considerable increase in fresh produce consumption is unquestionable. Supplies have been affected by the development and application of new technology, which has lowered marketing costs and improved the preservation of quality, for some products more than others.

A major decline has occurred in the market for several important items in recent years. Annual per capita use of whole milk and eggs has diminished (Fig. 1). Whole milk use dropped from 233 pounds per person in 1967 to 96 pounds in 1989, and

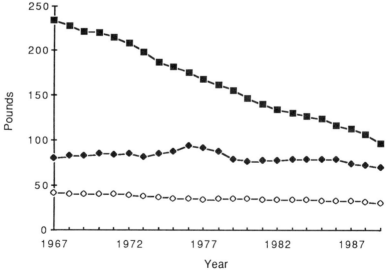

Figure 1   Changes in the annual per capita disappearance of whole milk (■), beef (♦), and eggs (◊), 1967–1989. (Adapted from Putnam and Allshouse, 1991.)

egg use from 41 pounds to 30 pounds over the same period. Beef consumption, influenced by the cattle cycle but also apparently benefitting initially from backyard barbecues and the growth of the fast food industry, increased from 80 pounds per person (retail cut equivalent) in 1967 to a peak of 94 pounds in 1976, but then declined to 70 pounds in 1988. Consumption of veal and lamb also declined over this period, whereas pork use varied from year to year without showing any major upward or downward trend (Putnam and Allshouse, 1991).

Fresh fruits and vegetables have been among the major food types whose use has increased because of recent changes in food demand and supply (Fig. 2). Per capita use of selected major fresh fruits and vegetables rose from 143 pounds in 1967 to 185 pounds in 1989. These data understate the change that occurred in this segment of the industry, since they fail to reflect the new items that have been introduced to the American public over this period, and the minor ones that have become much more important. Other changes include the expansion of lowfat milk use to compensate for the decline in whole milk consumption, increasing from 26 pounds in 1967 to 104 pounds in 1989. Consumers also have purchased more poultry and fish in recent years, increasing their consumption from 42 pounds to 77 pounds between 1967 and 1989 (Putnam and Allshouse, 1991).

There are other indications that the evolving patterns of food consumption are not entirely due to increased concern with nutrition and health, as some investigators would argue. Products that have experienced an increase in use also include cheese and cheese products, which doubled in per capita consumption from 1971 to 1989, and caloric sweeteners including sugar, corn syrup, and honey, which increased from 116 pounds in 1967 to 137 pounds in 1990 (Putnam and Allshouse, 1991).

The volume and variety of produce items consumed in the United States have increased substantially in recent years. Unfortunately, as we shall see later, official

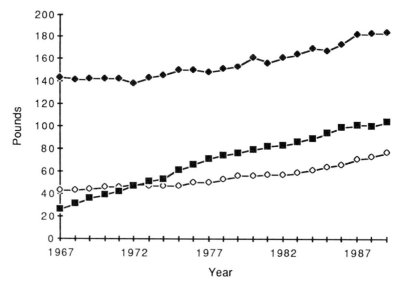

Figure 2 Changes in the annual per capita consumption of major fresh fruits and vegetables ($\blacklozenge$), lowfat milk ($\blacksquare$), and poultry and fish ($\diamond$), 1967–1989. (Adapted from Putnam and Allshouse, 1991.)

data are not available to document these changes fully. Published statistics do, however, give some indication of the growth that has occurred.

Among the major fresh fruits and vegetables, bananas, apples, and tomatoes have increased in popularity (Fig. 3). Most of this growth has occurred since the early to mid 1970s. Banana consumption hovered around 18 pounds per capita from 1967 until 1975, then increased to 25 pounds by the late 1980s. Twenty years ago, apple consumption averaged 15 pounds per person, but by 1987 exceeded 20 pounds. Fresh tomato consumption, according to these data, did not really increase until 1980, increasing from less than 11 pounds in 1979 to more than 15 pounds per person in 1989.

These data reflect not only the growth in consumption that has occurred but another very important characteristic of the industry. All three items, especially bananas and apples, vary widely in per capita consumption from year to year. These short-term changes in consumption, typical of many products, result mainly from changes in supply rather than changes in demand. For these perishable products, a market is found for the entire crop. Large apple crops in 1975, 1980, and 1987 brought higher per capita consumption but represented a real marketing challenge to the industry. Annual fluctuations in supply tend not to be as great for bananas and tomatoes, since shipments come from several different sources throughout the year, but supplies and prices can vary greatly over shorter periods.

Increases in consumption have been relatively greater for some lower volume crops such as grapes, broccoli, and pineapples (Fig. 4). Data for many other minor crops would be likely to show similar patterns, if they were available. Rapid expansion of the California wine industry contributed to the reduction in the sales of fresh grapes during the early 1970s. When increases in grape plantings in California and Chile became influential later in the decade, fresh consumption soared

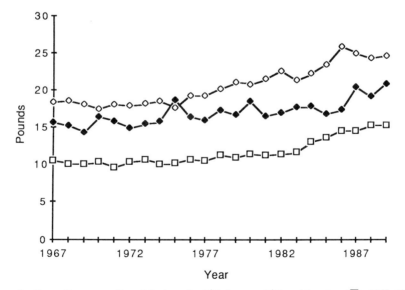

Figure 3 Per capita consumption of fresh apples (♦), bananas (◇), and tomatoes (□), 1967–1989. (Adapted from Putnam and Allshouse, 1991.)

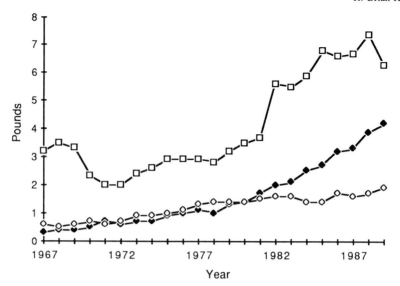

Figure 4 Per capita consumption of fresh grapes (□), pineapples (◇), and broccoli (♦), 1967–1989. (Adapted from Putnam and Allshouse, 1991.)

from under 3 pounds to almost 7 pounds per capita. Broccoli consumption also has expanded dramatically, especially in the last decade, increasing from 1 pound in 1978 to 4.2 pounds in 1989. Broccoli is a good example of a crop that is underreported, since official statistics on production for fresh sale are only gathered in four major states, although growers in many other areas are now undertaking commercial production of this crop. Consumption of pineapples also has risen since 1967, especially during the first half of this period. In the past few years, pineapple consumption has not shown any major upward trend despite new merchandising practices such as the use of in-store coring and peeling equipment.

Greater availability of many of these crops during the off season has been an important factor in increased consumption. In many cases, marketers have accessed sources in other countries to maintain supplies during seasons in which domestic production declines or ceases. New Zealand apples, Mexican tomatoes, and Chilean grapes are examples of sources of fresh items that regularly augment our supplies during the winter.

Perhaps even greater than the increase in consumption of traditional crops such as bananas, pineapples, and the like has been the acceptance of unusual tropical, oriental, or otherwise exotic items. Many fruits and vegetables almost unknown several years ago, for example, kiwifruit, are now familiar to shoppers. A major trade publication sets forth varieties, grades, availability, care, and trade sources for 71 different staple fresh produce items, and more abbreviated information on 88 specialties (The Packer, 1990). A sampling of the specialties includes belgian endive, bok choy, chayotes, fava beans, guavas, kumquats, lychees, pomegranates, sapotes, tamarillos, and water chestnuts. Data are available on imports and market arrivals for some of these items, but provide insufficient information to estimate per capita consumption.

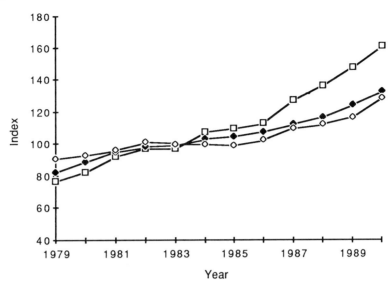

Figure 5 Consumer price index (1982–1984 = 100) for fresh fruits and vegetables (□), meat (◊), and all foods (♦) eaten at home, 1979–1990. (Adapted from Putnam and Allshouse, 1991.)

The popularity of fresh fruits and vegetables as a group is not because of the relatively low cost. Prices consumers have paid for fresh produce in retail stores have risen faster in recent years than have prices of many other products and product groups (Fig. 5). The index of prices for all foods eaten at home rose by about 50% from 1979 to 1990, whereas the prices for fresh fruits and vegetables almost doubled. In contrast, the index of prices for meats rose only about 30% during this period of declining consumption. Retail prices for other products also declining in per capita consumption, for example, eggs and milk, also have experienced very modest increases in recent years.

B. Marketing Response

Changes in food consumption are not, by any means, the only changes that have occurred in the United States food marketing system in recent years. Many changes have taken place in the methods and channels of production and distribution that have improved the consistency and quality of supplies reaching the market, as well as in merchandising practices that have appealed to consumers.

Probably the largest single change in food marketing in recent years has been the increase in the foodservice business. Total expenditures for food rose from $94 billion in 1967 to $546 billion in 1990. Over this period, however, the amount spent for food eaten away from home increased from $30 billion to $251 billion, or more than 8-fold, whereas that spent for food eaten at home increased from $63 billion to $295 billion, or less than 5-fold (Fig. 6). Foodservice requires major changes in the distribution system; preparing and serving meals generally costs 2 to 3 times what it costs to sell the same foods at retail. Increased marketing charges have reflected this fact.

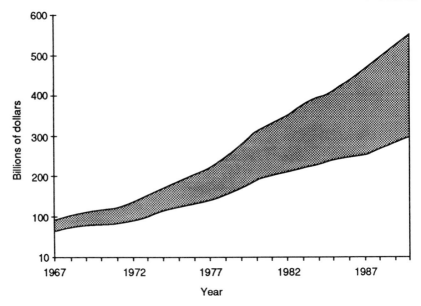

Figure 6 Changes in consumer expenditures for food eaten away from home (shaded) and for food for home use (not shaded), 1967–1990. (Adapted from Putnam and Allshouse, 1991.)

The increase in the marketing bill for domestically produced farm foods documents these changes in marketing services and in the channels of distribution (Fig. 7). Total consumer expenditures for United States farm-produced foods rose from $92 billion in 1967 to $440 billion in 1990. Two-thirds of this total, or $62 billion, covered marketing costs in 1967, leaving one-third for American farmers. By 1990, marketing charges had risen to $334 billion or more than three-fourths of the total, leaving less than one-fourth for the farmers. Much of this change took place during the last decade. Marketing charges for foods not originating on United States farms, for example, imports and United States-produced seafood and fish, which the U. S. Department of Agriculture (USDA) does not report, may be even higher. The USDA has estimated, however, that the total retail value of the foods we ate that did not originate on United States farms amounted to $106 billion in 1990, equal to almost 20% of total food expenditures (Dunham, 1991; Putnam and Allshouse, 1991).

In spite of these increased services, food actually has become a less important item in the budget of the typical consumer in recent years. Total food expenditures in relation to consumer disposable income declined from 14.2% in 1967 to 11.9% in 1990 (Fig. 8). The portion spent for food to eat at home dropped from 10.7 to 7.3% over this period, whereas that spent for food eaten away from home only increased from 3.5 to 4.6%.

Today, individuals and families differ more in their food consumption patterns than formerly. These differences are related to many varied socioeconomic and demographic factors. Household income has been found to be the most important factor associated with different food buying practices. The proportion of income

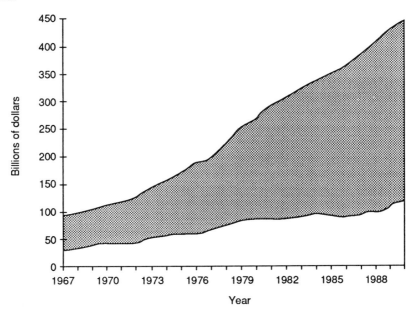

Figure 7   Changes in the farm value (not shaded) and marketing bill (shaded) for United States produced farm foods, 1967–1990. (Adapted from Dunham, 1991.)

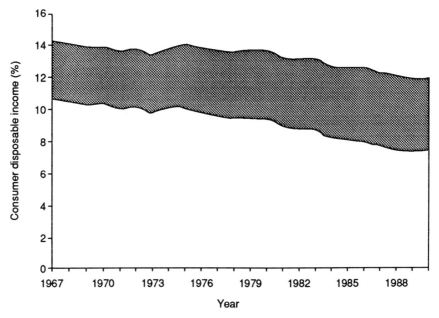

Figure 8   Changes in the percentage of United States consumer disposable income spent for food eaten away from home (shaded) and eaten at home (not shaded), 1967–1990. (Adapted from Putnam and Allshouse, 1991.)

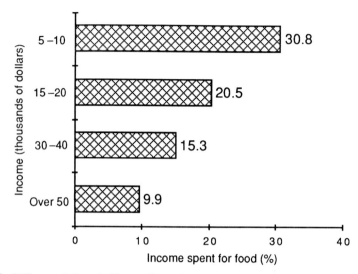

Figure 9 Differences in household expenditures for food relative to income before taxes, by selected income groups, 1989. (Adapted from Putnam and Allshouse, 1991.)

spent for food varies widely among households at different income levels (Fig. 9). Households with incomes over $50,000 annually, despite eating out more frequently and purchasing more unusual or exotic foods or foods requiring little preparation, spend a considerably smaller proportion of their income on food than do households with incomes under $10,000 a year.

Food marketers not only perform the services necessary to provide their customers with the basic food product, but perform other services that add value to the commodity. These value-added services, in the case of products intended for home consumption, include further preparation, processing, or packaging to make the item more table ready. Products to be consumed away from home require food-service preparation at the bare minimum, but also can have their value enhanced through additional marketing services. The increase in the proportion of food consumed away from home has caused a substantial expansion in total marketing charges.

C.   Sources of Supply of Fresh Produce

Increased domestic consumption of fresh fruits and vegetables has been made possible by increases in both domestic production and imports. Changes also have occurred in the sources of supply, within this country as well as in foreign countries. The volume and value of imports have increased much more rapidly than those of exports.

United States production of nine major vegetables for fresh market (asparagus, broccoli, carrots, cauliflower, celery, sweet corn, lettuce, onions, tomatoes, and honeydew melons) increased from 140 million hundredweight in 1970 to 240 million hundredweight in 1989 (Fig. 10). Large absolute increases in quantity were recorded for vegetables such as carrots, lettuce, onions, and tomatoes, but relatively greater

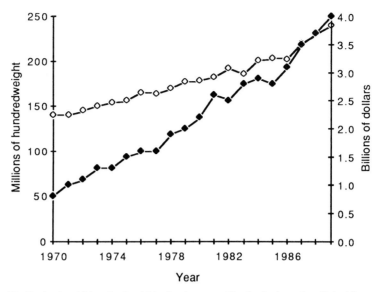

Figure 10 Production (◇) and value (◆) of major vegetables for fresh market, United States, 1970–1989. (Adapted from USDA/ERS, 1990b.)

expansion occurred in minor crops such as broccoli, cauliflower, and honeydew melons. During this period, the farm value of production of these crops rose from $800 million to $4 billion.

Fruit production and value showed a similar pattern over this period (Fig. 11). Utilized production of citrus and noncitrus fruits for fresh and processed consumption increased from 21.4 million short tons in 1970 to 29.5 million short tons in 1989, whereas the value of production rose from $1.8 billion to $7.8 billion. Greater increases were recorded for the noncitrus crops, led by apples and grapes, than for the citrus crops. Production and value of fruit has shown considerable year-to-year variation since 1970.

The quantity and value of fresh fruits and vegetables imported into this country since 1980 have increased markedly (Fig. 12). Imports rose from 3.8 million metric tons in 1979 to 5.5 million metric tons in 1988. Bananas constitute over half of these imports, and their consumption has increased substantially. Significant increases also have been recorded for other fresh produce items such as tomatoes, grapes, cantaloupes, and cucumbers. The value of our imports nearly doubled between 1979 and 1988, rising from less than $1 billion to almost $2 billion.

The quantity of our exports has remained relatively stable in recent years, but exports have increased in value (Fig. 12). Exports amounted to 2.1 million metric tons in 1979, rose to 2.3 million in 1983, then after a decline recovered to 2.5 million tons in 1988 (Buckley, 1990). Oranges contributed to the peak in 1983, and grapefruit and apples to the large volume in 1988. Exports almost equaled imports in value in 1979 and even exceeded them in 1981. Despite a substantial increase in the late 1980s, exports fell short of a trade balance by about $500 million in 1988 (Buckley, 1990).

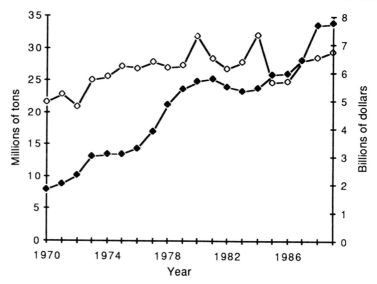

Figure 11   Utilized production (◊) and value of production (♦) of citrus and noncitrus fruit, United States, 1970–1989. (Adapted from USDA/ERS, 1990a.)

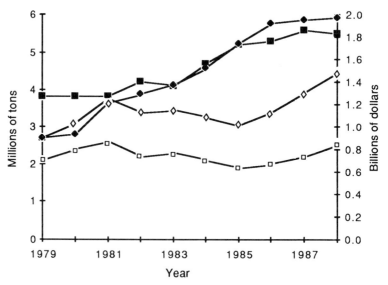

Figure 12   Quantity (squares) and value (diamonds) of selected fruit and vegetable imports (filled) and exports (open), United States, 1979–1988. (Adapted from Buckley, 1990.)

II. Marketing System for Fresh Fruits and Vegetables

A. Size and Scope of the United States System

Production and distribution of fresh fruits and vegetables in the United States is a complex operation; published statistics fail to describe the magnitude of this business fully. Many different crops are grown, most of them adapted to a wide range of climatic conditions and capable of being moved from production to retail sale with a minimum of processing and packaging. To obtain a complete and comprehensive measure of the industry would be prohibitively expensive. Consequently, there are significant gaps in our knowledge. Current USDA reports, for example, do not include the production and use of several major crops, such as cabbages and watermelons, nor most minor crops, and only include the output of major crops in major production areas. Until recently, exports to Canada were underestimated grossly; there still are deficiencies in trade data for fresh products. Available information would indicate, however, that total domestic use of fresh fruits and vegetables amounts to about 75 billion pounds annually (Table I). In addition, the industry exports more than 6 billion pounds of fresh produce each year. Per capita consumption is roughly equivalent to 300 pounds, of which about one-third consists of fresh fruits and the other two-thirds of fresh vegetables, melons of all kinds, and potatoes. Our exports of fresh produce constitute just under 10% of production, but imports supply 15–20% of domestic demand. These numbers can vary widely

Table I

Estimates of Fresh Fruit and Vegetable Use and Sources of Supply in the United States, Early 1990s[a]

Produce	United States			Domestic use	
	Production	Imports	Exports	Total	Per capita
	(millions of pounds)			(pounds)	
Fresh fruit					
Citrus	8,400	200	2,200	6,400	25.6
Bananas	10	6,200	—	6,210	24.8
Apples	6,000	250	800	5,450	21.8
Other fruit	6,800	1,350	500	7,650	29.4
Total	21,210	8,000	3,500	25,710	101.6
Fresh vegetables					
Major vegetables	24,000	2,800	1,750	25,050	100.2
Other vegetables	5,000	500	250	5,250	21.0
Melons, all kinds	5,000	1,000	300	5,700	22.8
Fresh potatoes					
Irish potatoes	12,300	600	400	12,500	50.0
Sweet potatoes	1,100	—	—	1,100	4.4
Total	68,610	12,900	6,200	75,310	300.0

[a]Author's estimates based on data in Putnam, 1990; USDA, AMS, 1990; USDA, ERS, 1990a,b; USDA, FAS, 1990.

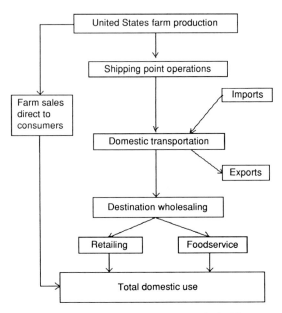

Figure 13 Marketing system for fresh produce in the United States: an overview.

from one year to the next; even well-established trends can change direction. Thus, it is important to monitor the data to detect emerging patterns of production, consumption, and trade.

The basic system for marketing fresh produce in the United States is straightforward and direct (Fig. 13). Most produce moves through several major stages. In areas distant from market, operations at the shipping point play an important role, as does long-distance transportation. Exports leave the system, and imports join it, at various stages but generally also require over-the-road transportation. Wholesaling operations at the destination provide a wide range of services for both the retail and the foodservice market. The growing volume of food eaten away from home has been accompanied by an increased diversity in foodservice institutions. Although supermarkets still dominate the retailing sector, their size and character are changing constantly. In practice, the system is complicated by many minor variations and frequent changes.

B. Economic Importance

The economic importance of this system is difficult to determine in the absence of official statistics. The USDA annually publishes an estimate of total United States expenditures for all foods, dividing this between the amount spent for food to eat at home and that spent for food eaten away from home (Table II). Another USDA source reports that expenditure for domestically produced farm foods amounts to about 83% of the total; domestically produced food constitutes a higher proportion of the foods eaten at home than of the foods eaten away from home. Expenditure for domestically produced fresh and processed fruits and vegetables was estimated

Table II
Total Expenditures for Food, for Unites States Farm Foods, and for Fresh Fruits and Vegetables, United States, 1989

	Total expenditure	At home use	Away from home use (billions of dollars)[a]	Marketing bill	Farm value
All foods[b]	507.2	284.3	222.9	NA	NA
	(100.0)	(56.1)	(43.9)	—	—
United States farm foods[c]	423.4	258.6	164.8	320.4	103.0
	(100.0)	(61.1)	(38.9)	(75.7)	(24.3)
United States fruits	95.1	NA	NA	77.5	17.6
and vegetables[c,d]	(100.0)	—	—	(81.5)	(18.5)
All fresh fruits	65.0	39.0	26.0	50.0	15.0[f]
and vegetables[e]	(100.0)	(60.0)	(40.0)	(76.9)	(23.1)[f]

[a]Numbers in parentheses represent percentages of total. NA, not available.
[b]Adapted from Putnam, 1990.
[c]Adapted from Dunham, 1990.
[d]Fresh and processed. Also includes soup, baby foods, condiments, dressings, spreads, and relishes.
[e]Author's estimate assuming fresh sales constitute 60% of total fresh and processed; imports valued at $8 billion.
[f]Author's estimate assuming United States farm value of $12 billion plus imports valued at $3 billion at point of entry.

at $95.1 billion in 1989. How this amount is divided among the at-home and away-from-home market is not specified; however, the farm value of the food was estimated at $17.6 billion and the marketing bill or value added at $77.5 billion.

Based on these data, it is estimated conservatively that total retail expenditures for fresh fruits and vegetables in the United States amounted to about $65 billion in 1989, of which about $26 billion or 40% was spent on away-from-home use and 60% or $39 billion was spent on at-home use (Table II). American farmers and foreign traders received about $15 billion, whereas the value-added or marketing charges earned by American marketers came to about $50 billion. United States marketers also earn marketing charges on domestically grown fresh produce that is exported from this country, which may amount to as much as $5 billion on foreign sales.

The marketing charges, marketing bill, or marketing costs incurred at various stages of the marketing system or for specific marketing functions are not reported for fresh fruits and vegetables in total. Data are available, however, on charges by marketing functions for total domestic farm foods sold both at retail and through foodservice operations. Marketing charges by function are also available for many individual fresh fruits and vegetables sold at retail (USDA/ERS, 1987; Pearrow 1988, 1989, 1990; Pearrow and Lofton, 1990). Similar information is not available on marketing charges through foodservice operations.

How charges by marketing function may vary from one item to another is illustrated by information on Northeast potatoes and on California oranges and lettuce sold at retail in 1990 (Fig. 14). The farm value of potatoes, for example, amounted to 22.4% of the retail price that year, compared with only 15.6% for

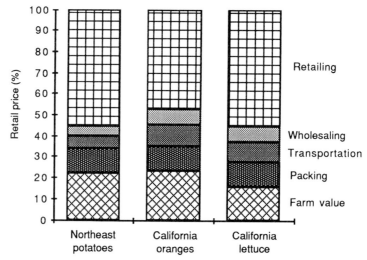

Figure 14 Farm value and marketing changes by major function as a percentage of the retail price of northeast potatoes, California oranges, and California lettuce, 1990. (Adapted from Dunham, 1991.)

lettuce. This difference might be explained partly by the fact that many potato growers grade, wash, and pack their potatoes, whereas California lettuce growers typically sell their lettuce in the field for the shippers to cut, pack, haul, cool, and ship. Charges for packing these three items varied from 11.5 to 12.2% of the retail price; charges for intercity transportation varied from 6 to 10%. Wholesaling charges constituted only about 5% of the retail price for potatoes, but over 7% for lettuce and oranges. Retailing charges or margins tend to vary widely, however; higher charges are seen for staple items such as potatoes, lettuce, and oranges, and lower percentage charges are seen for specialty items such as strawberries, asparagus, and nectarines (How, 1990). Supermarket retailers typically aim for a margin of 33% of the retail price on fresh produce, roughly equivalent to a 50% markup.

Marketing charges or costs separated into major marketing function components are published for all United States farm foods combined, separated by expenditures for foods purchased at foodstores and those for foods eaten away from home (Table III). These data are not disaggregated further by major food groups. For all United States foods in 1989, the charges for farm production and for processing and shipping each amounted to about one-fourth of consumer expenditures; retailing and foodservice charges together constituted more than one-third. Smaller proportions were applied to wholesaling and intercity transportation. For foods eaten at home, the retailing margin amounted to 23.4% of the total, whereas the foodservice margin constituted 60.3% of the total. The higher cost of foodservice can be illustrated by the fact that, in 1989, for each dollar United States farmers received for food sold at retail, the charge for retailing amounted to about $0.80, whereas for each dollar they received for food sold through foodservice, the charge for foodservice came to $3.75.

Table III

Marketing Function Components of United States Consumer Expenditures for United States Farm Foods and for United States Grown and Imported Fresh Fruits and Vegetables, 1989

Expenditure	Farm or import value	Processing or shipping cost	Intercity transportation cost	Wholesaling cost	Retailing/ foodservice cost	Total expenditure
			(billions of dollars)[a]			
Farm foods[b]						
At foodstores	76.6	81.3	14.3	26.0	60.4	258.6
	(29.6)	(31.4)	(5.5)	(10.1)	(23.4)	(100.0)
Eating away	26.5	24.7	4.3	9.9	99.5	164.8
from home	(16.1)	(15.0)	(2.6)	(6.0)	(60.3)	(100.0)
Total	103.0	106.0	18.6	35.9	159.9	423.4
	(24.3)	(25.0)	(4.4)	(8.5)	(37.8)	(100.0)
Fresh fruits and	15.0	8.5	6.0	5.5	30.0	65.0
vegetables[c]	(23.1)	(13.1)	(9.2)	(8.5)	(46.1)	(100.0)

[a] Numbers in parentheses indicate percentages of total.
[b] Adapted from Dunham, 1990.
[c] Author's estimates.

Based on published reports for individual fresh produce items and on aggregate data for all United States farm foods, an estimate has been made of the costs of major marketing function components for all fresh fruits and vegetables purchased by United States consumers (Table III). By far the largest cost is $30 billion for retailing and foodservice, equivalent to $2.00 for every $1.00 going to United States farmers or foreign suppliers. This figure represents about 46% of the total expenditure, greater than that for all other foods since retailing margins for fresh produce typically average about 33% of sales compared with 22 or 23% for all foods. Data are not available to compare foodservice margins for fresh produce with margins for all foods, so they have been assumed to be about the same. Given the lack of available data, the estimates for total costs of processing and shipping, for intercity transportation, and for wholesaling can be considered only approximate.

C. System Overview

Changing structure has dominated the food system during the 1980s (USDA/ERS, 1990c). Mergers, divestitures, and leveraged buyouts have been prevalent. In 1988, 573 acquisitions and 273 divestitures by food marketing processors, wholesalers, retailers, and foodservice operators were reported. The four largest leveraged buy-outs in history were all in the food marketing system. Much of this activity crossed international boundaries. Foreign buyers in 1988 bought United States food processing companies valued at $8.5 billion, whereas United States food marketers acquired foreign firms worth $1.6 billion (USDA/ERS, 1990c).

Aggregate concentration in food processing, wholesaling, and foodservice has risen sharply because of merger and acquisition transactions. Merger activity certainly has caused a decline in the number of firms. Intense competition also has forced many firms out of business. Farmers sell their products to fewer buyers, and consumers buy from fewer sellers. According to the USDA, however, the effect of this activity on consumers, farmers, and the food marketing system is unclear (USDA/ERS, 1990c).

Major differences exist among local markets and among separate industries. Some local shipping points or wholesale markets apparently are highly concentrated; most of the business is done by a few firms. Information is scarce on the specific nature of the structure of the fresh fruit and vegetable industry, and of recent changes in it. Indications are, however, that this industry reflects the general restructuring taking place in the entire food marketing system.

1. Farm production

In the United States, the bulk of farm production of fresh produce now occurs in specialized areas where favorable soils, climate, and other factors facilitate economical large scale production (How, 1990). Competition among growing areas is intense. Developments in long distance transportation have decreased the importance of location close to markets. Currently, Washington and a few other states grow most of the apples, Idaho and Washington most of the potatoes, Florida almost all grapefruit, and California most of the lettuce. In fact, California and Florida together account for over half the commercial shipments of fresh produce reported annually in the United States (How, 1990). California is the sole domestic source of many specialty crops. A few domestic production areas are solely responsible for some perishable crops during specific seasons of the year. Imports have increased in importance, not just for crops that could only be grown in volume domestically at great expense, for example bananas, mangoes, papayas, and other tropical fruits and vegetables. The resources of many foreign countries supplement the needs of the United States for variety and consistency of supply.

This concentration of production has not eliminated output in secondary growing areas entirely, although the diversified market garden farms that formerly encircled our cities are gone. These farms have been replaced by growers who have developed profitable operations selling particular items directly to consumers, to restaurants, or to specialty fruit and vegetable stores. Often their emphasis is on providing better internal qualities such as taste and texture rather than on external appearance.

Farming is still largely a family business, but farms are becoming fewer in number and larger. The Census Bureau tracks the number and size of United States fruit and vegetable farms, but makes no distinction between those producing for processing and those geared to the fresh market (USDC/BC, 1990). The Census Bureau also counts very small farms, and fails to consolidate the operations of large farms in noncontiguous counties or different states that operate under the same ownership. Census data indicate that, in 1987, 4.3% of the 60,800 farms harvesting vegetables for sale in the United States had 250 acres or more of these crops but accounted for 55.7% of the total acreage (How, 1990). Similar distribution exists in the lands used for growing Irish potatoes and in land used for orchards and vineyards. Although farms have become larger, these larger farms are still predominantly family owned and operated. With few exceptions, large corporations have had little success in growing and shipping fresh fruits and vegetables.

2. Shipping point operations

In addition to farm, ranch, and vineyard production, the specialized growing areas also have developed a complex infrastructure of shipping point operations (Kader *et al.*, 1985; Hardenburg *et al.*, 1986). Physical marketing services such as sorting, grading, storing, ripening, degreening, processing, and consumer packaging, many of which formerly were performed close to market, now are undertaken largely at the shipping point, where they often may be done more efficiently and economically. Marketers are able to exploit this opportunity because of improvements in transportation and handling practices. How these operations are performed and the sequence with which they are carried out, even for the same commodity, often differs from one firm or region to another, and changes over time. Apples may be run over a grading line both before and after storage, or only just before packing. Grapes may be graded and packed in the vineyard, or brought to a central packing house to prepare them for market. Fruit may be stored in common, refrigerated, or modified atmosphere storage. With all these possibilities, the potential handling combinations are almost infinite in number.

At the shipping point, grower–shippers generally provide a large share of the marketing services. Many growers still perform few or no marketing services and some shippers are not involved in growing, but firms at the shipping point that obtain part of their product from their own farms or ranches and part from independent growers seem to have an economic advantage. Cooperatives play an important role in packing, selling, and shipping some commodities such as citrus, nuts, and apples. Large federations of locally owned packing and shipping cooperatives, for example, Sunkist in California and Arizona and Sealdsweet in Florida, provide sales and merchandising services. The apparently favorable market prospects for fresh fruits and vegetables have attracted many large firms such as Campbell Soup, Castle and Cooke, Superior Oil, R. J. Reynolds, Procter & Gamble, and Tenneco to the shipping business, but in most cases these firms have soon withdrawn from fresh operations. Indications are that the firms were unable to cope with the instabilities and uncertainties of this phase of the business.

3. Transportation

The increased distance that a typical carton of fresh produce travels to market has been made possible by improved domestic transportation methods, but also continues to challenge those who use the transportation system (Ashby *et al.*, 1987; Beilock *et al.*, 1988). Trucks now handle about 90% of domestic shipments. Interstate highways and larger and more efficient trailers hauled by more powerful and reliable tractors provide shippers and receivers with the flexibility and dependability they need at relatively low cost. Deregulation has enabled railroads to operate more effectively; the trailer-on-flatcar system is competitive for certain products. Air shipments seldom amount to more than 1 or 2% of the total, but meet the needs of the shipper occasionally. Given the large number of produce items, many of which are incompatible with others and differ in optimum storage temperature and humidity requirements, maintaining quality over long distances still can be difficult, whatever the mode of transportation.

Deregulation of trucking in 1980 has led to considerable reorganization in that industry. Large fleet operations have developed that contract with receivers or shippers. Many independent owner–operators, however, still operate under lease arrangements with contract haulers. Few shippers or receivers have their own long distance transportation equipment.

Improvements in ocean transportation have had a major impact on both imports and exports (Risse *et al.*, 1982; Nicholas, 1985; McGregor, 1987). During the 1960s, the change from shipping bananas on the stalk to shipping them packed in cartons made a major improvement in the total system. Bananas now could be handled on pallets from the packing house in Central America to the grocery warehouse in the United States, with a great saving in cost and improvement in quality. When shipping containers were introduced a few years later, the potential for quality maintenance became even greater.

Marine or shipping containers—the large $8' \times 8' \times 20'$ or $40'$ boxes with individual systems to maintain constant temperature and humidity—revolutionized ocean transport by better maintaining quality, reducing labor in handling, and guarding against damage and pilferage. The full potential of these containers will be realized when railroads develop flatcars specially designed to carry containers stacked two deep.

4. International trading

Fresh produce imports have risen markedly in recent years. As in the case of domestic production, the impetus has come from changes in both demand and supply. Consumers have been receptive to a year-round supply of staple items, and to the introduction of exotic tropical and oriental novelties. In many cases, United States suppliers have taken the initiative and sought out sources of contraseasonal or unusual products in developing countries to the extent of becoming involved in production and packing operations abroad. Marketers in other countries also, at times, have introduced their specialties successfully. Foreign trade has been facilitated by greater ease in travel and communication and by the reduction in trade barriers (Paarlberg *et al.*, 1984).

The volume and value of exports of fresh produce have not kept pace with imports. Some markets for some products have enjoyed successful growth, but intense competition and some persistent barriers to trade have restricted or limited sales in other areas (Wright, 1988). Japan has barred the importation of apples and has restricted imports of other items. The European Community not only has raised barriers to that market but, through subsidized exports, also has engaged in unfair competition in other markets (Moulton, 1983). Efforts to liberalize agricultural trade, or at least provide equal opportunities through multinational agreements such as the General Agreement on Tariffs and Trade, have met with only limited success. The United States Trade Agreement with Canada that went into effect at the beginning of 1990 is expected to benefit both countries; a similar agreement with Mexico is under discussion.

5. Wholesale distribution

Recent changes in wholesaling at destination have been as great as those at any other stage in the marketing system (How, 1990). Supermarket chains have purchased most of their needs directly from shipping point sources for many years, bypassing traditional wholesalers and the bulk breaking, ripening, and repacking services these firms formerly performed. The major wholesale service still performed by integrated retailers, apart from physical distribution, is banana ripening. The independent or nonintegrated wholesalers still in operation owe their existence largely to the growth in foodservice and the proliferation in numbers of specialty items carried by supermarkets. Antiquated terminal market facilities are being replaced in some communities, another indication of the differences in wholesaler operations today (Overheim *et al.*, 1988). Some existing firms have taken advantage of the increased demand for salad mixes and precut packs by supermarkets and foodservice operations that have introduced salad bars and prepared salads. New businesses also have been formed to meet these needs.

The structure and organization of firms involved in wholesale operations at destination also has changed drastically in recent years. Retail chains continue to buy their major items directly from distant sources, but rely on other firms for specialty items or prepared products such as salad mixes or precut packs. Expansion of the foodservice business has provided opportunities for many independent wholesalers, despite the fact that some of the larger fast food and restaurant chains have set up their own wholesale distribution systems. The decision by dry grocery wholesalers many years ago to add fresh produce to their line has been made more recently by large foodservice wholesalers such as Sysco Corporation and Kraft Foodservice. Despite the apparent increase in size of wholesale firms, the 1987 census reported more merchant distributors of fresh fruits and vegetables than were recorded 5 years earlier (USDA/ERS, 1990c).

6. Retailing

Consumers in the United States purchase most of the food they eat at home in supermarkets. These stores provide an increasing number of services. As supermarkets have increased in size and as fresh produce has become relatively more important, retail merchandising has become much more effective (Leed and German, 1985). Whether bulk or prepackaged (the relative merits of these alternatives are still being debated), the foods offered are generally of high quality. Value-added services have become more important: stores have introduced salad bars, prepared salads, and precut packages. Concern with food safety has led some stores to carry organically grown fresh fruits and vegetables, but the prospects for this type of product are presently unclear (Morgan *et al.*, 1990). Variants of the typical supermarket have emerged, for example, hypermarkets in which customers can roam a vast expanse containing thousands of items and numerous brands, or wholesale clubs where members can economize by buying in quantity from a limited selection (Price and Newton, 1986).

As noted by the USDA, the structure and organization of food retailing changes little from year to year but, over a period of time, has changed drastically (USDA/ERS, 1990c). Supermarkets continue to account for about three-fourths of grocery store sales, but they vary in size, number of items, and services provided. The 20 largest grocery store chains continue to attract about 35% of the total business; however, within this group the share held by the largest firms has declined whereas the share held by the rest has increased. The smaller and medium-sized chains seem to have been able to respond more effectively to economic changes.

7. Foodservice

In 1989, about one-third by value of United States farm foods moved through foodservice, but this sector accounted for 39% of consumer expenditures. Additionally, the cost of providing foodservice that year amounted to $99.5 billion, two-thirds more than the cost of retailing (Dunham, 1990). Unlike food retailing, food-service is a highly diversified industry. Establishments providing foodservice can be divided broadly into two categories—commercial and noncommercial. Full service restaurants and fast food outlets are the major source of commercial sales; smaller business volume takes place at lodging places, retail stores, and others (USDA/ERS, 1990c). Restaurants nearly equal fast food outlets in number of establishments and in sales, but fast food outlets have shown more rapid growth in both categories in recent years (How, 1990). Commercial establishments account for almost three-fourths of total foodservice sales.

Many different types of noncommercial establishments serve food. Significant noncommercial foodservice expenditures are made at educational institutions, plants and office buildings, hospitals and extended care facilities, vending machines, military services, correctional facilities, and many other types of establishments.

Foodservice requirements for fresh produce vary widely, from the highest quality specialty products demanded by fancy restaurants to the standard quality staples required by correctional facilities (Van Dress, 1982). Purchases by fast food and restaurant chains are dominated by the need for uniformity as well as high quality, which is harder to achieve for fresh produce than for other foods. Also, prior preparation or value-added services are required in many instances by most food-service establishments, excepting some top quality restaurants. The best providers of these services have not been determined.

After many years of rapid growth, foodservice sales failed to rise in 1989 (USDA/ERS, 1990c). Sales are expected to resume their increase in the 1990s as the economy recovers from the recession. Take-out dinners continue to expand in popularity. Franchising is becoming even more important in the commercial food-service sector. Fast food operations are offering a wider selection of foods.

8. Direct marketing

A small but significant quantity of fresh produce, possibly 3–5% of the total, bypasses the usual marketing system in moving more directly from grower to retailer, foodservice operator, or consumer. Many factors have given rise to the revitalization of direct marketing in recent years (Henderson and Linstrom, 1980). The desire by some consumers, retailers, and restauranteurs for freshly harvested specialty crops such as herbs and leaf lettuce and the inability of the marketing system to provide them has encouraged enterprising farmers to grow and directly market these items (Dicks and Buckley, 1989). Growers marketing directly can be more responsive to the preferences of buyers for crop variety and quality. Some consumers apparently enjoy the opportunity to buy direct from growers at farm roadside markets, pick-your-own operations, and community farmers markets.

III.   Current Practices—A Critical Assessment

The conduct and performance of the food marketing system, especially of the fresh fruit and vegetable marketing sector, is difficult to evaluate (Marion, 1986). First, criteria must be agreed upon and then a means of measurement must be determined. Efficiency, equity, and innovation are doubtless relevant, but somewhat nebulous, concepts.

The United States fruit and vegetable marketing system provides consumers with a wide range of products of high quality all year long at a cost that appears to be reasonable relative to average incomes. The system appears to have responded well to changing consumer demand, and has grown rapidly in recent years. There has been considerable innovation in production, distribution, and merchandising practices. Food bought by or served to consumers is largely safe to eat, although it may have come from a distant country or may have been sprayed with toxic pesticides.

In spite of this admittedly superior conduct and performance, there are still deficiencies in the system that will be difficult to correct. The changes that lie ahead in technology, management science, and information sources will present even greater challenges than have been met in the past.

Losses of product still plague the fresh fruit and vegetable industry. The greatest loss for many crops is the product left in the field or discarded at the packing house because it fails to meet market requirements, but a significant quantity of fruits and vegetables of marketable quality still fails to reach the consumer in satisfactory condition and must be wasted.

External appearance plays a major role in the presentation and selection of fresh produce. Consumers have learned to make their selections on that basis, and government grades and trade practices reflect that fact. Internal quality attributes such as taste and nutritional value are recognized as increasingly important by some buyers but still play a minor role in marketing.

With the exception of a few minor incidents, our food supply has been safe, but such safety may be more difficult or costly to maintain in the future. Some pests are becoming harder to control, and greater restrictions are being placed on the use of existing chemical pesticides and the development of new materials. The withdrawal of approval for specific pesticide materials may lead to more problems than it would appear to solve. Organic farming, or forgoing the use of all synthetic chemicals, does not seem to be a practical alternative at this time. The adoption of a wide spectrum of control measures, as in integrated pest management, may still need to be coupled with a widespread testing system.

Supplies and prices of fresh produce items continue to be highly fluctuating. The extent of such fluctuations actually may be increasing as production becomes concentrated in fewer growing areas and consumer demand becomes less elastic. This presents a great challenge to the price system to transmit the proper messages between producers and consumers. How well the system is performing in this respect is an open question.

The farmers' share of the consumer food dollar has declined steadily. Currently, United States fruit and vegetable growers receive, on average, less than 20% of the consumer cost of their product (Table II). Marketers play a more important role in providing food and associated services. By itself this change is not necessarily detrimental, yet it has been accompanied by several apparent problems. Many United States fruit and vegetable growers now suffer low and variable incomes from their farming operations. Some farmers have been able to offset these losses by diversifying into other crops or extending their season by operating at other locations. A few have become involved in marketing, but many simply have been forced to adjust to lower net returns.

The scope and complexity of the marketing system has led to certain difficulties in responding to consumer preferences also. Fresh market tomatoes are a common example. Many consumers complain that, even during the local growing season, they are unable to buy juicy and flavorful tomatoes in the supermarket. Major efforts by marketers to instruct shippers, wholesalers, retailers, and consumers on how to handle tomatoes to maintain internal quality have been largely unsuccessful.

Only a few of the deficiencies that diminish the high standards of conduct and performance of the fresh fruit and vegetable industry have been presented. Such a diverse, complex, and rapidly changing industry must be prone to many other challenges for improvement. Continued study is necessary to enable this industry to maintain its proper role in the United States food marketing system.

Bibliography

Ashby, B. H., Hinsch, R. T., Risse, L. A., Kindya, W. G., Craig, W. L., Jr., and Turczyn, M. T. (1987). "Protecting Perishable Foods During Transport By Truck." USDA Agricultural Handbook No. 669. U.S. Govt. Printing Office, Washington, D.C.

Beilock, R., MacDonald, J., and Powers, N. (1988). "An Analysis of Produce Transportation: A Florida Case Study." USDA Agricultural Economic Report No. 597. U.S. Govt. Printing Office, Washington, D.C.

Buckley, K. C. (1990). "The World Market in Fresh Fruit and Vegetables, Wine, and Tropical Beverages—Government Intervention and Multilateral Policy Reform." USDA Staff Report No. AGES 9057. U.S. Govt. Printing Office, Washington, D.C.

Dicks, M. R., and Buckley, K. C., eds. (1989). "Alternative Opportunities in Agriculture: Expanding Output through Diversification." USDA Agricultural Economic Report No. 633. U.S. Govt. Printing Office, Washington, D.C.

Dunham, D. (1990). "Food Cost Review, 1989." USDA/ERS Agricultural Economic Report No. 636. U.S. Govt. Printing Office, Washington, D.C.

Dunham, D. (1991). "Food Cost Review, 1990." USDA U.S. Govt. Printing Office, Washington, D.C. ERS Agricultural Economic Report No. 651. U.S. Govt. Printing Office, Washington, D.C.

Hardenburg, R. E., Watada, A. E., and Wang, C. Y. (1986). "The Commercial Storage of Fruits, Vegetables, and Florist and Nursery Stocks." USDA Handbook 66, Rev. U.S. Govt. Printing Office, Washington, D.C.

Henderson, P. L., and Linstrom, H. R. (1980). "Farmer-to-Consumer Direct Marketing in Six States." USDA Agriculture Information Bulletin No. 36. U.S. Govt. Printing Office, Washington, D.C.

How, R. B. (1990). "Marketing Fresh Fruits and Vegetables." Van Nostrand Reinhold, New York.

Kader, A. A., Kasmire, R. F., Mitchell, F. G., Reid, M. S., Sommer, N.F., and Thompson, J. F. (1985). "Postharvest Technology of Horticultural Crops." Special Publication 3311. University of California, Division of Agriculture and Natural Resources, Davis, California.

Leed, T. W., and German, G. A. (1985). "Food Merchandising Principles and Practices." Lebhar-Friedman, New York.

McGregor, B. M. (1987). "Tropical Products Transport Handbook." USDA Agriculture Handbook No. 668. U.S. Govt. Printing Office, Washington, D.C.

Marion, B. W. (ed.) (1986). "The Organization and Performance of the U.S. Food System." D.C. Heath.

Morgan, J., Barbour, B., and Green, C. (1990). Expanding the organic produce niche: Issues and obstacles. In "Vegetables and Specialties Situation and Outlook Yearbook," pp. 55–60. USDA Publ. No. TVS-252. U.S. Govt. Printing Office, Washington, D.C.

Moulton, K. S. (1983). "The European Community's Horticultural Trade: Implications of EC Enlargement." USDA Foreign Agriculture Economic Report No. 191. U.S. Govt. Printing Office, Washington, D.C.

Nicholas, C. J. (1985). "Export Handbook for U.S. Agricultural Products." USDA Agriculture Handbook No. 593. U.S. Govt. Printing Office, Washington, D.C.

Overheim, R. K., Morris, J. N., Kessler, R. L., Bragg, E. R., Covey, E. S., Hunt, D. S., Meyer, G. W., Smalley, H. R., and Stewart, C. F. (1988). "Plans for Improved Food Distribution Facilities for San Diego, California." USDA Marketing Research Report No. 1150. U.S. Govt. Printing Office, Washington, D.C.

Paarlberg, P. L., Webb, A. J., Morey, A., and Sharples, J. A. (1984). "Impacts of Policy on U.S. Agricultural Trade." USDA/ERS Staff Report AGES 940802. U.S. Govt. Printing Office, Washington, D.C.

Pearrow, J. (1988). U.S. prices, costs, and spreads for California fresh oranges and Florida frozen concentrated orange juice, 1980–87. In "Fruit and Tree Nut Situation and Outlook Yearbook," pp. 73–77. USDA Publ. No. TFS-246. U.S. Govt. Printing Office, Washington, D.C.

Pearrow, J. (1989). Washington Red Delicious apples: Fresh market prices and spreads. In "Fruit and Tree Nut Situation and Outlook Yearbook," pp. 89–93. USDA Publ. No. TFS-250. U.S. Govt. Printing Office, Washington, D.C.

Pearrow, J. (1990). Fresh market and canned tomatoes: Prices and spreads, 1980–89. In "Vegetables and Specialties Situation and Outlook Report," pp. 25–31. USDA Publ. No. TVS-250. U.S. Govt. Printing Office, Washington, D.C.

Pearrow, J., and Lofton, N. (1990). "Bibliography of Economic Reports on Production and Marketing of Fruits and Vegetables, 1981–89." USDA Bibliography and Literature of Agriculture No. 102. U.S. Govt. Printing Office, Washington, D.C.

Price, C. D., and Newton, J. (1986). "U.S. Supermarkets: Characteristics and Services." USDA Agricultural Information Bulletin No. 502. U.S. Govt. Printing Office, Washington, D.C.

Putnam, J. J. (1990). "Food Consumption, Prices, and Expenditures, 1967–88." USDA/ERS Statistical Bulletin No. 804. U.S. Govt. Printing Office, Washington, D.C.

Putnam, J. J., and Allshouse, J. E. (1991). "Food Consumption, Prices, and Expenditures, 1968–89." USDA/ERS Statistical Bulletin No. 825. U.S. Govt. Printing Office, Washington, D.C.

Risse, L. A., Miller, W. R., and Moffitt, T. (1982). "Shipping Fresh Fruits and Vegetables in Mixed Loads to the Caribbean." USDA Agriculture Research Service Publ. No. AAT-S-27. U.S. Govt. Printing Office, Washington, D.C.

The Packer (1990). "The Packer 1990 Produce Availability & Merchandising Guide." Vance, Overland Park, Kansas.

Thompson, G. D., Conklin, N.C., and Dono, G. (1990). The demand for fresh fruit. In "Fruit and Tree Nuts Situation and Outlook Report," pp. 39–44. USDA Publ. No. TFS-256. U.S. Govt. Printing Office, Washington, D.C.

U.S. Dept. of Agriculture, Agricultural Marketing Service (1990). "Fresh Fruits and Vegetable Shipments by Commodities, States, and Months." FVAS-4 Calendar Year 1989. U.S. Govt. Printing Office, Washington, D.C.

U.S. Dept. of Agriculture, Economic Research Service (1987). "Fresh Fruits and Vegetables Prices and Spreads in Selected Markets, 1975–84." Statistical Bulletin No. 752. U.S. Govt. Printing Office, Washington, D.C.

U.S. Dept. of Agriculture, Economic Research Service (1990a). "Fruit and Tree Nuts Situation and Outlook Report Yearbook." Publ. No. TFS-254. U.S. Govt. Printing Office, Washington, D.C.

U.S. Dept. of Agriculture, Economic Research Service (1990b). "Vegetables and Specialties Situation and Outlook Yearbook." Publ. No. TVS-252. U.S. Govt. Printing Office, Washington, D.C.

U.S. Dept. of Agriculture, Economic Research Service (1990c). "Food Marketing Review, 1989–90." Agricultural Economic Report No. 639. U.S. Govt. Printing Office, Washington, D.C.

U.S. Dept. of Agriculture, Foreign Agricultural Service (1990). "Horticultural Products Review." Publ. No. FHO RT 8-89. U.S. Govt. Printing Office, Washington, D.C.

U.S. Dept. of Commerce, Bureau of the Census (1990). "Census of Agriculture 1987." U.S. Govt. Printing Office, Washington, D.C.

Van Dress, M. G. (1982). "The Foodservice Industry: Structure, Organization, and Use of Food, Equipment, and Supplies." USDA Statistical Bulletin No. 690. U.S. Govt. Printing Office, Washington, D.C.

Wright, B. (1988). "Trade Barriers and Other Factors Affecting Exports of California Specialty Crops." Agricultural Issues Center, University of California, Davis, California.

CHAPTER 2

CHALLENGES IN HANDLING FRESH FRUITS AND VEGETABLES

Robert L. Shewfelt and Stanley E. Prussia

American consumers continue to increase their per capita consumption of fresh fruits and vegetables (Figs. 1, 2; Chapter 1). Health concerns of consumers are a driving force for this increase. Virtually every governmental report on the American diet in the 1980s [U.S. Department of Agriculture and Health and Human Services (USDA/HHS), 1985; U.S. Department of Health and Human Services (USDHHS), 1988, National Academy of Sciences and National Research Council (NAS/NRC), 1989] recommends greater consumption of fruits and vegetables to promote good health. Fruits and vegetables are low in calories and fats and are rich sources of nutrients including dietary fiber, vitamins, and minerals. Part of the increase in the consumption of fresh fruits and vegetables comes at the expense of processed products because of real and perceived losses in appearance, flavor, texture, and nutritional quality during processing.

Fresh fruits and vegetables displayed on supermarket shelves are unlike a fresh tomato grown in a backyard garden, picked at the peak of flavor, and eaten minutes after harvest. Before appearing in the supermarket display, a fruit or vegetable has taken a long and sometimes circuitous journey. Part of each shipment becomes unacceptable and is discarded at culling points within the system because of the development of physiological disorders, handling damage, visible decay, or other causes. Typical losses range from 10 to 25%, but it is estimated that at least 5% and as much as 100% of a given crop can be lost between the field and the consumer (NAS, 1978; Coursey, 1983).

Since fresh produce is living, respiring tissue, physiological processes between harvest and consumption can result in the loss of quality characteristics. In some cases, quality degradation leads to a discarded product whereas in others it reduces consumer acceptability. Such losses are minimized by proper and efficient handling of the fruit or vegetable from farm to market. More subtle losses of quality result

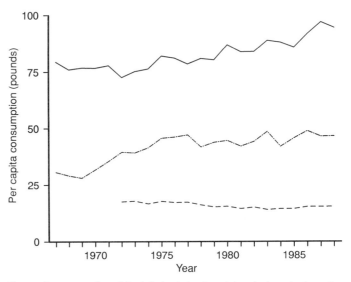

Figure 1 Changes in consumption of fresh fruits (—), citrus juices (-··-), and other selected processed fruit products (----) 1966–1988. (Adapted from Putnam, 1990.)

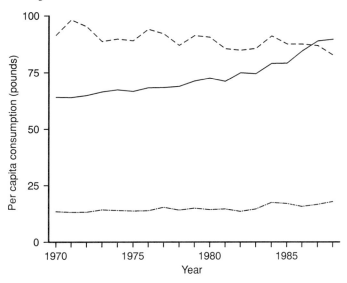

Figure 2 Changes in consumption of fresh (—), canned (---), and frozen (-··-) vegetables, 1970–1988. (Adapted from Putnam, 1990.)

from production, harvesting, and handling strategies that emphasize appearance at the expense of flavor and nutritional quality.

The previous chapter provides a strong rationale for the need for a better understanding of postharvest handling to reduce the losses of fresh fruits and vegetables from farm to consumer, as well as to improve quality and shelf life of these items when they reach the consumer. To achieve this goal, two disciplines have developed. Postharvest physiology seeks to understand the basic physiological changes and underlying mechanisms that occur in a detached plant organ during handling and storage. Postharvest technology seeks to understand the handling and storage conditions that lead to extended shelf life and improved quality of harvested produce. This chapter

- describes current approaches to postharvest handling
- suggests a need for a more integrated approach to postharvest research
- introduces the concept of a systems approach
- illustrates challenges in postharvest handling amenable to systems solutions

I. Handling of Fruits and Vegetables from Farm to Consumer

Scientific research usually is directed at narrowly defined problems using hypothesis testing or empirical observation to draw conclusions. Efficient handling and distribution of fresh fruits and vegetables is the direct result of the current understanding of postharvest physiology and the development of new technologies from narrowly focused studies. Future progress may require a broader-based program. Before studying the handling system, the component handling steps must be understood.

A.   Production Phase Operations

Although the emphasis of this book is postharvest handling, conditions in the field before harvest influence quality and shelf life after harvest. Genetic potential, growing conditions, and cultural practices all influence quality at harvest as well as shipping and storage stability (see Chapter 4). The relationship between preharvest factors and postharvest quality is complex and not well understood. For example, Hobson and Bedford (1989) conclude that preharvest cultural factors are more important than cultivar in acceptability of cherry tomatoes, but Bottcher (1988a,b) suggests that postharvest conditions are more important than growing conditions in the quality of fresh asparagus.

Plant breeders must satisfy many requirements in the breeding and selection of commercial cultivars. Most importantly, a cultivar must produce high yields under a wide range of growing conditions. More emphasis is being placed on greater resistance to stress, disease, and insects because of increasing consumer concern about the safety of agricultural chemicals. Uniformity of maturity at harvest permits the use of once-over harvest techniques. Resistance to mechanical damage during harvesting or subsequent handling operations improves shipping and storage stability. Flavor and nutrient composition are important to the consumer, but maintenance of acceptable appearance and firmness or turgor is more important to other buyers within the handling system. Achieving all these desirable characteristics in a single genotype is a difficult task; thus, a cultivar usually is judged by its most limiting characteristic.

Most commercial cultivars are selected primarily on the basis of potential yield over a range of growing conditions with the idea of maintaining an acceptable level of shipping quality. Biotechnological techniques such as cell culture and genetic engineering greatly accelerate the breeding and selection process. Cell culture techniques have the potential to provide a means to screen large numbers of genotypes for specific traits, but the journey from culture tube to stable, commercial cultivar is a long difficult one. Advances in genetics and genetic engineering offer potential for improved quality, but further advances will be limited by a lack of understanding of many basic physiological processes and unexpected modification of unrelated traits.

Growing conditions play an important role in postharvest performance of harvested crops. Preharvest stress conditions can affect the flavor, microbial quality, and composition of a fruit or vegetable. Cultural practices are chosen for other reasons including maximizing yield, minimizing visual damage, and improving efficiency of farm operations. Row spacing and training regimes facilitate field operations such as harvest or the application of agricultural chemicals. Growth regulators promote common growth patterns of crops, resulting in greater uniformity of maturity at harvest. The pressure to reduce the use of agricultural chemicals resulted in development of a strategy of Integrated Pest Management (IPM), which seeks to apply chemicals only when required to prevent economic damage (NRC, 1989). IPM helps reduce pesticide use, but requires close monitoring and a good understanding of the biology of the crop and the pests.

B. Harvest

By definition, postharvest handling begins at harvest. Numerous reviews point to the importance of the maturity of the crop at harvest on subsequent postharvest quality and shelf life (Dull and Hulme, 1971; Kader, 1985; Shewfelt, 1987; IFT, 1990; Morris, 1990). Determination of the harvest date is based on yield, visual appearance, anticipated prices, estimated culling losses to achieve shipping quality, and field conditions. Harvesting is accomplished by hand, by mechanically assisted picking devices, or by mechanical harvesters (O'Brien et al., 1983; Prussia and Woodroof, 1986). Robotics offers the long-term potential of combining the efficiency of machines with the selectivity of humans, but greater sophistication and cost reductions are needed before widespread adoption by commercial growers will be achieved (Isaacs, 1986; Harrell, 1987). Factors during harvesting operations that can influence postharvest quality include the degree of severity of mechanical damage induced by machine or human, the accuracy of selecting acceptable and unacceptable fruit, the time of day of harvest, and the pulp temperature at harvest (O'Brien et al., 1983; Prussia and Woodroof, 1986; Morris, 1990).

C. Packing

Placement of the harvested crop into shipping containers is one of many activities described as packing operations. Packing may occur directly in the field or in specially designed facilities called packinghouses. Most packing operations include a means of removing foreign objects, sorting to remove substandard items, sorting into selected size categories, inspecting samples to insure that the fruit or vegetable lot meets a specified standard of quality, and packing into a shipping container. Some commodities are washed to remove soil and decrease microbial load. Many commodities are precooled to remove field heat and slow physiological processes (Hurst, 1984; Kasmire, 1985a; Mitchell, 1985). Some special functions such as the removal of trichomes (fuzz) from peaches are also part of packing operations (Kays, 1991). Each operation is designed to achieve a product of uniform quality, but each handling step provides the opportunity to induce damage or disease.

D. Transportation

The wide availability of fresh fruits and vegetables year-round and the availability of items for sale where they cannot be grown is a triumph of modern transportation systems. The primary transportation step carries the crop from the growing region to the selling region. This trip may be cross-continent by truck or rail, overseas by ship or plane, or across the county line in a pickup truck. Minimizing mechanical damage, maintaining proper temperatures, and insuring commodity compatibility are the most important considerations in transportation operations. Mechanical damage occurs during loading, unloading, and stacking operations or from shock and vibration during transport (Peleg, 1985). Shipment of a load at or near its optimal temperature is affected by the initial temperature, refrigeration capacity, condition of refrigeration equipment, and degree of airflow around the product. Construction of the shipping container, proper alignment of the vent holes in the containers, and use of approved and appropriate stacking patterns insures adequate airflow (Kasmire, 1985b).

Attention must be given to commodity compatibility within a load. Ethylene-sensitive commodities such as lettuce should not be shipped with ethylene generators such as apples. A complete description of compatible and incompatible commodities is available (Ashby et al., 1987). The most common cause of freight claims is load shifting and crushing, but the costliest claims are the result of inadequate temperature control (Beilock, 1988).

Other transportation steps are also important in quality maintenance, for example, from field to packing facility and from wholesale distribution point to retail outlet. The same principles that apply to long-distance ones, apply to short-distance shipments, but handling practices tend to receive less attention when the shipping distance is short. Fields and rural roads are usually bumpier than highways; thus, vehicles hauling the harvested crop from field to packinghouse are generally not as capable of preventing shock and vibration damage as are tractor-trailer rigs. The delay of cooling of a crop is affected by the time required to load a vehicle in the field, the distance from field to packinghouse, the speed of the vehicle, and the number of vehicles waiting to be unloaded at the packinghouse (Garner et al., 1987). The trip from wholesale warehouse to retail outlet brings together a wide range of commodities arranged by store. Mechanical damage results from shifting of loads in transport or crushing of cartons due to the unconventional stacking of containers with differing sizes, shapes, and strengths. Quality losses also can result from inadequate temperature control or product incompatibility. Even the most careful attention to proper stacking methods and proper temperature management can be defeated on loading docks by rough handling or long delays in unrefrigerated conditions.

E. Storage

Within the handling system, fruits and vegetables are placed in storage for a few hours up to several months, depending on the commodity and storage conditions. Storage of a commodity serves as a means to extend the season, to delay marketing until prices rise, to provide a reserve for more uniform retail distribution, or to reduce the frequency of purchase by the consumer or food service establishment. The commodity must have sufficient shelf life to remain acceptable from harvest to consumption.

Shelf life of a fruit or vegetable during storage is dependent on its initial quality, its storage stability, the external conditions, and the handling methods. Shelf life can be extended by maintaining a commodity at its optimal temperature, relative humidity (RH), and environmental conditions as well as by use of chemical preservatives or gamma irradiation treatment (Shewfelt, 1986). An extensive list of optimal storage temperatures and RHs with anticipated shelf life is available (Hardenburg *et al.*, 1986). Controlled atmosphere storage is a commercially effective means of extending the season of apples (Lidster, 1984). Atmosphere modification within wholesale or retail packages is a further extension of this technology. Modification of the atmosphere is being achieved by setting initial conditions and using absorbent compounds to limit carbon dioxide (CO_2) and ethylene (C_2H_4) concentrations (Kader *et al.*, 1989; Labuza and Breene, 1989). Use of gamma irradiation extends the shelf life of some commodities, particularly strawberries (Morris, 1987).

Physiological disorders that reduce the acceptability of susceptible commodities can develop during storage. Chilling injury, damage incurred at low temperatures above the freezing point, leads to a wide range of quality defects (Jackman *et al.*, 1988). Crops also may be sensitive to high levels of CO_2 (Herner, 1987) or C_2H_4 (Lougheed *et al.*, 1987), low levels of oxygen (Weichmann, 1987), water stress due to high transpiration (Ben-Yehoshua, 1987), high temperatures (Maxie *et al.*, 1974; Buescher, 1979), and irradiation (Morris, 1987; Thayer, 1990).

F. Retail Distribution

The ultimate destination of most fresh fruits and vegetables is the retail market where a consumer makes the final decision to accept or reject the product. Retail distribution is the most visible of all handling steps, and frequently the least controlled. Merchandising displays are designed to enhance quick, impulsive purchases, not necessarily to maintain quality. Conditions within the outlet (temperature, RH, lighting), close display of incompatible commodities, length of exposure to conditions or incompatible commodities, and the degree and severity of handling by store personnel or consumers all affect quality and acceptability. Addition of ice to lower temperatures and maintain high RH and timed water misting are examples of techniques used to maintain quality. The most effective way to prevent quality losses at retail, however, is a rapid turnover of stock on the shelves. Because it is the only part of the process most consumers see, retail distribution provides an excellent opportunity to communicate with the consumer.

II. Toward a More Integrated Approach to Handling

As a result of physiological and technological studies, guidelines for the efficient management of fresh fruits and vegetables are available for each handling step described earlier. Although these guidelines are not always followed, postharvest technologists do have the knowledge of how to handle these produce items properly at each step. A basic premise of this book, however, is that many handlers of produce within the postharvest system do not have a good understanding of the interaction between these handling steps. Optimization of each handling step does not necessarily result in the best handling system. In extreme cases, an emphasis on individual handling steps results in poorer final quality. Questions that need to be answered to improve postharvest handling that have not been adequately studied by conventional approaches include

- How do preharvest cultural factors affect consumer acceptance?
- How does storage at nonoptimal conditions affect quality and consumer acceptance?
- Are handlers who adopt new methods that result in better consumer acceptance properly rewarded for their improvements?

To answer these and other questions that require an understanding of the interaction of various handling steps, a greater integration of specialized expertise and research perspectives is needed. We propose a greater emphasis on integrated studies between

- postharvest technologists and postharvest physiologists
- crop production (horticulture, entomology, pathology) and utilization (economics, engineering, food science) disciplines
- university laboratories and commercial establishments
- field and quality assurance departments within food processing companies

QUALITY

Figure 3   An integration of handling steps from farm to retail is the key to quality.

Such studies require a better definition of commercially relevant goals (economics, quality, shelf life) within the confines of environmental and economic constraints. Successful interaction of "basic" and "applied" research is synergistic. Technological problems require immediate attention, which stimulates basic inquiry into underlying physiological mechanisms. New basic knowledge suggests, in turn, new approaches and solutions to old problems.

With an improved knowledge of interactions between handling steps and a clearer understanding of the ultimate goals, integrated handling systems can be developed that incorporate answers to the questions posed earlier (Fig. 3). Traditional postharvest studies alone are not capable of answering these questions. The adoption of a systems approach provides a context for future advances in postharvest science and its commercial application.

Operations research is the scientific discipline that emerged from the need to provide troops with necessary supplies at appropriate times in World War II (Karnopp and Rosenberg, 1975). A systems approach, derived from operations research, seeks to provide a means of studying broader issues than those addressed by the typical, narrowly focused approaches employed by most scientists (Churchman, 1968; Checkland, 1981; Wilson, 1986).

III.   Challenges Amenable to Systems Solutions

Research with selected fruits and vegetables (Prussia and Shewfelt, 1985; Campbell *et al.*, 1986; Shewfelt *et al.*, 1986, 1987; Ott *et al.*, 1989; Jordan *et al.*, 1990) reveals several critical problems that require systems studies to provide meaningful solutions. Particular attention is required to identify conditions encountered in postharvest handling that affect consumer acceptance as well as preharvest factors that influence postharvest quality. Research challenges that are particularly amenable to systems solutions include latent damage, stress physiology, quality management, marketing, and food safety.

A.   Latent Damage and Its Detection

Latent damage (Peleg, 1985) is "damage incurred at one step but not apparent until a later step" (Shewfelt, 1986). Bruising is the most obvious example of latent damage, since the impact that causes a bruise can occur during rough handling but the discoloration and textural breakdown might not become visible for up to 12 hr later (J.D. Klein, 1987). Other types of latent damage not requiring mechanical impact include quiescent infections that are incurred preharvest but become evident in postharvest storage (Ben-Arie and Lurie, 1986), physiological disorders resulting from inadequate nutrition or preharvest stress conditions (Bangerth, 1983; Poovaiah, 1985), or postharvest stress disorders such as chilling injury, which is not usually evident until the item is returned to ambient temperatures (Jackman *et al.*, 1988).

Latent damage lowers quality and increases the cost of produce, since the cost of transportation must be incurred for fruits and vegetables that are discarded eventually. Early detection of latent damage can reduce costs, help identify causes of damage, and lead to new techniques that limit losses or eliminate the damage source. An understanding of the extent of latent damage and its economic impact can provide important clues to whether techniques developed to detect and prevent latent damage are economically feasible. Latent damage is discussed from a perspective of visual inspection (Chapter 9), physiology (Chapter 10), and nondestructive evaluation (Chapter 11).

B. Stress Physiology

An "aberrant change in physiological processes brought about by one or a combination of environmental biological factors" is known as the stress response (Hale and Orcutt, 1987). Almost any handling technique used to keep harvested crops fresh for an extended period of time causes some stress to that tissue. Temperature extremes, desiccation, microbial invasion, gaseous atmosphere, light, and mechanical handling all can induce stress in a harvested fruit or vegetable. Certain fruits and vegetables are more susceptible to disorders such as chilling, freezing, and CO_2 injury than others. Many factors are implicated in the syndromes associated with stress response, but the physiological mechanisms of these responses remain elusive (Chapter 12). Advances in molecular biology promise to provide techniques that will help unravel the physiological basis of quality degradation (Nevins and Jones, 1987; Romig and Orton, 1989).

C. Quality Management

Quality Assurance is an integral part of most manufacturing industries, including food processing. There is less motivation to develop quality management programs for fresh produce than for other food items, partly because of the lack of foodborne pathogens associated with fresh fruits and vegetables, the generic nature of produce marketing, and the difficulty of applying principles developed for processed foods to living, respiring tissue. The primary differences between fresh and processed foods that affect quality management factors include the following:

- Fresh fruits and vegetables are maintained in recognizable form whereas processed products are modified.
- Variability in response to storage conditions among different items in the same lot is much greater in fresh fruits and vegetables than in processed products.
- The relationship between physiological processes and food quality has not been defined clearly in many fresh fruits and vegetables.
- Latent damage is a greater factor in quality losses in fresh produce than in processed products.

The fruit and vegetable processing industry is able to avoid these problems by (1) treating the crop as raw material, thus mixing lots of varying composition to produce a product that meets uniform product specifications and (2) inactivating physiological processes during food processing operations. Despite these drawbacks, frameworks from Australia (Holt and Schoorl, 1981) and Israel (Lidror and

Prussia, 1990) provide a basis for quality management of fresh produce (Chapter 13).

D. Marketing

Fresh produce is a major profit center for supermarket food chains. Fierce competition among chains is changing the merchandising of fresh items (Chapter 14). With the exception of a few commodities, most fresh fruits and vegetables are marketed at retail in bulk displays without brand identification. Brands are used in marketing schemes of shippers directed at wholesale distributors, but whether brands will have an impact at retail distribution is still uncertain (Zind, 1990).

Display of consumer information, including nutritional composition, handling and preparation suggestions, point of origin, and "best if consumed by" dates are part of the merchandising process in many outlets. Use of universal price codes (UPCs) is not being exploited fully in the management of fresh produce, particularly in providing feedback to earlier steps in the system. Retail distribution is arguably the most important step of the entire postharvest system for determining consumer acceptance, yet this step may be the least understood in physiological and technological terms.

E. Food Safety

The growing demand for fresh fruits and vegetables by health-conscious consumers also results in an increased concern about food safety. Media attention to the use of agricultural chemicals to maintain "cosmetic" quality of fresh produce has heightened this concern. A prestigious American committee advocates the use of alternative production techniques to reduce pesticide use dramatically (NRC, 1989). A survey of United States consumers suggests that they would be willing to pay more for certified, residue-free produce but are not willing to sacrifice quality (Ott, 1989). It is not clear how much pesticide use can be reduced without loss of visual quality of fresh fruits and vegetables, nor is it clear how reduced visual quality would affect consumption [Institute of Food Technologists (IFT), 1990].

Lost in the public debate of this issue are the greater dangers of microbial pathogens associated with fresh produce. New techniques in handling fresh produce, particularly those that result in atmosphere modification, change the environment, which can favor the growth of organisms pathogenic to humans. Refrigeration temperatures, once thought to guarantee the safety of fresh fruits and vegetables, do not protect fresh produce from psychotrophic pathogens such as *Listeria* and *Aeromonas* (Chapter 15).

F. Working at the Interfaces of the Postharvest System

When we initiated research on the application of the systems approach to the handling of fresh fruits and vegetables, we tended to study the postharvest system in isolation and ignore what happened before harvest (production) or after retail sale (home storage and consumption). We soon learned the limitations of this perspective. Much of the variation observed during postharvest storage was

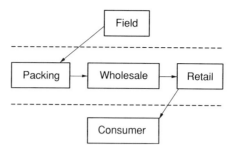

Figure 4 A generalized postharvest handling system and its interfaces with production systems and the consumer.

attributed to preharvest factors. In addition, the key to increasing the amount of an item consumed and the economic value of the item lies in understanding consumer desires. Progress in quality improvement of fresh fruits and vegetables will be made possible by working at the interfaces of the system (Fig. 4) and providing

- a clearer specification of quality and value of an item from the consumer perspective
- an ability to understand preharvest factors that contribute to sample variability and predetermine storage stability
- a means to predict mathematically the period of optimum marketability under a specified set of handling conditions

The remainder of this book places postharvest handling in a systems context. We propose a systems approach as a new paradigm for postharvest research. The ideas of numerous individuals and teams influenced the thoughts that formed the basis of this book. Critical areas of thought include dynamic systems (Karnopp and Rosenberg, 1975), systems thinking (Checkland, 1981; Wilson, 1986), application of a systems approach to processing tomatoes (O'Brien *et al.*, 1983), identification of preharvest factors on quality of processed fruit and vegetable products (Morris *et al.*, 1978, 1983, 1989), development of systems frameworks for fresh fruits and vegetables (Holt and Schoorl, 1981; Schoorl and Holt, 1982), a comprehensive approach to postharvest technology (Kader, 1985), introduction of engineering principles into handling and distribution (Peleg, 1985), an innovative approach to quality assurance (Groocock, 1986), a perspective on fresh fruit and vegetable quality (Dull and Hulme, 1971), a cellular approach to quality (Hultin, 1981), and the application of hedonic price functions to fruit and vegetable quality (Jordan *et al.*, 1985).

Bibliography

Ashby, B. H., Hinsch, R. T., Risse, L. A., Kindya, W. G., Craig, W. L., and Turczyn, M. T. (1987). "Protecting Perishable Foods during Transport by Motor Truck." USDA Handbook No. 669. U.S. Govt. Printing Office, Washington, D.C.

Bangerth, F. (1983). Hormonal and chemical preharvest treatments whch influence quality, maturity and storeability of fruit. *In* "Postharvest Physiology and Crop Preservation" (M. Lieberman, ed.), pp. 331–354. Plenum Press, New York.

Beilock, R. (1988). Losses in the logistical system: The case of perishables. *J. Food Dist. Res.* **19(2),** 20–28.

Ben-Arie, R., and Lurie, S. (1986). Prolongation of fruit life after harvest. *In* "Handbook of Fruit Set and Development" (S. Monseliese, ed.), pp. 493–520. CRC Press, Boca Raton, Florida.

Ben-Yehoshua, S. (1987). Transpiration, water stress, and gas exchange. *In* "Postharvest Physiology of Vegetables" (J. Weichmann, ed.), pp. 113–170. Marcel Dekker, New York.

Bottcher, H. (1988a). Quality changes of asparagus (*Asparagus officinalis* L.) during storage. I. External and sensory quality traits. *Nahrung* **32,** 179–187.

Bottcher, H. (1988b). Quality changes of asparagus (*Asparagus officinalis* L.) during storage. II. Nutritional quality. *Nahrung* **32,** 189–200.

Brackett, R. E. (1987). Microbiological consequences of minimally processed fruits and vegetables. *J. Food Qual.* **10,** 195–206.

Buescher, R. (1979). Influence of high temperature on physiological and compositional characteristics of tomato fruit. *Lebensm.-Wiss. Technol.* **12,** 162–164.

Campbell, D. T., Prussia, S. E., and Shewfelt, R. L. (1986). Evaluating postharvest injury to fresh market tomatoes. *J. Food Dist. Res.* **17(2),** 16–25.

Checkland, P. (1981). "Systems Thinking, Systems Practice." John Wiley & Sons, New York.

Churchman, C. W. (1968). "The Systems Approach." Dell, New York.

Coursey, D. G. (1983). Postharvest losses in perishable foods of the developing world. *In* "Postharvest Physiology and Crop Preservation" (M. Lieberman, ed.), pp. 485–514. Plenum Press, New York.

Dull, G. G., and Hulme, A. C. (1971). Quality. *In* "The Biochemistry of Fruits and Their Products" (A. C. Hulme, ed.), Vol. 2, pp. 721–725. Academic Press, Orlando, Florida.

Garner, J. C., Prussia, S. E., Shewfelt, R. L., and Jordan, J. L. (1987). Peach quality after delays in cooling. American Society of Agricultural Engineers Tech. Paper # 87-6501 St. Joseph, Michigan.

Groocock, J. M. (1986). "The Chain of Quality: Market Dominance through Product Superiority." John Wiley & Sons, New York.

Hale, M. G., and Orcutt, D. (1987). "The Physiology of Plants under Stress." John Wiley & Sons, New York.

Hardenburg, R. E., Watada, A. E., and Wang, C. Y. (1986). "The Commercial Storage of Fruits, Vegetables, and Florist and Nursery Stocks." USDA Agriculture Handbook No. 66. U.S. Govt. Printing Office, Washington, D.C.

Harrell, R. (1987). Economic analysis of robotic citrus harvesting in Florida. *Trans. ASAE* **30,** 298–304.

Herner, R. C. (1987). High CO_2 effects on plant organs. *In* "Postharvest Physiology of Vegetables" (J. Weichmann, ed.), pp. 239–253. Marcel Dekker, New York.

Hobson, G. E., and Bedford, L. (1989). The composition of cherry tomatoes and its relation to consumer acceptability. *J. Hort. Sci.* **64,** 321–329.

Holt, J. E., and Schoorl, D. (1981). Fruit packaging and handling distribution systems: An evaluation method. *Agric. Systems* **7,** 209–218.

Hultin, H. O. (1981). Food quality control at the cellular level. *In* "Recent Advances in Food Science and Technology" (C. C. Tsen and C.-Y. Lii, eds.), Vol. 2, pp. 393–405 Hua Shiang Yuan, Taipei, Taiwan.

Hurst, W. C. (1984). "Building and Operating a Vegetable Packingshed on the Farm." University of Georgia Cooperative Extension Bulletin No. 899. Athens, Georgia.

Huxsoll, C. C., and Bolin, H. R. (1989). Processing and distribution alternatives for minimally processed fruits and vegetables. *Food Technol.* **43(2),** 124–128.

Institute of Food Technologists (1990). Quality of fruits and vegetables. A scientific status summary by the Institute of Food Technologists' Expert Panel on Food Safety and Nutrition. *Food Technol.* **44(6),** 99–106.

Isaacs, G. W. (1986). Robotic applications in agriculture. *Acta Hort.* **187,** 123–128.

Jackman, R. L., Yada, R. Y., Marangoni, A., Parkin, K. L., and Stanley, D. W. (1988). Chilling injury, a review of quality aspects. *J. Food Qual.* **11,** 253–278.

Jordan, J. L., Shewfelt, R. L., Prussia, S. E., and Hurst, W. C. (1985). Estimating implicit marginal prices of quality characteristics of tomatoes. *South. J. Agric. Econ.* **17,** 139–146.

Jordan, J. L., Prussia, S. E., Shewfelt, R. L., Thai, C. N., and Mongelli, R. (1990). "Transportation Management, Postharvest Quality, and Shelf-Life Extension of Lettuce." University of Georgia Agricultural Bulletin No. 386. Athens, Georgia.

Kader, A. A. (1985). "Postharvest Technology of Horticultural Crops." Agriculture and Natural Resources Publication, University of California, Berkeley.

Kader, A. A., Zagory, D., and Kerbel, E. L. (1989). Modified atmosphere packaging of fruit and vegetables. *Crit. Rev. Food Sci. Nutr.* **28,** 1–30.

Karnopp, D., and Rosenberg, R. (1975). "Systemic Dynamics: A Unified Approach." John Wiley & Sons, New York.

Kasmire, R. F. (1985a). Cooling operations: Evaluation of efficiency. *In* "Postharvest Technology of Horticultural Crops" (A. A. Kader, ed.), pp. 44–48. Agriculture and Natural Resources Publications, University of California, Berkeley.

Kasmire, R. F. (1985b). Transportation of horticultural commodities. *In* "Postharvest Technology of Horticultural Crops" (A. A. Kader, ed.), pp. 104–110. Agriculture and Natural Resources Publications, University of California, Berkeley.

Kays, S. J. (1991). "Postharvest Physiology of Perishable Plant Products." AVI/Van Nostrand Reinhold, New York.

Klein, B. P. (1987). Nutritional consequences of minimal processing of fruits and vegetables. *J. Food Qual.* **10,** 179–193.

Klein, J. D. (1987). Relationship of harvest date, storage conditions, and fruit characteristics to bruise susceptibility of apple. *J. Amer. Soc. Hort. Sci.* **112,** 113–118.

Labuza, T. P., and Breene, W. M. (1989). Applications of "active packaging" for improvement of shelf-life and nutritional quality of fresh and extended shelf-life foods. *J. Food Proc. Pres.* **13,** 1–69.

Lidror, A., and Prussia, S. E. (1990). Improving quality assurance techniques for producing and handling agricultural crops. *J. Food Qual.* **13,** 171–184.

Lidster, P. D. (1984). Storage of apples. *J. Inst. Can. Sci. Technol. Aliment.* **17(3),** xii–xiii.

Lougheed, E. C., Murr, D. P., and Toivonen, P. M. A. (1987). Ethylene and nonethylene volatiles. *In* "Postharvest Physiology of Vegetables" (J. Weichmann, ed.), pp. 255–276. Marcel Dekker, New York.

Maxie, E. C., Mitchell, F. G., and Sommer, N. F. (1974). Effect of elevated temperature on ripening of 'Bartlett' pear, *Pyrus communis.* L. *J. Amer. Soc. Hort. Sci.* **99,** 344–349.

Mitchell, F. G. (1985). Cooling horticultural commodities. *In* "Postharvest Technology of Horticultural Crops" (A. A. Kader, ed.), pp. 35–43. Agriculture and Natural Resources Publications, University of California, Berkeley.

Morris, J. R. (1990). Fruit and vegetable harvest mechanization. *Food Technol.* **44(2),** 97–101.

Morris, J. R., Ray, L. D., and Cawthon, D. L. (1978). Quality and postharvest behavior of once-over harvested Clingstone peaches treated with daminozide. *J. Amer. Soc. Hort. Soc.* **103,** 716–722.

Morris, J. R., Spayd, S. E., and Cawthon, D. L. (1983). Effects of irrigation, pruning severity, and nitrogen levels on yield and juice quality of Concord grapes. *Amer. J. Enol. Vitic.* **32,** 229–233.

Morris, J. R., Sims, C. A., and Cawthon, D. L. (1989). Effects of production systems, plant populations, and harvest dates on yield and quality of machine-harvested strawberries. *J. Amer. Soc. Hort. Sci.* **110,** 718–721.

Morris, S. C. (1987). The practical and economic benefits of ionising radiation for the postharvest treatment of fruits and vegetables: An evaluation. *Food Technol. Aust.* **39,** 336–342.

National Academy of Sciences (1978). "Postharvest Food Losses in Developing Countries." National Academy Press, Washington, D.C.

National Academy of Sciences and National Research Council (1989). "Diet and Health: Implications for Reducing Chronic Disease Risk." National Academy Press, Washington, D.C.

National Research Council (1989). "Alternative Agriculture." National Academy Press, Washington, D.C.

Nevins, D., and Jones, R. A. (1987). "Tomato Biotechnology." Liss, New York.

O'Brien, M., Cargill, B. F., and Fridley, R. B. (1983). "Principles and Practices for Harvesting and Handling Fruits and Nuts." AVI/Van Nostrand Reinhold, New York.

Ott, S. L. (1989). Pesticide residues: Consumer concerns and direct marketing opportunities. University of Georgia Agricultural Experimental Station Research Report No. 574. Athens, Georgia.

Ott, S. L., Jordan, J. L., Shewfelt, R. L., Prussia, S. E., Beverly, R. B., and Mongelli, R. (1989). An economic analysis of postharvest handling systems for fresh broccoli. University of Georgia Agricultural Research Bulletin No. 387. Athens, Georgia.

Peleg, K. (1985). "Produce Handling, Packaging, and Distribution." AVI/Van Nostrand Reinhold, New York.

Poovaiah, B. W. (1985). Role of calcium and calmodulin in plant growth development. *HortScience* **20**, 347–352.

Prussia, S. E., and Shewfelt, R. L. (1985). Ventilation cooling of bulk shipments of shelled southern peas. *Trans. ASAE* **28**, 1704–1708.

Prussia, S. E., and Woodroof, J. G. (1986). Harvesting, handling, and holding fruit. *In* "Commercial Fruit Processing" (J. G. Woodroof and B. S. Luh, eds.), pp. 25–97. AVI/Van Nostrand Reinhold, New York.

Putnam, J. J. (1990). "Food Consumption, Prices, and Expenditures, 1967–88." USDA/ERS Statistical Bulletin No. 804. U.S. Govt. Printing Office, Washington, D.C.

Rolle, R. S., and Chism, G. W. (1987). Physiological consequences of minimally processed fruits and vegetables. *J. Food Qual.* **10**, 157–177.

Romig, W. R., and Orton, T. J. (1989). Applications of biotechnology to the improvement of quality of fruits and vegetables. *In* "Quality Factors of Fruits and Vegetables—Chemistry and Technology" (J. Jen, ed.). ACS Symposium Series 405. American Chemical Society, pp. 381–393. Washington, D.C.

Schoorl, D., and Holt, J. E. (1982). Fresh fruit and vegetable distribution—Management of quality. *Sci. Hort.* **17**, 1–8.

Shewfelt, R. L. (1986). Postharvest treatment for extending the shelf life of fruits and vegetables. *Food Technol.* **40(5)**, 70–89.

Shewfelt, R. L. (1987). Quality of minimally processed fruits and vegetables. *J. Food Qual.* **10**, 143–156.

Shewfelt, R. L., Prussia, S. E., Jordan, J. L., Hurst, W. C., and Resurreccion, A. V. A. (1986). A systems analysis of postharvest handling of fresh snap beans. *HortScience* **21**, 470–472.

Shewfelt, R. L., Myers, S. C., Prussia, S. E., and Jordan, J. L. (1987). Quality of fresh-market peaches within the postharvest handling system. *J. Food Sci.* **52**, 361–364.

Thayer, D. W. (1990). Food irradiation: Benefits and concerns. *J. Food Qual.* **13**, 147–169.

U.S. Dept. of Agriculture and U.S. Dept of Health and Human Services (1985). "Dietary Guidelines for Americans." U.S. Govt. Printing Office, Washington, D.C.

U.S. Dept. of Health and Human Services (1988). "The Surgeon General's Report on Nutrition and Health." U.S. Govt. Printing Office, Washington, D.C.

Weichmann, J. (1987). "Postharvest Physiology of Vegetables." Marcel Dekker, New York.

Wilson, B. (1986). "Systems: Concepts, Methodologies, and Applications." John Wiley & Sons, New York.

Zind, T. (1990). Fresh trends '90—A profile of fresh produce consumers. *The Packer Focus* **96(54)**, 37–68.

CHAPTER 3

SYSTEMS APPROACH TO POSTHARVEST HANDLING

Stanley E. Prussia and Robert L. Shewfelt

The systems approach provides a set of tools for use in cases with well defined objectives (hard systems) or when it is difficult to define perceived problems (soft systems). A few of these tools have been applied to postharvest systems. Ideally, a situation is studied by combining the systems approach with scientific and engineering methods. Most postharvest studies have treated produce handling as a hard system, but a soft-system approach may be more appropriate. The biological variability of fruits and vegetables, combined with the frequent change of handlers from farm to consumer, suggests that postharvest systems are soft. This chapter describes postharvest handling systems, explains the systems approach, presents applications of the systems approach to postharvest handling, and suggests the implications of such an application in the modification of handling operations.

I. Postharvest Systems

Terminology relating to postharvest systems ranges from the broad and inclusive, encompassing all the traditional food processing operations, to the more narrow (Table I). We prefer the narrow interpretation, which includes the period during which the harvested item is living and respiring. This period typically ends during a blanch stage in processed products and with cooking or consumption of fresh items. "Fresh," however, is a vague term with many incongruous connotations, including the distinction from "processed" and being recently harvested (Breene, 1983).

Actually, there is no one postharvest system. Rather numerous interacting operations can be considered systems that have been designed to handle and distribute

Table I

Terminology Relating to Postharvest Systems Indicating Wide Variations in Perceptions

Term	Definition (Reference)
System	A set of parts that behave in a way that an observer has chosen to view as coordinated to accomplish one or more purposes (Wilson and Morren, 1990)
Postharvest losses	Losses occurring between the completion of harvest and the moment of consumption (Bourne, 1977)
Postharvest period	Period beginning at separation of food item from medium of immediate growth or production and ending when food enters process of preparation for final consumption (National Academy of Science, 1978)
Postharvest technology	Functions of assembly, processing, packaging, warehousing, storing, transportation, and distribution of agricultural products through the institutional food trade and wholesale and retail outlets (Garner and Dennison, 1979)
Postharvest functions	On-farm operations after harvesting and before processing the crop (Monroe and O'Brien, 1983) [to include] handling and transport . . . after harvest (O'Brien and Gaffney, 1983)

various perishable fruits and vegetables. These systems vary by product, end use of that product, and the level of technology available, affordable, and acceptable. In a systems approach, the boundaries of each system must be delineated clearly and the steps must be documented.

If living, respiring tissue is used as the criterion for the postharvest period, then the boundaries of a system for a given commodity are delineated on one end by harvest and on the other end by death or loss of commercial viability (Variyam *et al.*, 1988). In processed products, postharvest handling ends when most or all enzymes have been inactivated, frequently during blanching. In fresh-market items, postharvest handling ends at consumption or discard. Documentation of a system is accomplished by carefully tracing the flow of commercial handling from farm

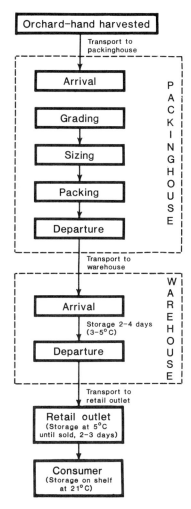

Figure 1 Operations in a typical postharvest system. [Reprinted by permission of the Institute of Food Technologists; from Shewfelt, R. L., *et al.* (1987a). *J. Food Sci.* **52,** 361–364.]

to market, listing unit operations performed on an item, and identifying branch points. Schematics then are drawn to permit a visual perspective. Such schematics are available for many fruits and vegetables (Kader, 1985); an example is provided in Fig. 1. The primary unit operations in postharvest handling are harvesting, packing, transportation, storage, and retail distribution (Chapter 2). Production phase operations exert such a critical impact on postharvest handling that they also must be considered (Chapters 2, 4). Variation in the importance of these operations by commodity, end use, and technology is described in the following sections.

A. Commodity

As a general rule, differences in the treatment of specific fruits and vegetables are more apparent during production phase and early postharvest operations and less apparent during wholesale and retail distribution. Thus, flexibility for change in a system to accommodate a specific commodity tends to decrease as it approaches consumer purchase. An appreciation of this tendency is beneficial prior to any attempt to modify a handling system for a specific commodity. A critical point within a system tends to exist in the packing or transportation stage of operations. Prior to this point, technology can be adapted to meet the specific requirements of a commodity, but after this point the commodity begins to lose its identity within the larger wholesale and retail distribution system. Thus, specific requirements of a commodity are superseded by the limited options of existing technology.

Differences in production phase operations among commodities and within a commodity are discussed in Chapter 2. These differences, in turn, are reflected in the state of physiological maturity of the crop at harvest. Maturity at harvest affects color, flavor, texture, and nutritional quality of a commodity at harvest and also can affect subsequent storage stability (Morris *et al.,* 1979; Shewfelt, 1987, 1990). Most maturity indices are based on differences in color, textural properties, solids or moisture content, or carbohydrate composition. A listing of some maturity indices is provided in Chapter 5.

Differences in harvesting operations are widespread among commodities. Some crops such as sweet corn, beans, processing tomatoes, cane fruits, and grapes are amenable to mechanical harvesting (Morris, 1990). Many other crops, including small fruits and fresh-market tomatoes, are successfully harvested with mechanically assisted picking operations (O'Brien and Berlage, 1983). Field packing is a third variation widely used for lettuce and broccoli, for example, and involves hand picking and placement on a conveyor for transport to a mobile trimming and packing operation. Most commodities around the world are picked primarily by hand (Prussia and Woodroof, 1986).

As mentioned in Chapter 2, the packing stage encompasses several unit operations such as transport to the packing facility, precooling, washing, sanitizing, grading, and packing into a shipping container. The primary limitations to field packing are the specificity and complexity of packaging steps, control of quality, and the required scale of harvesting operations.

Rapidly respiring items such as green leafy vegetables and broccoli, as well as soft fruits such as strawberries and peaches, all of which are susceptible to decay microorganisms, must be cooled rapidly and shortly after harvest (Wills *et al.,*

1989). Rapid cooling can be achieved by vacuum cooling, hydrocooling, package icing, or forced-air cooling. Vacuum cooling requires large capital investments but is very effective on commodities such as lettuce that have large surface-area-to-mass ratios but ineffective on tightly packed spherical items. Hydrocooling permits rapid transfer of heat from the product to chilled water and is used widely to cool peaches. However, water can enhance the chances for decay. Package icing, used widely for broccoli and greens, exhibits advantages and disadvantages similar to those of hydrocooling.

Forced-air cooling can be achieved in the shipping container and is an effective technique that has ready application to many harvested products such as cauliflower and grapes. Merely placing a freshly harvested crop into a cold-storage facility or refrigerated truck will not allow sufficient cooling. In addition to removing the field heat (heat present in the item at harvest), the refrigeration must remove the heat of respiration which increases at a logarithmic rate with increasing pulp temperature (Monroe and O'Brien, 1983; Prussia and Woodroof, 1986). Certain commodities including many tropical fruits, snap beans, and tomatoes are sensitive to low temperature storage, called chilling susceptibility (Jackman *et al.*, 1988). These items should not be cooled or stored below the temperature of susceptibility (see Hardenburg *et al.*, 1986).

Items that are in contact with the soil, either as root crops (potatoes, carrots, radishes) or grown on the plant in such a way that they might contact soil (tomatoes, bush beans), are prime candidates for washing. Other crops, including soft fruits that may harbor decay organisms and may be penetrated easily by nicks and scratches, also can benefit from washing and application of fungicides. Problems are engendered in these steps since brushing to clean can cause abrasions and use of water can enhance decay.

In addition, application of fungicides is coming under increased scrutiny with growing concerns about agricultural chemicals. During the same series of operations, wax is added to certain commodities such as citrus fruits, cucumbers, and apples. In citrus fruits, the wax is essential to prevent dehydration whereas in other products it enhances moisture retention and visual appeal by providing greater gloss. This enhanced visual appeal may be counteracted, however, by consumer skepticism of the "unnatural" wax. Other special operations performed at the packing stage include "defuzzing" of peaches, the removal of trichomes from the outside of the fruit (Monroe and O'Brien, 1983).

Sorting and grade determination operations are part of most postharvest systems. In field-pack operations that involve a person picking and placing the item directly into the shipping container, the picker/packer also must serve as a sorter. Most fruits and vegetables are bought and sold on the basis of meeting predetermined specifications. These specifications may be set by government grades and standards, industry-wide or grower's cooperative standards, or buyer specifications. Many sorting operations function by passing the commodity over a sorting belt at which workers remove defective items. Defects may include visual damage but also can include misshapen items or products outside the range of acceptable maturity (see Chapter 9).

Quality standards and levels of acceptable maturity vary widely among commodities (Chapter 5). Studies have shown that sorters are more efficient at removing defective material when items are coming directly toward them than from a traditional side-to-side orientation (Meyers *et al.*, 1990). Rapid nondestructive sorting techniques have been developed for certain crops, including cherries and tomatoes (Prussia and Woodroof, 1986). These techniques tend to be more effective if an optical property such as color, not pattern recognition, is the criterion for elimination. Uniformity of size is also a desirable attribute. Thus, many commodities are sized after sorting operations.

Packing into the shipping container is another operation that is not as simple as it might first appear. Premium packs of bruising-susceptible fruit may be separated on fiberboard trays to minimize fruit-to-fruit impact. Wholesalers and retailers frequently demand uniformity, either of count or weight per container, by commodity. Standardization of cartons for unitization in transport and further distribution is desirable, but such a task can be difficult for large commodities such as watermelons. Direct stacking of the shipping containers onto a pallet to be placed in the transport vehicle saves time and expense later and improves airflow necessary for temperature control during transport.

Most crops are transported several times from one postharvest operation to the next, but there is usually one major transportation step from the growing locale to the selling or processing locale. Many factors other than commodity affect this step, but it is important to keep noncompatible commodities separated when transporting mixed loads (Chapter 2). Ethylene-generating equipment such as propane-powered forklifts must be kept away from ethylene-sensitive commodities (Kasmire, 1985). Maintaining the proper temperature of the item(s) shipped is also important. Most harvested crops are stored at least once during handling. At wholesale or retail facilities, the commodity must fit into one of a narrow range of storage options, usually based on product tolerance to ice and high humidity conditions or the degree of chilling susceptibility. The growing location tends to be more flexible in maintaining optimum temperatures, since only one or a few crops may be harvested during the same season. Controlled atmosphere storage of apples is an example of sophisticated storage of a commodity at the growing location.

Optimal storage conditions normally are achieved during distribution for high-value items such as tomatoes, bananas, and selected tropical fruits.

B. End Use

Although a cucumber may be merely a cucumber to a consumer, the end use of that cucumber is important in determining the varietal selection, cultural practices, harvesting methods, and handling system. Different attributes are important in a fresh cucumber destined for the supermarket shelf than in one headed for the pickle factory (Salunkhe and Desai, 1984). Likewise, attributes vary for many fresh commodities. Before entering a store, consumers usually have decided that certain products they purchase will be fresh, frozen, or canned (Resurreccion, 1986). However, consumers do select which product to purchase within a category depending on appearance, price, and other factors (e.g., fresh peas or fresh corn).

The primary difference in postharvest systems between fresh-market and processed fruits and vegetables is shelf life requirement. Fresh-market produce has a limited shelf life, and the system should be designed to permit distribution at an acceptable level of quality. Food processing sacrifices freshness for extended shelf life. Minimal processing has the purpose of extending shelf life without losing freshness. Fresh items must be able to withstand the handling system while still alive and respiring. Processed items must be able to maintain certain attributes during processing operations. As mentioned earlier, the attributes of importance to the consumer vary for fresh and processed items.

Differences in systems for fresh and processed crops begin at the production phase. Different quality specifications lead to the selection of different cultivars for fresh-market and processed crops, as demonstrated by peaches (Sistrunk, 1985), tomatoes (Stevens, 1985), pears (Quamme and Gray, 1985), and strawberries (Sistrunk and Morris, 1985). Production systems are affected in addition to the harvesting systems. Mechanical harvesting has been adapted more widely to processed items than to fresh items (Morris, 1990). Mechanical harvesting can lead to bruising, which can become visible over a period of hours or days in an intact fruit, making it unacceptable for retail distribution. The same bruise may not be harmful to a processed product, because the bruise may not become visible prior to processing or it may be removed in operations such as peeling or trimming. Crops amenable to mechanical harvesting tend to be amenable to mechanization of other production practices such as pruning.

Decisions about the timing of harvest are dependent on many factors, including yield at harvesting or processing, maturity stage, field conditions, quality characteristics, expected storage or shipping requirements, and economic considerations. Again, optimal harvest maturities can vary greatly depending on whether a crop is destined for the fresh market or the processing plant. Once-over harvests usually are attempted for mechanical operations, whereas hand harvesting provides a greater opportunity for multiple harvests. Optimal maturity at harvest may vary within each commodity, for example, between conventional and vine-ripened or between crops grown for freezing or for canning.

Likewise, transportation considerations for fresh and processed items vary in postharvest systems. Fresh produce usually is packed into a shipping container in the field or in a packing facility and subsequently transported to the point of wholesale distribution. Transport from the field to the packing facility, when necessary, is usually a short distance (< 75 km) but may be a major source of damage, depending on the amount of vibration and the size of the bulk shipment (Prussia and Woodroof, 1986). Since the processing plant is the final postharvest destination for items to be processed, the critical transportation step in these systems is from the field to the processing plant. Ideally this distance is short, but it is not uncommon to see crops shipped hundreds of kilometers in bulk containers. In cases of long shipments, some operations, including sorting, grading, and precooling, may be performed in the field or in small centralized facilities that collect material from several fields and prepare the crop for shipment.

Most postharvest handling systems require one or more storage periods. The critical time period for most commodities is between harvest and the first storage period. Since quality deterioration is associated so closely with respiration and transpiration, control of these two processes is achieved primarily by temperature and humidity control, respectively. Expeditious handling from field to consumer for fresh-market items or from harvest to blanch in their processed counterparts is also an effective tool to maintain harvested quality. Notable exceptions to this rule exist in both fresh-market and processed systems. Controlled atmosphere storage permits year-round availability of fresh (unprocessed) apples (Smock, 1979). Tomatoes can be held in the mature-green state for several weeks at 12.5–15°C in the absence of ethylene to provide a more uniform distribution of fruit to markets.

C. Technology

The availability, appropriateness, and economic feasibility of technology affects the design of handling systems. Industrialized societies have infrastructures that can provide long-distance transportation and continuous refrigeration, permitting mass distribution of fresh-market items when not available from local growers. Developing countries tend to lack the infrastructure necessary to supply consumers with a wide range of products during all seasons. On the other hand, it may not be appropriate nor economically feasible to develop technologically intensive systems when selected perishable crops can be grown year-round near the major population centers.

Unavailability of labor or inability to pay the cost of available workers tends to be the motivation for mechanization of production and harvesting practices. Highly mechanized operations are capital intensive and typically require large growing areas to be effective. Mechanization is more common on large farms than on small ones, although mechanical equipment may be available to smaller operations through growers' cooperatives or rental agreements. Mechanical expertise and the availability of spare parts are additional considerations for mechanized production and harvesting systems.

In industrialized nations, the fiberboard box is becoming the container of choice for packing and shipping of most commodities. By adding wax, fiberboard boxes can retain their strength when used in high humidity environments, with ice in the box, and even when submerged in hydrocooler water. Use of fiberboard boxes has aided the adoption of standardization, unitization, and palletization practices for handling produce shipments (Hardenburg *et al.*, 1986). Consumer packs in over-wrapped foam trays can be packed in the fiberboard container, and metal straps, glue, or plastic overwrap can be used to hold a pallet-load of containers together. Fiberboard containers tend to be used once and then discarded. Unwaxed boxes are readily recyclable whereas waxed ones are more difficult to recycle.

In developing countries, a greater emphasis is placed on packing with lower cost or reusable containers. For example, in India mangoes are wrapped in the leaves of the tree and shipped in bulk to market. Elsewhere, baskets made from cane are used and reused to ship tomatoes (Labios *et al.*, 1984) and other commodities (Musa, 1984). Such practices may result in increased bruising during shipment or inoculation with decay microorganisms from one shipment to the next when containers are reused.

Long-distance transportation permits a grower to serve markets not accessible otherwise. To compete with locally grown produce, items shipped long distances must be out-of-season or have a competitive advantage sufficient to recover the cost of transportation. Losses that occur during shipment can be reduced by refrigeration. The decision to refrigerate can be made by balancing the cost of losses against the cost of refrigeration. When refrigeration is not available, the movement of the truck can be used to create airflow across the product to lower temperatures. When not in transit, parking vehicles in shaded areas can lower temperatures and reduce losses. Shorter distances, such as those traveled by periurban farmers who grow crops in or near metropolitan centers, may permit animal- or human-powered conveyances to transport products to market.

Maintenance of storage conditions at or near optimal temperatures and relative humidity (RH) can extend shelf life of an item and permit greater flexibility during marketing and distribution. In the later phases of wholesale and retail distribution only a limited choice of refrigerated conditions is available. Some warehouses keep nonperishables at ambient or "room" temperatures ranging from 15–25°C and perishables at 4–8°C. Perishables usually are stored in either wet (high RH and/or ice) or dry refrigerated rooms. Other facilities with separate storage areas at 1–2°C and 8–10°C permit the segregation of chilling-tolerant and chilling-sensitive items. In locations where refrigerated storage is not available or economical, expeditious handling of the product is the most effective means of maintaining quality.

II.   Systems Approaches

The reasons for implementing a systems approach for postharvest research are similar to the reasons that systems approaches were developed. There is a need for understanding the interactions among the many operations necessary for delivering horticultural crops to consumers. It is beneficial to predict the impact of actions or changes at one point on other operations without actually changing existing systems. Considering the complete system helps to identify any gaps in knowledge and to prioritize research projects. Thus, research results are more likely to be compatible with actual systems.

This section presents a historical perspective of scientific, engineering, and systems approaches for learning, discusses some systems techniques and methodologies, and considers the utility of a systems approach to postharvest systems.

A.   Historical Perspective

The study of a complex system is more complete when the accumulated knowledge from several scientific disciplines is combined. Thus, a study based on a systems approach benefits from the formation of an interdisciplinary team. For example, postharvest systems have operations that require expertise from disciplines such as economics, horticulture, plant pathology, entomology, food science, and engineering. However, team members often have difficulty working together. They face the challenge of accommodating differences in the paradigms of each discipline that predispose team members to have conflicting views on how to define the problem and the best approach to find answers. Cooperation among team members can be improved when team members understand the historical developments that led to the current scientific and engineering methods and systems approaches accepted by their discipline.

1.   Scientific approach

The scientific revolution of the seventeeth century resulted in the scientific method as the accepted standard for generating knowledge. The process of producing publicly testable knowledge involves reductionism, repeatability, and refutation (Checkland, 1981). A complex situation is reduced to an understandable level that enables the design of experiments for testing hypotheses. Results from the experiments are considered valid when they are repeatable. Knowledge is built by the refutation of competing hypotheses. Both basic and applied research depend on the scientific method.

The purpose of basic research is to explain why, to find cause and effect relationships, or to describe how things interact. A scientific problem is defined when an observation is not explained adequately by existing conceptual frameworks or contradicts the existing explanatory and meaningful body of knowledge (paradigm). After a problem is identified, the approach of basic research is to reduce the problem conceptually to a collection of facts, generate hypotheses, experimentally test the hypotheses several times, and refute competing hypotheses.

The major objective of applied research is to gain the knowledge necessary to alter critical components of a real situation. A frequent aim is to optimize actions to be taken in a particular situation. Applied research shares with basic research the same procedures of reductionism, repeatability, and refutation. The scientific approach has been tremendously successful in learning about the world and making changes when desired. However, the scientific method is less suitable for dynamic and complex situations, social issues, and management applications. When complex situations are reduced to controllable experiments it is not possible to include characteristic properties at each level that are irreducible. After obtaining valid evidence, the scientific method does not provide techniques for reassembling the knowledge to higher levels in the system where decisions must be made.

2.   Engineering approach

An underlying aim of engineering is to generate technical changes thought to improve a situation. Such improvements are based on efforts to understand natural and designed systems sufficiently to control them for beneficial uses. Engineering efforts depend on scientific and mathematical knowledge. However, engineering and science have different aims. Science achieves understanding of a structure whereas engineering achieves development of a new structure. Science studies what *is* whereas engineering considers what *could be*. Science attaches high value to advancement of knowledge whereas engineering values the efficient accomplishment of some defined purpose.

Typically, an improvement needed is made known to a design engineer who tries to find ingenious, low cost, and efficient technical alternatives, that is, the problem is structured in terms of "what" is needed and the engineering task is to determine "how" to meet the need. Examples of the engineering technologies that benefit postharvest operations include equipment for materials handling, cooling, sorting, packing, refrigerating, transporting, and displaying fresh fruits and vegetables at retail.

Important limitations of the engineering approach are the lack of a method for selecting which problem to solve and the inability to accommodate unpredictable human actions.

3. Systems approach

Over the past 50 years, several systems approaches have been developed, ranging from formalized mathematical procedures for optimizing a system to broad guidelines for thinking about situations involving both technical and human components. Systems approaches became necessary as designed systems became more complex and as our understanding of natural systems expanded. The scientific method continues to be an essential tool for reducing complexity to knowledge. In a complementary manner, systems approaches enable specific knowledge to be integrated into a complete system.

B.   Techniques and Methodologies

Systems approaches can be described as either "hard" or "soft." Hard systems approaches have goals or end states that can be defined readily and optimized or maximized through quantitative models. Soft systems approaches enable the study of problem situations, usually involving people, that are difficult to define and quantify because there is no general agreement on the problems, goals, or purpose of the real-world systems.

1.   Hard systems approaches

An urgent need to allocate scarce resources to various military operations during World War II stimulated the formation of teams of scientists to conduct research on developing methods to optimize military operations. A result of this work was the development of quantitative approaches to decision making, called "operations research." After the war operations research (OR) approaches were adopted rapidly by private businesses and management consultants.

Linear programming is a popular OR tool. A common application of linear programming is the optimum location of warehouses to minimize transportation costs when multiple manufacturing and distribution locations exist. Other tools are dynamic programming, which can help evaluate the best policies needed to reach a desired state, gaming theory, which helps identify the best strategies, queuing theory, which can reduce waiting times, inventory theory, which helps minimize storage space requirements, and simulation models, which help evaluate alternative systems.

Systems engineering (SE) and systems analysis (SA) are two approaches that are very similar to each other and to OR techniques. Both SE and SA are based on the premise that the real-world problems can be formulated as the difference between a desired state and the present state. When using SE, emphasis is placed on carefully considering the steps necessary to complete an engineering project successfully. Use of SA emphasizes appraisals appropriate for decision-making about projects.

A limitation of hard systems approach is revealed by many unsuccessful attempts to apply OR, SE, and SA to the needs related to criminal justice, public policy, and other social and political issues. The difficulty is the wide range of opinions that exists among the people affected by or involved in the situations.

2.   Soft systems methodologies

Soft systems methodologies (SSM) were developed by researchers at the University of Lancaster in the United Kingdom. Initially, the researchers applied the previously described hard systems engineering approaches to ill-structured problem situations encountered when consulting with industry and public agencies. Through active learning, they developed a methodology based on real-world management situations considered "human activity systems." SSM commonly is described as a seven-stage approach for working on "messy" problem situations to bring about improvements. More recently, Checkland and Scholes (1990, p. 300) state, "It is simply an organized version of doing purposeful 'thinking'!"

The first stage identifies a situation considered problematic. Stage two expresses the elements of structure and process and the climate surrounding them. In stage three, the effort moves from real world to systems thinking about real world by writing root definitions (RD) to describe several viewpoints of the situation. Stage four develops a conceptual model (CM) for each RD in the form of a diagram showing activities and interactions among them. Stage five compares the diagrams with the real-world situation. Any differences stimulate debate, which leads to the definition of feasible and desirable changes in stage six. Finally, in stage seven, action is taken to improve the situation. The cycle can be repeated after the situation originally described in stage one has changed as a result of the process.

The RDs are evaluated for completeness by using the mnemonic CATWOE, in which C is customer, A is actors, T is transformation, W is weltanschauung or worldview, O is owner, and E is environment (see Table II). Conceptual models typically contain five to nine verbs that represent the activities that are the minimum necessary for the model to function as the system described in the RD. Normally, one or more of the activities is expanded into a hierarchy of systems by asking how the activity is accomplished. The resulting subsystem consists of verbs describing how one of the activities in the system is done. Thus, the "hows" for an activity in the system become the "whats" for the subsystem. Careful attention to a hierarchy helps to keep activities at similar levels.

Many people with a background in OR, SE, or SA, consider one weakness of SSM to be a lack of mathematical or scientific rigor. When evaluating a methodology such as SSM, Checkland (1981, p. 242) states,

> In dealing with human activities as perceived problems, the best we can hope for is that in the eyes of concerned people former problems are now rated as "solved" or that problem situations are rated as "improved."

Table II
Ingredients for a Checkland CATWOE Analysis of a System as Applicable to Postharvest Systems

Abbreviation	Term	Definition
C	Customer (beneficiary of system)	Ultimately the consumer, but changes between subsystems
A	Actors	Includes growers, packers, shippers, and distributors
T	Transformation	Raw harvested crops become items offered for consumer purchase
W	Weltanschauung (worldview)	Not always clearly defined, but usually to generate a profit by providing an acceptable product
O	Owner (possesses power to make policy decisions)	With the exception of vertically-integrated systems, ownership is diffuse and identifiable only within subsystems
E	Environmental constraints	Includes biological nature of fresh produce, availability and capability of labor and technology, and economics

C. Utility of a Systems Approach

A major thesis of this book is that a more rigorous application of the systems approach to postharvest handling of fresh fruits and vegetables will serve to develop systems that are more responsive to consumer needs and desires and to provide information that will lead to more efficient and effective production of crops for the postharvest system. Although a formal application of the Checkland methodology to postharvest systems has not been published yet, one such study has been conducted on the wholesale distribution in the United Kingdom (Prussia and Hubbert, 1991). Completing the formal methodology is not essential to having a positive impact on existing systems. Examples are provided on the application of systems thinking to analysis of a system, developing and evaluating alternatives, and improving actual postharvest systems.

The systems approach has many advantages as a tool to improve handling systems. First, it forces a consideration of the goals and objectives within a handling system. Also, it helps identify areas of need so resources can be reapportioned to be most effective. Additionally, the analysis can serve as a basis for structuring management decisions.

The systems approach is merely a tool, and one with limitations. Solutions proposed are only as effective as the information generated and the accuracy with which the investigator perceives reality. The temptation to conduct a quick analysis or to confuse written documentation with reality should be avoided. Likewise, caution must be exercised in the general application of one system to other seemingly similar systems. The RD of a system provides the key, and RDs may vary widely for tomatoes destined for fresh market or processing in the United States, Japan, or India.

III. Applications of a Systems Approach to Postharvest Handling

---

No records have been found showing that investigations have used the systems approach in a complete sense, as outlined in the preceding discussion, to modify postharvest handling of fresh fruits and vegetables. Rather, several studies have employed systems thinking to view handling from a different perspective, leading to improvements of specific handling systems. Mention of a systems approach to production or handling systems in the literature often can be considered systems thinking.

A. General Principles

1. Systems thinking

Systems thinking is described as a holistic way of viewing a process or problem using a structured methodology to place the process or problem in a larger context. The entire process from soil preparation and planting through consumer purchase and consumption can be divided into two distinct subsystems—production and postharvest. Harvest provides the intervening boundary.

A noteworthy application of systems thinking in production phase operations is integrated pest management (IPM), which was developed "to maintain damage below an economic threshold with a minimal input of externally applied chemicals" (Pedigo et al., 1986; Institute of Food Technologists, 1990). To be effective, IPM requires a basic understanding of the host–pest relationship, clearly defined levels of economic injury, a means for assessing in-field damage on a periodic basis, and careful day-to-day management of the crop. Use of this tool is a cornerstone of a detailed report calling for reduced dependence on agricultural chemicals in the United States (National Research Council, 1989). IPM programs have been described for many crops, including apples (Whalon and Croft, 1984).

Systems thinking was employed in the development of an integrated production and harvest scheme for the management of peach orchards (Horsfield et al., 1971). Preharvest management for fruits to be processed has been described from the systems perspective for grapes (Morris et al., 1980, 1986), blackberries (Sims and Morris, 1982; Morris and Sims, 1985), and strawberries (Morris et al., 1985a,b; Main et al., 1986). The systems perspective is apparent in a description of loss reduction techniques in postharvest management (Bourne, 1977, 1984; National Academy of Science, 1978; O'Brien et al., 1978).

The most clearly delineated description of systems thinking applied to postharvest handling has been provided by Australian researchers. A method for evaluating handling systems was introduced by Holt and Schoorl (1981). Subsequent articles described a time–temperature framework for evaluating disease control measures (Schoorl and Holt, 1982a) and schemes for managing quality (Schoorl and Holt, 1982b, 1985). Economic components were incorporated into systems thinking about postharvest handling of fresh products (Jordan et al., 1985; Prussia et al., 1986).

Primary considerations in systems thinking applied to postharvest handling include the following.

- Handling steps are not isolated events within a process but may be influenced by preceding steps and can exert influence on succeeding steps.
- Output of the system must be responsive to consumer needs and desires, not merely to handler convenience.
- Factors occurring in the production system can influence postharvest performance, and a clearer understanding of the interaction of these two systems is needed if production is to be more responsive to postharvest needs.
- Studies intended to improve systems need full integration of disciplinary perspective at the problem definition and experimental design stage.
- Social factors should be studied in addition to technical issues by using approaches such as soft systems methodologies.

2. Analyzing current systems

In applying any systems approach, an investigator must learn more about the existing system. Through a series of interviews with the "actors" in the system, a pattern begins to emerge. This pattern then is documented in the form of a schematic, as shown for peaches in Fig. 1. Such a system for a specific commodity with a single end use may be reasonably consistent for growers, packers, and shippers in a growing locale or it may vary at certain steps. An understanding of current practices and the range of variability in handling is essential to any analysis of the system.

Once the structure of the system is established, the function of each handling step and the rationale for the sequential order of steps should be delineated. Within a postharvest system may be subsystems that merit more detailed analysis. Packing is an example of a subsystem within the fresh peach system outlined in Fig. 1. Depth of analysis of a system and component subsystems will depend on the interests and resources of the investigator. Postharvest systems that have been described include those for broccoli (Brennan and Shewfelt, 1989), lettuce (Jordan *et al.,* 1990), mangoes (Hubbert *et al.,* 1987), peaches (Shewfelt *et al.,* 1987a), snap beans (Shewfelt *et al.,* 1986), southern peas (Shewfelt *et al.,* 1984), and tomatoes (Shewfelt *et al.,* 1987b).

3. Developing and implementing alternative systems

Once existing systems have been analyzed carefully, changes can be considered. Certain handling steps may have evolved to meet earlier needs that are no longer relevant or may have been adopted for handler convenience. Each step is evaluated for its contribution to the purpose of the system. Idealized systems might be slight modifications of existing systems, might include rearrangement of handling steps, or might involve radical restructuring. At this stage, the investigator is not bound by the existing conditions but seeks to find the pathway that will achieve the purpose most effectively.

Once an idealized system has been constructed, it can be compared with the existing one to reveal differences. Realistic alternatives to the existing handling system then can be posed to improve it. These alternatives must be evaluated within the context of the entire system. A progression of analyses could include (1) computer simulation modeling of several alternatives to screen out unwieldy possibilities, (2) pilot scale evaluation of 3 to 5 of the most promising alternatives, using physical simulation of a defined system under controlled conditions, and (3) evaluation of the best alternative(s) in actual commercial operations.

Systems thinking has been used to evaluate in-field handling practices for southern peas (Shewfelt *et al.*, 1984) and broccoli (Shewfelt *et al.*, 1988), new marketing and distribution systems for broccoli (Ott *et al.*, 1989) and mangoes (Schoorl and Holt, 1986; Hubbert *et al.*, 1987), and orchard packing of fresh peaches (Garner *et al.*, 1987). Marketing and distribution of fresh-market items has benefitted from systems thinking, particularly in Australia and the United Kingdom.

Improvement is achieved by first viewing the entire system and determining which subsystems or handling steps are most likely to limit desired output. Resources then are devoted to improving output at these points. Modification at one point may have unexpected consequences elsewhere in the system. Thus, all modifications must be evaluated in the context of the entire system. The new system or process can serve as a basis for a quality management program.

Quality management programs have been proposed for the handling of fresh produce (Schoorl and Holt, 1982b,1985; Lidror and Prussia, 1990) and have been adopted for inspection of freshly harvested produce in Israel and evaluation of items for wholesale distribution in Australia and the United Kingdom.

As indicated previously, the use of a systems approach to address postharvest problems is a relatively recent development. Few successful examples are available and we know of no examples of a total systems approach. Five examples of the use of systems thinking to solve postharvest problems are provided.

B. Mechanization of Strawberry Harvest Operations for Processing

Unavailability and escalating labor costs led United States strawberry processors to consider mechanical harvesting to maintain competitiveness (Morris, 1990). Achieving consistently high quality frozen strawberries and strawberry jam required an integrated effort. Cultivars were identified that could produce a high percentage of fruit that would ripen within a 5-day optimum harvest period with acceptable color, firmness, soluble solids, and acidity in the harvested raw fruit. Fruit firmness was the most critical quality attribute for raw fruit (Morris *et al.*, 1979). These factors also needed to be coordinated carefully with production phase operations to insure that the plants and production practices were compatible with the harvesting equipment (Morris *et al.*, 1978,1985). Likewise, the limit of green fruit in puree (Sistrunk *et al.*, 1983), the length of storage permissible prior to processing (Morris *et al.*, 1985), and effects of added calcium on firmness (Morris *et al.*, 1985; Main *et al.*, 1986) have been established.

C. Distribution and Marketing of Fresh Mangoes

The traditional marketing of Queensland mangoes permitted a 10-day distribution life consisting of 5 days from harvest to wholesale and 5 days from wholesale to consumer. This market life was insufficient for export and the fragmentation of decision-making within the system did not permit the growers to meet expected increases in demand (Schoorl and Holt, 1986; Schoorl et al., 1988). Improved production phase practices, cool-chain management, and temperature monitoring increased harvest-to-wholesale life to 10 days and, thus, shelf life to 15 days. Implementation of full quality assurance programs, attention to harvest maturity, additional fungicide treatment, and product sourcing from cooperating producers added 6 more days from harvest to wholesale, permitting a 21-day shelf life. A 26-day shelf life could be achieved by stricter sorting operations, hiring of consultants or staff trained in postharvest handling, and use of controlled-atmosphere during transport. Each increase in the level of control provided greater access to markets at increased operating costs and reduced losses (Hubbert et al., 1987; Hubbert, 1989).

The Queensland projects emphasize (1) identification of priorities from a marketing perspective, (2) cooperation with growers and handlers to promote greater mutual understanding, (3) quantification of the benefits of technology from a price and quality perspective, (4) identification of all factors that may affect price, and (5) cultivation of an appreciation for consumer demand and marketing structure among growers (Hubbert, 1989).

D. Orchard Packing of Fresh Peaches

A limiting factor in fresh peach quality is the maturity at harvest (Shewfelt et al., 1987a). Harvesting too early results in fruit that will not develop the characteristic sweetness and volatile components to accompany the desirable postharvest changes in color and softness. Harvesting peaches too late results in premature ripening (soft fruit), increased bruise damage, and increased incidence of decay prior to purchase. Baumgartner (1985) recommends harvesting at a color stage 3–4, based on comparison of ground color with prepared chips, but a survey at one location showed that as much as 65% of the fruit was being harvested and shipped at an earlier stage (Jordan et al., 1986).

It was proposed that orchard packing of peaches directly from the tree into shipping containers could save 1 to 2 days in the system and reduce handling by eliminating the packinghouse operation. Thus, fruit could be harvested at a later stage of maturity and still arrive at the market at the same time (Garner et al., 1984). However, such a change would require a shift of workers from the packinghouse to the less favorable conditions in the field. Also, field packing eliminated operations that removed peach fuzz, which wholesale dealers considered an objectionable quality attribute. A later consumer evaluation study of peaches indicated that "fuzzy" peaches were not objectionable and that orchard packing had potential advantages for the early-season cultivars that do not develop flavor as full as that of later-season fruit. Orchard-packing operations did not seem to be meaningful for the later-season cultivars (Shewfelt et al., 1989).

E. Tomato Loss Simulation Model

Postharvest loss is a topic of global interest. However, standardized procedures are not available for measuring losses as produce progresses from farm to consumer. The ability to predict losses would help identify steps in the system that cause loss at later steps. Proposed changes in a handling step can be evaluated readily for their capability of reducing losses when the system is represented by a simulation model.

The physical loss of tomatoes occurring at the packinghouse, repacker, and the retail store was modeled (Campbell *et al.*, 1986) by representing the probability of loss at the three steps. Data was collected from commercial shipments for obtaining loss probabilities. The network option in the SIMAN computer simulation package (Systems Modeling Corporation, Sewickley, Pennsylvania), was used for developing a model that simulated the commercial system. Additional resolution was obtained by including probabilities of damage at steps additional to those at which losses are removed. Finally, alternative systems were evaluated by changing the probabilities of loss.

The model was run for several alternative systems such as perfect sorting at a packinghouse, 50% reduction in damage for fruit entering the packinghouse, field-pack of fruit with no damage or loss at the packinghouse, and no repacking steps. The best reduction in losses was for the field-pack system. An interesting result was evidence that previously damaged fruit provided cushioning for good fruit. Thus, some alternatives with reduced initial damage had less benefit than expected. Probability data from other commercial operations, from other regions, and from other seasons is needed to make the model more representative of the fresh tomato industry. Additionally, an unexpected result of the modeling effort was the sharply defined need to understand latent damage.

F. Soft Systems for Studying Import Companies

Growers, packers, and exporters could enhance their international marketing operations and opportunities of fresh horticultural crops through an improved understanding of the role of importers in the receiving country. Likewise, researchers could benefit from a systematic approach to learning about the activities of importing companies and other business in the postharvest system. Several new approaches have been developed for making decisions about problems that exist in messy ill-structured situations typical of real life (Rosenhead, 1989). SSM emphasizes the learning process and the need for multiple viewpoints of a problem situation (Checkland, 1981; Checkland and Scholes, 1990; Wilson, 1990). A textbook by Wilson and Morren (1990) presents agricultural applications of SSM.

The objectives of a study by Prussia and Hubbert (1991) were

- to learn the operations of fresh produce importing companies
- to evaluate the applicability of SSM for modeling multibusiness postharvest systems

Two British companies that specialize in fresh produce importing agreed to participate in a study that included using SSM to describe their operations. Both companies were major suppliers to supermarket chain stores. The corporate directors and other top level managers in both companies were interviewed. They identified five to seven of the main activities they controlled by giving verbs representing each activity. Questions were asked that provided input for each element of the CATWOE representation of their system and subsystems.

Over 15 CMs at various levels of detail were developed from interviews with managers in the two companies. The CM in Fig. 2 represents the total system as developed from input by all managers at one company. One possible RD for the total system is

> a system owned by the Board of Directors and administered by a Managing Director with the purpose of converting supermarket requirements into the decisions and actions necessary to deliver fresh produce consistently at the desired times, quantities, and qualities and at margins lower than the costs for supermarkets to internalize the operations given that profit margins are limited and it is difficult to buy consistently acceptable fresh produce.

The trading system shown in Fig. 3 expands the "trade" activity in Fig. 2 into a subsystem consisting of the activities buy, sell, and manage, which are expanded into subsystems. A possible RD for the trading system is

> a system owned and operated by the trade managers for the purpose of establishing and maintaining long-term relationships with senders and supermarkets while constrained by the need to make a profit.

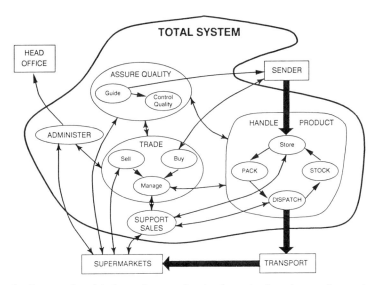

Figure 2 Conceptual model of a total system showing the system boundary, environment, and major activities in the system. [Reprinted by permission from Prussia, S. E., and Hubbert, C. A. (1991). *Acta. Hort.* **297**, 649–654.]

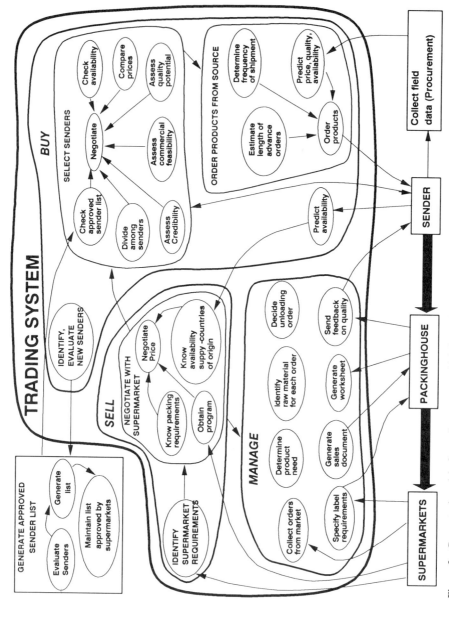

Figure 3 Conceptual model of a trading system, which is a subsystem of the total system. [Reprinted by permission from Prussia, S. E. and Hubbert, C. A. (1991). *Acta. Hort.* **297**, 649–654.]

SSM provided a framework for learning about the operations of importing companies through the development of RDs and CMs. The ability to combine the input from several managers into a CM of the total system was a notable contribution of SSM. Working as a team was critical to document learning about import companies successfully.

IV. Implications of a Systems Approach to Postharvest Handling

Use of a systems approach to evaluate postharvest handling systems will result in changes in handling practices. Impetus for change probably will come in the marketplace, with a single entrepreneur leading the way to a better understanding of the interactions of their system with other systems that make the postharvest continuum. Competitors then will be forced to adapt to stay in business. Even with the limited applications of systems thinking to postharvest handling to date, the following specific implications are evident: importance of latent damage, a perspective on quality perception, marketing and merchandising during retail distribution, management of crops in the production phase, and the consideration of "soft" issues.

A. Latent Damage

In some studies, systems thinking was used to develop an accounting system for quality loss during postharvest handling (Shewfelt et al., 1984,1986,1987a,b). The purpose was to identify the step(s) resulting in greatest losses so resources could be focused where they could be most effective. In work with southern peas (Shewfelt et al., 1984), this approach was effective, but with other commodities such as peaches (Shewfelt et al., 1987a) and tomatoes (Shewfelt et al., 1987b) it was less useful in identifying limiting steps. One problem was that latent damage was incurred at one step but not evident until a later step (Chapter 10, 11). In the accounting system being used, quality losses due to latent damage were being assessed at the step at which they became evident and not where they were incurred. Thus, without a knowledge of latent damage, false conclusions about the source of problems result.

An assessment of the existence and economic consequences of latent damage should be part of any analysis of a postharvest system. Identification of the causes of latent damage represents a potential to reduce total damage and, thus, improve quality, provides a point for monitoring within a quality management program, and suggests potential for nondestructive screening during packing to prevent shipment of material that eventually will be discarded (Chapter 13).

B. Quality

The primary value of systems thinking about the quality of fruits and vegetables is a greater emphasis on satisfying consumer needs and desires. Present systems appear to have developed primarily to serve the distributors and handlers, not to satisfy consumers. Left with little choice of a particular item on the retail shelf consumers can either select what is available or do without. Innovation to meet the desires of consumers tends to be discouraged within the primary distribution system because of lack of economic incentive. Specialty markets, roadside stands, and pick-your-own outlets offer more economic advantages in serving the market of consumers who are willing to seek out a higher quality item.

Systems thinking suggests that a greater emphasis on meeting consumer needs and desires will provide economic rewards within an integrated system. A greater responsiveness to these needs requires

- an understanding of quality at the time of both purchase and consumption
- increased emphasis on improved quality at sale, even when it may come in conflict with shelf life extension
- movement away from traditional grades and standards to programs of quality management

Food quality is defined as the "composite of those characteristics that differentiate individual units of a product and have significance in determining the degree of acceptability of that unit by the buyer" (Kramer and Twigg, 1970; see also Chapters 5 and 7 for a more detailed discussion). This definition emphasizes the importance of the buyer in determining quality, that quality is not the absolute but relative to what is available, and that quality may have a different meaning to different people.

Consumers evaluate quality of a selected fresh fruit or vegetable twice. An evaluation is made at the time of purchase using one set of criteria; another evaluation is made at the time of consumption using a different set of criteria. Purchase quality attributes are used by handlers within a system to evaluate grade and condition, to make "accept or reject" trading decisions, to help determine wholesale and retail prices, and to determine shelf life, as well as by retail store customers. Consumption quality attributes are associated most closely with "liking" and "acceptance" as the product is eaten by the consumer, and are thought to be important in repurchase decisions by consumers. Purchase quality attributes tend to be the ones that can be measured nondestructively whereas consumption quality attributes such as flavor and mouthfeel are necessarily destructive.

Potential effects of modifying handling conditions on purchase quality (Q_P) and consumption quality (Q_C) are shown for hypothetical nonclimacteric products in Fig. 4 and hypothetical climacteric fruits in Fig. 5. Such modifications, in turn, affect the purchase life (t_P) and consumption life (t_C) of the product. The quality scale in Fig. 4 is normalized by assigning a value of 1.0 to all quality attributes at harvest. In Fig. 5, the quality scale for Q_P and Q_{PM} is normalized by giving maximum quality value of 1.0. The Q_C curve also is normalized by giving a value of 1.0 to its maximum point whereas the Q_{CM} curves are scaled relative to the Q_C curve. In Figs. 4 and 5, quality is acceptable only if above the threshold value, q.

To capture full market price of a product, retail distributors must sell that product before Q_P reaches its minimal acceptable level, q. Thus, a driving force for maintenance of appropriate conditions and efficient distribution of products is an effort to sell the item at the retail outlet before the end of t_p. When making a purchase

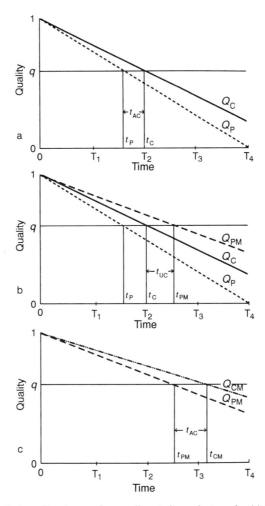

Figure 4 Hypothetical quality changes for nonclimacteric products under (a) typical handling and storage conditions and (b and c) modified conditions that extend both purchase life and consumption life, where Q_P is purchase quality, Q_C is consumption quality, q is lowest acceptable quality, Q_{PM} is Q_P under modified conditions, Q_{CM} is Q_C under modified conditions, t_P is purchase life, t_C is consumption life, t_{PM} is t_P under modified conditions, t_{CM} is t_C under modified condition, t_{AC} is acceptable home storage life, and t_{UC} is inferior quality zone.

decision, the consumer is likely to project a period of time (t_{AC}) after which the product would still be consumable, although it would not be acceptable for purchase.

Modification of the handling system has consequences for both quality and shelf life. For a nonclimacteric product (Fig. 4), modifications could result in maintenance of appearance characteristics to increase t_P to t_{PM} with little or no change in the development of off flavors (Fig. 4b). In this case, the consumer becomes vulnerable to a zone of inferior quality (t_{UC}) in which the product looks good but the consumption quality is unacceptable. Such experiences would be expected to affect

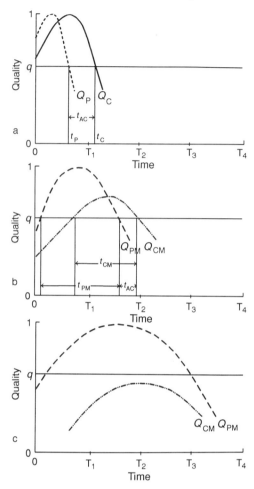

Figure 5 Hypothetical quality changes for climacteric fruits harvested at (a) advanced stage of ripening, (b) an earlier stage of ripening to extend purchase life, and (c) an immature stage. See Fig. 4 for description of symbols.

future decisions. Thus, it is important to modify handling systems to extend consumption as well as purchase life (Fig. 4c).

Climacteric fruits generate much more complex relationships (Fig. 5). When harvested after climacteric has been initiated (Fig. 5a), optimal color and flavor can be achieved at the cost of short t_P and t_C. Harvesting prior to onset of climacteric but at a fully mature state (Fig. 5b) may extend the life of the fruit at a cost of achieving full Q_C. If the fruits are picked in an immature state, acceptable Q_C may never be reached (Fig. 5c).

Differences between the changes in Q_P and in Q_C possibly explain the lack of flavor noted in many fresh-market tomatoes, peaches, and other products. Thus, when considering any proposed change in the handling system, it is essential to consider the effects on consumption quality in addition to those on shelf life (purchase life).

Systems thinking considers fresh fruits or vegetables to be food products, not raw agricultural commodities. As food products, these items should have consistent quality. Product specifications can be developed and shipments can be monitored for compliance. Because these products are senescing and, thus, perishable, management of quality is essential (Chapter 13). If expected handling conditions of the product are known, then predictive equations can be developed for quality degradation (Chapter 8) and "use by" dates can be provided.

C. Economics and Marketing

Systems thinking eventually leads to the conclusion that all product handling must be market driven, that is, all decisions within the handling system should be aimed at satisfying consumer desires. Thus, product quality becomes part of the market acceptability of that product. Market acceptability is influenced by many other factors including price, availability, merchandising techniques, product image, and consumer attitudes.

Supply and demand dictate that the price of an item is lowest when the quantity (and usually the quality) is highest. When shelf life extension can be managed, the quantity supplied can be distributed more evenly, thus allowing smaller fluctuations in prices.

D. Preharvest Factors

Systems thinking suggests that production phase decisions also should be market driven. Traditionally, growers have been independent in their decisions of which crops they will plant, cultivate, and harvest. If too many growers produce too much of an item, the quantity supplied increases and prices decrease. If prices decrease below the cost of production, the grower loses money and is less likely to grow that crop in subsequent seasons. If too many growers stop production, the quantity supplied decreases, forcing higher prices for the consumer.

Grower's cooperatives and marketing orders help maintain a healthy balance between quantity supplied and reasonable price. Market-driven decisions that surpass what, when, and where to plant include

- which cultivar to plant
- irrigation, fertilization, and pesticide application schedules, to maximize quality and shelf life
- cultural practices to adopt to provide both short- and long-term returns
- physiological maturity for optimal harvest time

These and other factors are covered in more detail in Chapter 4.

E. Soft Systems Methodologies

The use of soft system methodologies (SSM) provides a recognized approach for addressing the wide range of both technical and social issues that normally is involved in real situations. Typical applications of SSM are for improvements in a single company or one department in a company. However, Prussia and Hubbert (1991) proposed a method for modeling complete postharvest systems using SSM. Growers, packers, shippers, and others interacting with importers could develop individual CMs of importers. Comparisons with the CMs presented in this chapter could result in discussions leading to desirable changes. Likewise, importers could develop CMs for ideal exporting systems which would be compared with CMs developed for existing activities. Similar comparisons could be made for grocery chain stores and importers. Such comparisons would identify activities that might need to be changed or developed. How to accomplish these changes would be decided by the owner of the system.

V. Summary

Application of the systems approach to postharvest systems requires clarification of the starting and ending points for both fresh-market produce and produce delivered to food processing plants. No individual system exists; differences depend on the commodity and technologies that are available or appropriate. A systems approach must be placed in the context of the historical perspective of scientific, engineering, and systems approaches. The strength of a systems approach is the ability to integrate diverse functions into a functioning system. Soft systems meet the need for a methodology to address poorly defined problem situations in which many viewpoints are possible. The usefulness of systems approaches is their ability to identify critical steps in a system at which resources can be used best.

Bibliography

Baumgartner, R. A. (1985). Harvesting quality peaches. *Proc. Nat. Peach Council* p. 17.

Breene, W. M. (1983). Raw product selection: Factors to be considered. In "A New Market Nutrition and Quality," pp. 1–14. Food Proc. Inst., Washington, D.C.

Brennan, P. S., and Shewfelt, R. L (1989). Effect of cooling delay at harvest on broccoli quality during postharvest storage. *J. Food Qual.* **12,** 13–22.

Bourne, M. C. (1977). Postharvest food losses—the neglected dimension in increasing the world food supply. Cornell Institute of Agriculture. Mimeo 53. New York State College of Agriculture and Life Science, Cornell University, Ithaca.

Bourne, M. C. (1984). Proper care of food after harvest. *In* "World Food Issues," 2nd edition, (M. Drusdoff, ed.). pp. S1–S4. Center for Analysis of World Food, Issues, Ithaca, New York.

Campbell, D., Prussia, S. E., Shewfelt, R. L, Hurst, W. C., and Jordan, J. L. (1986). Postharvest handling systems analysis of fresh produce. *In "Engineering for Potatoes,"* (B. F. Cargill, ed.), pp. 350–392. American Society of Agricultural Engineers, St. Joseph, Michigan.

Checkland, P. (1981). "Systems Thinking, Systems Practice." John Wiley & Sons, New York.

Checkland, P. B., and Scholes, J. (1990). "Soft Systems Methodology in Action." John Wiley & Sons, Chichester.

Garner, R. G., and Dennison, R. A. (1979). "Cooperative Research in Postharvest Technology." USDA Publ. No. ARM-H-3. U.S. Govt. Printing Office, Washington, D.C.

Garner, J. C., Prussia, S. E. Shewfelt, R. L., and Jordan, J. L. (1987). "Peach Quality after Delays in Cooling." American Society of Agricultural Engineers, St. Joseph, Michigan. Tech. Paper #87-6501.

Hardenburg, R. E., Watada, A. E, and Wang, C. Y. (1986). "The Commercial Storage of Fruits, Vegetables, and Florist and Nursery Stocks." USDA Agriculture Handbook No. 66. U.S. Govt. Printing Office, Washington, D.C.

Holt, J. E., and Schoorl, D. (1981). Fruit packaging and handling distribution systems: An evaluation method. *Agric. Systems* **7**, 209–218.

Horsfield, B. C, Fridley, R. B., and Claypool, L. L. (1971). Systems analysis of postharvest handling of mechanically harvested peaches. *Trans ASAE* **14**, 1040–1050.

Hubbert, C. A. (1989). Technical aspects of marketing fresh fruits and vegetables—The Queensland experience. *Aspects Appl. Biol.* **20**, 23–31.

Hubbert, C. A. Johnson, G. I., Muirhead, I. F., and Ledger, S. N. (1987). Postharvest handling of mangoes—Resource booklet. Queensland Department of Primary Industries Publ. No. RQ 87001. Brisbane, Queensland.

Institute of Food Technologists (1990). Quality of fruits and vegetables. A scientific status summary by the Institute of Food Technologists' Expert Panel on Food Safety and Nutrition. *Food Technol.* **44(6)**, 99–106.

Jackman, R. L., Yada, R. Y., Marangoni, A., Parkin, K. L., and Stanley, D. W. (1988). Chilling injury, a review of quality aspects. *J. Food Qual.* **11**, 253–278.

Jordan, J. L. Shewfelt, R. L., Prussia, S. E., and Hurst, W. C. (1985). Estimating implicit marginal prices of quality characteristics of tomatoes. *South. J. Agric. Econ.* **17**, 139–146.

Jordan, J. L., Shewfelt, R. L., and Prussia, S. E. (1986). Postharvest peach research at the University of Georgia. *Proc. 7th Ann. S.E. Peach Conv.* 49–50.

Jordan, J. L., Prussia, S. E., Shewfelt, R. L., Thai, C. N., and Mongelli, R. (1990). "Transportation Management, Postharvest Quality, and Shelf-Life Extension of Lettuce." University of Georgia Agricultural Research Bulletin No. 386. Athens, Georgia.

Kader, A. A. (1985). "Postharvest Technology of Horticultural Crops." Agriculture and Natural Resources Publications, University of California, Berkeley.

Kasmire, R. F. (1985). Handling of horticultural crops at destination markets. *In* "Postharvest Technology of Horticultural Crops" (A. A. Kader ed.), pp. 111–117. Agriculture and Natural Resources Publications, University of California, Berkeley.

Kramer, A., and Twigg, B. A. (1970). "Quality Control for the Food Industry," 3d Ed. Vol. 1. AVI/Van Nostrand Reinhold, New York.

Labios, E. V., Esguerra, E. G., and Bautista, O. K. (1984). Extent and nature of postharvest losses in tomato cv. "Improved Pope" at a major handling route. *Postharvest Res. Notes* **1(4)**, 104–105.

Lidror, A., and Prussia, S. E. (1990). Improving quality assurance techniques for producing and handling agricultural crops. *J. Food Qual.* **13**, 171–184.

Main, G. L., Morris, J. R., and Wehunt, E. J. (1986). Effect of reprocessing treatments on the firmness and quality characteristics of whole and sliced strawberries after freezing and thermal processing. *J. Food Sci.* **51**, 291–394.

Meyers, J. B., Prussia, S. E., Thai, C. N., Sadosky, T. L., and Campbell, D. T. (1990). Visual inspection of agricultural products moving along sorting tables. *Trans. ASAE* **33**, 367–372.

Monroe, G. E., and O'Brien, M. (1983). Postharvest functions. *In* "Principles and Practices for Harvesting and Handling Fruits and Nuts" (M. O'Brien, B. F. Cargill, and R. B. Fridley, eds.), pp. 307–375. AVI/Van Nostrand, New York.

Morris, J. R. (1990). Fruit and vegetable harvest mechanization. *Food Technol.* **44(2)**, 97–101.

Morris, J. R., and Sims, C. A. (1985). Effects of cultivar, soil moisture, and hedge height on yield and quality of machine-harvested erect blackberries. *J. Amer. Soc. Hort. Sci.* **110**, 722–725.

Morris, J. R., Kattan, A. A., Nelson, G. S., and Cawthon, P. L. (1978). Developing a mechanized system for production, harvesting, and handling of strawberries. *HortScience* **13**, 413–422.

Morris, J. R., Sistrunk, W. A., Kattan, A. A., Nelson, G. S., and Cawthon, D. L. (1979). Quality of mechanically harvested strawberries for processing. *Food Technol.* **33(5)**, 92–98.

Morris, J. R., Cawthon, D. L., and Fleming, J. W. (1980). Effects of high rates of potassium fertilization on raw product quality and changes in pH and acidity during storage of Concord grape juice. *Amer. J. Enol. Vitic.* **31**, 323–328.

Morris, J. R., Sims, C. A., and Cawthon, D. L. (1985a). Effects of production systems, plant populations, and harvest dates on yield and quality of machine-harvested strawberries. *J. Amer. Soc. Hort. Sci.* **110**, 718–721.

Morris, J. R., Sistrunk, W. A., Sims, C. A., Main, G. L., and Wehunt, E. J. (1985b). Effects of cultivar, postharvest storage, preprocessing dip treatments, and style of pack on the processing quality of strawberries. *J. Amer. Soc. Hort Sci.* **110**, 172–177.

Morris, J. R., Sistrunk, W. A., Junek, J., and Sims, C. A. (1986). Effects of fruit maturity, juice storage, and juice extraction temperature on quality of 'Concord' grape juice. *J. Amer. Soc. Hort. Sci.* **111**, 742–746.

Musa, S. K. (1984). Reduction of post-harvest losses in vegetables and fruits in a developing country. *In* "Food Science and Technology: Present Status and Future Direction" (J. V. McLaughlin and B. M. McKenna, eds.), Vol. 5, pp. 165–174. Boole Press, Dublin.

National Academy of Science (1978). "Postharvest Food Losses in Developing Countries." National Academy Press, Washington, D.C.

National Research Council (1989). "Alternative Agriculture." National Academy Press, Washington, D.C.

O'Brien, M., and Berlage, A. G. (1983). Manual harvesting of fruits and nuts and use of mechanical aids. *In* "Principles and Practices for Harvesting and Handling Fruits and Nuts" (M. O'Brien, B. F. Cargill, and R. B. Fridley, eds.), p. 73–117. AVI/Van Nostrand, New York.

O'Brien, M., and Gaffney, J. J. (1983). Postharvest handling and transport operations. *In* "Principles and Practices for Harvesting and Handling Fruits and Nuts" (M. O'Brien, B. F. Cargill, and R. B. Fridley, eds.), pp. 413–470. AVI/Van Nostrand, New York.

O'Brien, M., Fridley, R. B., and Claypool, L. L. (1978). Food losses in harvest and handling system for fruits and vegetables. *Trans. ASAE* **21**, 386–390.

Ott, S., Jordan, J. L., Shewfelt, R. L., Prussia, S. E. Beverly, R. B. , and Mongelli, R. (1989). "An Economic Analysis of Postharvest Handling Systems for Fresh Broccoli." University of Georgia Agricultural Research Bulletin No. 387. Athens, Georgia.

Pedigo, L. P., Hutchins, S. H., and Higley, L. G. (1986). Economic injury levels in theory and practice. *Ann. Rev. Entomol.* **31**, 341–368.

Prussia, E. S., and Hubbert, C. (1991). "Soft System Methodologies for Investigating International Postharvest Systems." ASAE Techn. Paper 91-7050.

Prussia, S. E., and Woodroof, J. G. (1986). Harvesting, handling, and holding fruit. *In* "Commercial Fruit Processing" (J. G. Woodroof and B. S. Luh, eds.). pp. 25–97. AVI/Van Nostrand Reinhold, New York.

Prussia, S. E., Jordan, J. L., Shewfelt, R. L., and Beverly, R. B. (1986). "A Systems Approach for Interdisciplinary Postharvest Research on Horticulture Crops." Georgia Agricultural Experimental Station Research Report No. 514. Athens, Georgia.

Quamme, H. A., and Gray, J. I. (1985). Pear fruit quality and factors that condition it. *In* "Evaluation of Quality of Fruits and Vegetables" (H. E. Pattee, ed.), p. 47–61. AVI/Van Nostrand, New York.

Rosenhead, J. (ed.) (1989). "Rational Analysis for a Problematic World." John Wiley & Sons, Chichester.

Resurreccion, A. V. A. (1986). Consumer use patterns for fresh and processed vegetable products. *J. Cons. Studies Home Econ.* **10**, 317–332.

Salunkhe, D. K., and Desai, B. B. (1984). "Postharvest Biotechnology of Fruits," Vol. 1 CRC Press, Boca Raton, Florida.

Schoorl, D., and Holt, J. E. (1982a). Fresh fruit and vegetable distribution—Management of quality. *Sci. Hort.* **17**, 1–8.

Schoorl, D., and Holt, J. E. (1982b). The evaluation of post-harvest disease control measures using a time–temperature framework. *Agric. Systems* **13**, 97–111.

Schoorl, D., and Holt, J. E. (1985). A methodology for the management of quality of horticultural distribution. *Agric. Systems* **16**, 199–216.

Schoorl, D., and Holt, J. E. (1986). An analysis of mango production and marketing in Australia. *Agric. Systems* **21**, 171–188.

Schoorl, D., and Holt, J. E., and Meyer, D. G. (1988). Marketing models to aid decision making in horticulture. *Acta Hort.* **247**, 284.

Shewfelt, R. L. (1987). Quality of minimally processed fruits and vegetables. *J. Food Qual.* **10**, 143–156.

Shewfelt, R. L. (1990). Sources of variation in the nutrient content of agricultural commodities from the farm to the consumer. *J. Food Qual.* **13**, 37–54.

Shewfelt, R. L., Prussia, S. E., Hurst, W. C, and Jordan, J. L. (1984). A systems approach to the evaluation of changes in quality during postharvest handling of southern peas. *J. Food Sci.* **50,** 769–772.

Shewfelt, R. L., Prussia, S. E., Jordan, J. L., Hurst, W. C., and Resurreccion, A. V. A. (1986). A systems analysis of postharvest handling of fresh snapbeans. *HortScience* **21,** 470–472.

Shewfelt, R. L., Myers, S. C., Prussia, S. E., and Jordan, J. L. (1987a). Quality of fresh-market peaches within the postharvest handling system. *J. Food Sci.* **52,** 361–364.

Shewfelt, R. L., Prussia, S. E., Resurreccion, A. V. A., Hurst, W. C., and Campbell, D. T. (1987b). Quality of vine-ripened tomatoes within the postharvest handling system. *J. Food Sci.* **52,** 661–672.

Shewfelt, R. L., Garner, J. C., Beverly, R. B., Prussia, S. E., Jordan, J. L., and Miller, J. (1988). Effect of field handling methods on broccoli quality within the postharvest handling system. Presented at the Annual Meeting of the Institute of Food Technologists. June, 1988, New Orleans.

Shewfelt, R. L., Garner, J. C., Resurreccion, A. V. A., and Prussia, S. E. (1989). Consumer acceptability of 'fuzzy' peaches. Presented at the Annual Meeting of the Institute of Food Technologists. June, 1989, Chicago.

Sims, C. A., and Morris, J. R. (1982). Effects of cultivar, irrigation, and ethephon on the yield, harvest distribution, and quality of machine-harvested blackberries. *J. Amer. Soc. Hort. Sci.* **107,** 542–547.

Sistrunk, W. A. (1985). Peach quality assessment: Fresh and processed. *In* "Evaluation of Quality of Fruits and Vegetables" (H. E. Pattee, ed.), pp. 1–46. AVI/Van Nostrand, New York.

Sistrunk, W. A., and Morris, J. R. (1985). Strawberry quality influence of cultural and environmental factors. *In* "Evaluation of Quality of Fruits and Vegetables" (H. E. Pattee, ed.), pp. 217–256. AVI/Van Nostrand, New York.

Sistrunk, W. A., Wang. R. C., and Morris, J. R. (1983). Effect of combining mechanically harvested green and ripe puree and sliced fruit, processing methodology and frozen storage on quality of strawberries. *J. Food Sci.* **48,** 1609–1612, 1616.

Smock, R. M. (1979). Controlled atmosphere storage of fruits. *Hort. Rev.* **1,** 301–336.

Stevens, M. A. (1985). Tomato flavor: Effects of genotype, cultural practices, and maturity at picking. *In* "Evaluation of Quality of Fruits and Vegetables" (H. E. Pattee, ed.), pp. 367–386. AVI/Van Nostrand, New York.

Variyam, J. N., Jordan, J. L., Beverly, R. B., Shewfelt, R. L., and Prussia, S. E. (1988). An application of models for survival data to postharvest systems evaluation. *Proc. Fla. State Hort. Soc.* **101,** 200–202.

Whalon, M. E., and Croft, B. A. (1984). Apple IPM implementation in North America. *Ann. Rev. Entomol.* **29,** 435–470.

Wills, R. B. H., McGlasson, W. D., Graham, D., Lee, T. H., and Hall, E. G. (1989). "Postharvest: An Introduction to the Physiology and Handling of Fruits and Vegetables." AVI/Van Nostrand, New York.

Wilson, B. (1990). "Systems: Concepts, Methodologies, and Applications," 2d ed. John Wiley & Sons, New York.

Wilson, K., and Morren, E. B. G., Jr. (1990). "Systems Approaches for Improvement in Agriculture and Resource Management." Macmillan, New York.

CHAPTER 4

PREHARVEST PHYSIOLOGICAL AND CULTURAL EFFECTS ON POSTHARVEST QUALITY

R. B. Beverly, J. G. Latimer, and D. A. Smittle

I. Introduction

The quality of fresh fruits and vegetables offered to consumers is constrained by the level of quality achieved at harvest, and generally cannot be improved by postharvest handling. Myriad preharvest genetic and environmental factors affect the growth, development, and final quality of fresh fruits and vegetables (Pantastico, 1975; Mengel, 1979; Shewfelt et al., 1986; Monselise and Goren, 1987; Lill et al., 1989; Shewfelt, 1990). Many of these factors are beyond the control of production managers. Understanding the complexity of quality production and the inherent variability in living fruits and vegetables will afford the postharvest worker a greater appreciation for the challenges and limitations growers face in delivering consistently high-quality produce to the marketplace. At the same time, understanding the quality needs of consumers and postharvest handlers will help growers produce fruits and vegetables that are acceptable in the marketplace. The objective of the grower of maximizing marketability or profitability depends on both productivity and quality, as defined by initial buyers. Since quality and yield may be related inversely, growers must balance these factors to maximize net income.

The effects of preharvest factors on postharvest quality operate on both individual plants and plant communities. On the individual plant scale, genetic and environmental factors combine to affect plant physiology in such processes as energy fixation and metabolism, water and nutrient uptake, and carbohydrate partitioning. However, growers manage entire populations of plants rather than individuals, and cultural practices on the field scale affect physiological processes at the plant level. We will show how physiological processes, mitigated by field management, determine quality of plant products.

II. Whole Plant Model: Physiological Responses to Environmental Effects

The production of quality produce requires the adequate growth, development, and yield of fruits and vegetables. Therefore, much of this discussion will address factors affecting plant growth and development, with specific attention paid to factors that directly affect quality. A detailed review of the processes involved is beyond the scope of this chapter; interested readers should refer to general plant physiology texts for more information.

To understand whole-plant processes that affect crop yield and postharvest quality, consider plant growth and development as the integration of the flow of energy, water, and nutrients through the plant. For simplicity, consider a plant to consist of photosynthetic, conductive, root, and fruit tissues (Fig. 1). (This discussion will assume that the harvested plant product is a fruit, although the same principles apply whether the plant product of interest is a reproductive structure, a vegetative structure, or a storage organ, that is, a fruit, seed, root, tuber, stem, or leaf.)

Energy, water, and nutrient movements through the plant are relatively simple and generally unidirectional prior to the formation of fruits or storage organs (Fig. 1). Energy is fixed by the plant through photosynthesis, then is transported in the form of photosynthates (sugars or organic acids) down the phloem to the roots. Energy exits the plant from all tissues in the form of heat of respiration. Water and nutrients flow through the soil and are taken up by roots, then transported through the xylem primarily to the leaves. In leaves, nutrients are incorporated into structural organs, remain in the cytoplasm, or accumulate in the vacuole, whereas water is lost by transpiration.

As the harvested plant product begins to develop, flows become more complex and partitioning becomes important. The fruit competes with roots for photosynthates, nutrients, and water. Nutrients may be remobilized from leaves and stems into the developing fruit. The fruit also loses energy through respiration and water through transpiration.

Fruit quality at harvest and at consumption depends on the combined net effects of energy, water, and nutrient flows into and out of the fruit. The same principles apply to postharvest quality changes, since continued respiration and transpiration constitute net losses of energy and water from the harvested fruit that result in quality loss.

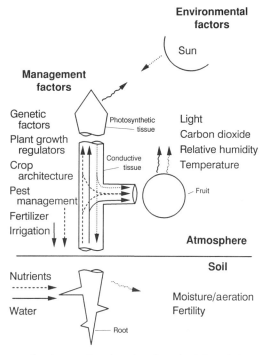

Figure 1 Flows of energy ($\cdots$), water (——), and nutrients (----) through the environment and plant. Management factors are designed to optimize yield and quality within constraints imposed by environmental conditions.

Environmental factors affecting energy, water, and nutrient flows within the plant (Fig. 1) include atmospheric and edaphic (soil) factors. Atmospheric environmental conditions include solar radiation, carbon dioxide concentration, relative humidity, and temperature, which are virtually unmanageable in field production. Soil factors include soil physical conditions, which determine water and oxygen supply to roots, and soil fertility, which determines the supply of nutrients to roots. The soil environment is more manageable than the atmospheric environment, albeit indirectly. The environmental conditions establish constraints within which growers must operate.

A. Energy Flows

Through photosynthesis, green plants convert atmospheric carbon dioxide to reduced carbon compounds, using light energy collected by chlorophyll. Accordingly, the amounts of light and carbon dioxide available influence the rate of photosynthesis. Temperatures higher or lower than the optimum reduce the assimilation of carbon dioxide by photosynthesis (Björkman et al., 1978; Farquhar and Sharkey, 1982).

In order to affect yield and quality, photosynthates must be transported from leaves to developing fruits (Daie, 1988), since photosynthesis in the fruit is generally negligible. Jones (1981) found that photosynthesis in apple fruits never exceeded respiration, resulting in a net loss of photosynthates in the fruit. Similarly, Ho (1988) reported that photosynthesis in tomato fruit accounted for less than 10% of its dry weight. This allocation of energy to fruits comes at the expense of the rest of the plant (especially roots), which continues to require energy to function. This complicated issue of carbohydrate partitioning and long-distance transport has been reviewed in detail (Beevers, 1985; Daie, 1985; Patrick, 1988). Understanding and manipulating carbohydrate partitioning may provide the key to enhancing crop yields and quality.

Energy transported to fruits in the form of photosynthates can be lost from the fruit by respiration, which is highly dependent on temperature (Gent and Enoch, 1983). Since temperature generally remains higher at night in humid regions than in arid regions, higher nighttime respiration rates could account for the lower accumulation of soluble solids and yields in humid regions. For example, sunny days and cool nights improve the flavor of strawberries (Sistrunk and Morris, 1985).

B. Water Flows

The flow of water into fruits provides the force for the cell expansion necessary for fruit growth (Lee, 1990). As important as this process is, however, the plant exerts relatively little control over water movement into and through the plant. Water flows from areas of high potential energy (high water potential) to areas of low potential energy (low water potential), or from "wetter" to "dryer" areas. Water moves from the bulk soil into plant roots, then through the stem to leaves and fruits, and finally into the atmosphere in response to decreasing water potential (Jones et al., 1985). The flow of water from the soil and through the plant is driven by the low water potential in the atmosphere (Tibbitts, 1979; Grange and Hand, 1987).

At low relative humidity, a large water potential gradient exists between wet leaf tissues and relatively dry air. The difference between the vapor pressure of the air and the vapor pressure of the leaf (the vapor pressure deficit) causes the loss of water from the leaves that is known as transpiration. Water loss from leaves creates an attraction for more water from the stems, which is transmitted to the roots, and ultimately drives water uptake from the soil. As the soil dries or salts accumulate in the soil, the plant must lower the water potential in other ways to absorb water into tissues. One such method is the accumulation of solutes such as sugars or inorganic ions, a process known as osmotic adjustment. This process requires energy either to accumulate inorganic ions against a potential energy gradient, or in the formation of photosynthates that serve as osmotica.

When the vapor pressure deficit becomes excessive, due to either low relative humidity or limited water supply to leaves, stomata in the leaf surface close and limit further water loss. This stomatal closure also limits the exchange of carbon dioxide between the atmosphere and the leaf tissue, thereby limiting photosynthesis. Since leaves have more stomata per unit area than fruits, more water tends to flow to leaves. This effect is exacerbated under conditions of high vapor pressure deficits, and can limit the supply of water and important nutrients to fruits.

C. Nutrient Flows

In addition to carbon, hydrogen, and oxygen, which the plant derives in the form of atmospheric carbon dioxide, oxygen, and soil water, the plant requires 13 essential inorganic nutrients to function properly. Plant nutrients affect energy flow in the plant since they are essential for photosynthesis and for regulating energy metabolism and carbohydrate transport. Nutrients regulate water flow by affecting the opening of stomata and by osmotic adjustment. Some nutrients are essential to the structural integrity of the plant by affecting protein formation or the integrity of cell walls and membranes. Many of the nutrients are required for the formation and proper function of enzymes.

As for energy and water, the movement of nutrients into fruits depends on the supply of available nutrients and their partitioning among various sinks within the plant (Patrick, 1988). Nutrient uptake from the soil is generally against a potential energy gradient, so it requires the expenditure of energy by the roots. Thus, a trade off develops between the supply of energy and the supply of nutrients to developing fruits. Since respiration is required for nutrient uptake, the process is quite sensitive to soil temperature and oxygen supply.

The supply of nutrients to fruits includes both current uptake by roots and remobilization of nutrients previously assimilated into other plant tissues. Some nutrients, most notably nitrogen, can be remobilized from older leaf tissue to younger developing leaves and fruits under conditions of limited supply. However, this depletion of leaf nutrients can lead to leaf senescence, thereby limiting the ultimate supply of energy to developing fruits (Sinclair and deWit, 1975). In contrast, many nutrients cannot be remobilized from older to younger tissues. Nutrients that are not mobile in the phloem must be supplied to developing fruits from current uptake, so partitioning becomes an important issue. Calcium and boron move to leaves and fruits in the xylem, their movement determined by the movement of water in the transpiration stream (Tibbitts, 1979), but they are not redistributed in the phloem. Hence, factors such as low relative humidity that limit the partitioning of water to fruits also affect partitioning of calcium and boron, and can cause nutritional disorders such as blossom end rot (Tibbitts, 1979; Grange and Hand, 1987).

D. Environmental Effects on the Rate of Growth and Development

The growth and development of many fruits and vegetables progresses according to cumulative temperature effects (Kinet *et al.*, 1985). High temperatures during the vegetative stage hasten growth and reproductive maturity. However, excessive growth rates, as occur in cool-season vegetables under high temperatures, can result in misshapen produce and unpleasant flavors. Since the stage of maturity at harvest profoundly affects the quality of most fruits and vegetables, careful attention to this rate of development is crucial.

The most profound effect of growth and development on produce quality is the initiation and completion of the reproductive cycle. The effect of flower initiation and fruiting on produce quality depends on whether the harvested plant product is a vegetative or reproductive structure. Many of the cool-season vegetables, such as lettuce, cabbage, carrot, celery, and onion, are grown for vegetative or storage organs. For these crops, the initiation of flowering represents a serious loss of quality because of flower stalk elongation, called bolting. These plants require a period of cool weather before flowering occurs, a process called vernalization. After being vernalized, these plants flower in response to lengthening days (or shortening nights), so they are classified as long-day plants. Preventing quality deterioration as a result of bolting in these crops requires planting late enough in the spring to avoid sufficient cool weather to complete vernalization (Kinet *et al.*, 1985) or harvesting before the critical day length triggers flowering. A related quality defect, buttoning in cauliflower, occurs when seedlings are exposed to sufficient cool weather to initiate curd development. The curds may remain small or develop to marketable size, but insufficient leaf growth prevents proper blanching (Skapski and Oyer, 1964).

On the other hand, flowering and fruiting must occur if the harvested product is a fruit or seed. Temperature can have chronic (cumulative) or acute (short-term) effects on flower initiation and viability in fruit and vegetable crops. For example, peaches require a minimum number of chill units, or hours below 6°C, to initiate budbreak (Kinet *et al.*, 1985). The chilling requirement differs among cultivars, so only low chill-requiring cultivars will produce fruit in warmer climates. Acute temperature effects also can reduce flower viability. Freezing temperatures after budbreak severely limit yield or damage the quality of spring-flowering fruits (Kinet *et al.*, 1985). Similarly, extremely high temperatures can result in poor pollen production or viability and flower abortion in many vegetable crops (Picken, 1984). Determinate crops such as corn, which have only a short reproductive period in which to set viable fruits, are especially sensitive to acute high-temperature stress during the reproductive period. Indeterminate crops such as tomato may have a temporary lapse in fruit set due to high temperatures, but can continue to set additional fruits on subsequent blooms after the stress passes. This type of fruiting pattern may have only slight impact on the quality of selectively hand-harvested fruits such as fresh-market tomatoes. However, quality of indeterminate crops such as snap beans, which are once-over harvested at an immature stage of development, may be reduced drastically.

The amount of light striking leaves of fruiting branches affects flower initiation and development in apple, apricot, and several other woody fruit species (Kinet *et al.*, 1985). Similarly, high light intensity is required for flower initiation and fruit set in tomato (Picken, 1984; Kinet *et al.*, 1985). Flower initiation also responds to drought stress, mineral nutrition, carbon dioxide, previous fruit load, and growth regulators such as ethephon and daminozide (Picken, 1984; Kinet *et al.*, 1985; Forshey and Elfving, 1989).

E.   Implications of Energy, Water, and Nutrient Imbalances

Since the integrated effects of energy, water, and nutrient flows into the harvested plant product determine fruit and vegetable quality, how do imbalances among these processes develop, and how do they reduce quality?

Excess energy relative to water in fruits can result from high photosynthetic rates or low respiration rates, combined with dry or saline soil (low water supply) or low relative humidity (high transpiration rate). In mild cases, this condition causes the accumulation of photosynthates in fruit, leading to high total soluble solids, which usually improves quality. However, insufficient water can reduce fruit enlargement, and ultimately can lead to wilting, which is a serious defect in most fruits.

Excess energy relative to nutrients in fruits can result from high net photosynthesis combined with limited nutrient uptake or partitioning to fruits. This situation leads to an insufficient supply of nutrients to support the fruit growth that would be possible with the available photosynthate supply. The most obvious example of fruits outgrowing their nutrient supply is the breakdown of developing tissue observed in calcium deficiency.

Excess water with respect to the energy supply in fruits results from high water availability or low transpiration rates coupled with low net photosynthetic rates. A high moisture content in fruit tends to dilute the soluble solids present, leading to lower flavor intensity. An extreme form of quality defect occurs when water pressure increases rapidly inside plant structures. For example, during extended dry periods, soluble solids accumulate within plant structures such as fruits. When the water supply increases suddenly, for example, by irrigation or rainfall, water quickly moves into the plant cells in response to the osmotic gradient. The rapid rise in hydrostatic pressure can rupture cell membranes and walls, leading to splitting of fruits and other structures.

An excessive supply of nutrients relative to photosynthates in fruits can develop when the rate of nutrient assimilation is high relative to net photosynthesis. In this case, product quality can suffer because of the accumulation of nutrients in the fruits to levels that are toxic either to the plant or to humans consuming the fruits. One example is the accumulation of excess nitrate nitrogen when excess nitrogen fertilizer is applied. Although not strictly nutrients, many heavy metals can accumulate to potentially harmful levels when materials such as municipal sludge, mine tailings, or fly ash (Wadge and Hutton, 1986) have been applied to the soil.

Nutrient excesses relative to water can occur when excess nutrient availability coincides with limited supplies of water or high transpiration rates. Salinity stress results from excess application of fertilizers, especially forms containing sodium and chloride. These solutes can accumulate in plant tissue, especially at the extremes of the transpiration stream, and cause tissue damage such as tip and marginal leaf necrosis. When the harvested plant product is leaf tissue, for example, leafy greens, the detrimental effect on quality can be severe.

Finally, it is important to appreciate how interdependent the environmental factors and whole plant responses are. Light is intercepted and used only when water and nutrient supplies are sufficient to prevent stomatal closure and support metabolic functions. Relative humidity drives the uptake and partitioning of water within the plant, and coincidentally determines the partitioning of calcium and boron. Since nutrient uptake depends on respiration, it diminishes under conditions of low soil temperature, low oxygen supply caused by excessive soil moisture, or an inadequate energy supply due to limited photosynthesis. These few examples serve to illustrate that even understanding the processes that determine fruit and vegetable quality is difficult; managing those processes is even more challenging.

III. Whole Field Model: Cultural Practices to Optimize Produce Quality

Whereas plants integrate their individual environmental conditions into physiological responses as dictated by their particular genetic codes, farmers must manage entire populations of plants. This field-scale management requires the integration of information and uncertainty regarding legal, socioeconomic, environmental, and biological factors that affect yield, quality, and profitability (Table I). In contrast to the unmanageable environmental factors listed in Fig. 1, farmers attempt to optimize yield and quality indirectly through genetic management, cultural practices, pest management, and soil management. Temperature regimes can be managed passively to a certain extent by varying planting dates to alter the temperature conditions to which crops are exposed (Smittle, 1986). Also, effects of temperature extremes can be mitigated by protective strategies such as irrigation, wind machines, or floating row covers to prevent frost damage. Similarly, overhead irrigation decreases air temperature slightly to reduce high-temperature effects. Still, growers cannot increase or decrease temperature greatly on a field scale, so temperature is considered an unmanageable environmental factor. Further, light interception and capture depend on plant canopy architecture. Nonetheless, even pruning is considered an indirect influence: pruning enhances light penetration into the canopy, but does not increase the light supply to the tree directly. Similarly, applying fertilizer increases the supply of nutrients in the soil, but chemical transformations and soil conditions affect root uptake of those nutrients. Thus, it is important to realize that growers use only indirect interventions in attempts to optimize the quality of an

Table I

Examples of Considerations Affecting Crop Production Management Decisions on a Field Scale

Decision domain	Considerations		
	Legal and socioeconomic	Environmental	Biological
Crop and cultivar selection	Profitable markets; government support programs	Temperature extremes; length of growing season; soil conditions	Crop adaptation; pest resistance
Land preparation	Cost of tillage equipment and energy	Risk of erosion; risk of decline in soil productivity	Soil biology
Crop establishment timing and method: direct seed or transplant	Price expectations at harvest; cost of seeds versus transplants; value of earlier harvest	Risk of frost; soil temperature and moisture; soil strength	Germination; emergence; growth rate
Irrigation management: surface, sprinkler, or drip	Costs and efficiency of alternative methods; availability of water and competition by other users; risk of non-point-source pollution	Temperature; humidity; rainfall amount and distribution; light intensity; soil moisture characteristics	Water required by crop; water available to crop; crop water use efficiency
Fertility management: preplant or sidedress	Cost of material and application; risk of non-point-source pollution; risk of reduced yield or quality	Soil chemical conditions; soil moisture and aeration; soil temperature	Soil nutrient transformations; plant nutrient uptake; growth rate; crop residue contribution to subsequent crops
Pest management: pesticides, biological control, or cultural practices	Cost and availability of materials and application; risk of reduced yield or quality; risk of non-point source pollution; risk of adverse public perception	Temperature; humidity; rainfall	Weed, insect, and disease populations; populations of pest predators or parasites
Harvest timing and method: hand harvest or mechanical harvest	Market price and expectations; market demands for quality throughout marketing; cost of harvest machinery; cost and availability of harvest labor; product recovery by hand and mechanical harvest; quality and value of hand versus mechanically harvested product	Temperature; rainfall; light intensity; humidity	Rate of maturation; risk of loss due to overmaturity or pest damage

entire population of biological individuals that are inherently variable because of their differing genetic and environmental histories.

The following sections will integrate general principles of physiology with practical aspects of plant growth and development, yield, and product quality from the standpoint of energy, water, and nutrient flows within the whole plant. By following the crop production sequence from cultivar selection and stand establishment through field culture and harvest, we will discuss the challenges inherent in producing high-quality fruits and vegetables from the whole-field perspective. Finally, we will suggest potential areas for improving the production system to enhance ultimate product quality for fresh fruits and vegetables.

A.   Genetic Management

Within a given crop, different cultivars have varied genetic makeups or genotypes. Therefore, cultivars vary in size, color, flavor, firmness, nutrition, pest resistance, and productivity. Genetic management for quality production includes both farmer selection from available cultivars and the development of new cultivars with desirable characteristics through traditional plant breeding or biotechnology. The challenge is to find cultivars that satisfy the need of the farmer for profitability and the demand of the consumer for quality at purchase and at consumption.

From a quality standpoint, cultivar selection may be the single greatest management decision, especially in perennial crops such as tree fruits. Pear fruit flavor is determined primarily by genetic makeup, irrespective of environmental conditions (Quamme and Gray, 1985). Eight peach cultivars evaluated for harvest quality and processed fruit quality showed significant variations in almost every quality attribute measured (Kader et al., 1982). Late-maturing peach and nectarine cultivars tend to have a shorter storage life than early-maturing cultivars (Lill et al., 1989). These and other cultivar evaluations (Sistrunk, 1985) indicate that quality attributes vary greatly among cultivars. To further complicate the issue of cultivar effects on postharvest quality, many fruit trees consist of fruiting wood (scion) of one cultivar grafted onto a genetically different rootstock. Rootstocks are selected largely for their adaptation to soil conditions, their effect in controlling tree size, or other production advantages. Nonetheless, rootstocks exert considerable influence on the quality of citrus (Fellers, 1985; Monselise and Goren, 1987) and pear (Fallahi and Larsen, 1981) products. These rootstock effects on fruit parameters such as total soluble solids (TSS) and acidity may be caused by differences in nutrient and water uptake and translocation, or differences in photosynthate partitioning.

In contrast to that of tree fruits, quality of annual crops may be more sensitive to environmental conditions and cultural practices than to genetic makeup. Increasing the TSS concentration generally increases fruit quality, but TSS in tomatoes (Gull et al., 1989) and strawberries (Shaw, 1990) appears to depend more on environmental conditions than on genetic control. Increasing the energy supply and decreasing the water content of fruits increases TSS in tomatoes (Stevens and Rudich, 1978; Ho, 1988). Breeding tomatoes specifically for higher TSS generally has not been successful because of the lack of understanding of the genetic mechanisms involved (Ho, 1988; Bartels et al., 1989). In addition, TSS exemplifies the trade off between yield and quality, since yield generally decreases with increasing TSS (Stevens and Rudich, 1978).

Human nutritional quality of horticultural products also varies significantly within a species (Shewfelt, 1990). For example, the vitamin A content of grapefruit cultivars may vary as much as 20-fold. Dry bean cultivars differ significantly in calcium, iron, thiamine, and riboflavin contents. Thus, the opportunities for improving nutritional quality of horticultural crops through plant breeding are very good (Munger, 1979). However, increasing consumption of fresh fruits and vegetables may improve human nutrition more effectively than increasing the nutritional value of any single fruit or vegetable (White, 1979).

Traditional cultivar development based on continual selection of desirable genotypes and the intentional development of inbred lines to produce hybrids of superior quality has produced an almost endless variety of crop quality characteristics. Two examples of particular interest with respect to fruit and vegetable quality are supersweet corn and nonripening tomatoes.

In "supersweet" (sh_2) corn cultivars, the sugar levels are four or more times greater than those of normal cultivars (Laughnan, 1953; Boyer and Shannon, 1983). Kernel tenderness, succulence, and creaminess also are improved; these characteristics are stable for up to 9 days after harvest under refrigeration (Wolf and Showalter, 1974). In addition, kernels lose less water during postharvest storage and, therefore, are less susceptible to "denting" (Risse and McDonald, 1990). Since these characteristics also affect seed quality and viability, additional research is necessary to improve production practices affecting yield and economics of these high quality cultivars (Guzman et al., 1983).

Nonripening tomato mutants provide valuable physiological tools for the study and management of quality development in tomatoes (Tigchelaar et al., 1978). The flavor qualities of fruits of these mutants are reduced substantially, but flavor and sweetness may be improved by crossing the mutants with cultivars known for high soluble solids in the fruit (Jones, 1986). These genes have been incorporated into the commercial cultivars 'Moneymaker' and 'Alisa Craig' by traditional breeding programs. The resulting hybrids produce fruits that ripen almost normally, with acceptable flavor but a much longer shelf life (Hobson, 1988).

Traditional plant breeding offers the opportunity to reduce susceptibility to environmentally induced quality deterioration. For example, solar injury of exposed cantaloupe fruits could be reduced by the development of cultivars that are netted completely and heavily (Lipton, 1977). Another complex problem is a tomato fruit quality defect known as fine-net cracking. Affected tomato fruits (generally greenhouse-grown) are covered partially or completely with a network of fine hair-like cracks that reduce shelf life because of rapid postharvest water loss (Hayman, 1988). Cultivars range from very high to absolutely no expression of the disorder. Therefore, excellent opportunities may exist to control or eliminate this disorder genetically (Hayman, 1988). Breeding for improved postharvest quality also has excellent potential in other crops, such as nectarines and peaches, which exhibit genetic differences in susceptibility to flesh browning during postharvest storage (Lill et al., 1989).

Biotechnology can reduce the time and space required to improve crop quality compared with traditional plant breeding, provided the necessary genes have been located adequately. A specific genetic change without the addition of associated detrimental genes can be achieved in one step in a test tube, whereas traditional breeding programs would require many generations and many acres of land (Meredith, 1982; Qualset, 1982). However, as Wasserman (1990) points out, the ability to manipulate specific gene targets such as enzymes does not insure that the intended and desired physiological benefits will follow.

Genetic engineering also increases the diversity of genes and germ plasm available for development of new crop varieties by allowing the incorporation of "foreign" genes (Gasser and Fraley, 1989). To date, genetic engineering has improved crop performance primarily through manipulation of single gene traits in crop species in which the genetics are understood fairly well. One example of the commercial application of genetic engineering is the incorporation of the genes for the insect control proteins of *Bacillus thuringiensis* into tomato plants, resulting in tomato plants resistant to lepidopteran caterpillar pests (Bartels *et al.*, 1989; Gasser and Fraley, 1989). Under field conditions, transgenic tomato plants (with incorporated *B. thuringiensis* genes) suffered no damage whereas control plants were defoliated completely by caterpillar pests. Other examples of breeding progress through genetic engineering include virus resistance of transgenic tomato plants carrying the tobacco mosaic virus (TMV) coat protein gene (Gasser and Fraley, 1989) and herbicide resistance in several agronomic crops (Bartels *et al.*, 1989). Some improvement of protein quality of cereal grains and legumes has been realized (Bartels *et al.*, 1989). There is also progress in regulating gene expression, for example, of the genes encoding polygalacturonase, an enzyme active in softening of tomato fruit cell walls (Gasser and Fraley, 1989). However, most quality and yield characteristics are under multiple gene control. Genetic engineering of quantitatively controlled traits is much more difficult, but new technology using restriction fragment length polymorphisms (RFLPs) is promising. For example, using this technology, genes controlling soluble solids content of tomato fruits have been identified, although not characterized fully yet (Bartels *et al.*, 1989).

Although biotechnology offers much hope for improvements in crop production practices and product quality, the acceptance of these advances in the marketplace will depend on many nontechnical issues such as regulatory approval in the United States, proprietary protection for companies investing in the research and development of these new crops, and, perhaps most importantly, public perception of the risks and benefits of this new technology (Gasser and Fraley, 1989; Wasserman, 1990).

B. Crop Establishment

In addition to carrying the genetic material that determines the quality potential of a crop, seeds provide the energy for initial growth and stand establishment. For vegetable crops in particular, uniform emergence is vital to insure even growth and development and allow a concentrated harvest of consistent quality. Strategies to enhance uniform stand establishment include seed grading, seed coating, and seed treatments such as osmotic conditioning and pregermination. Although these strategies generally are intended to facilitate production and increase yields, selection for seed quality and pregermination also improves quality in the form of increased cauliflower curd weight (Finch-Savage and McKee, 1990).

Many vegetable crops are started as seedlings in plant beds or greenhouse flats, then transplanted to the field for production. Proper handling of plants prior to and during transplanting is vital to good stand establishment. Various methods are available to prepare seedlings to withstand transplant shock, for example, withholding water, exposing to cool temperatures, or brushing. Research on the relationship between transplant handling and final quality is limited. Improper handling of cauliflower transplants can lead to buttoning, which reduces marketable quality (Skapski and Oyer, 1964). Also, conditioning tomato (Melton and Dufault, 1991) and cauliflower (McGrady, 1990) transplants with nutritional treatments improves yield and earliness. Further, nutritional conditioning of transplants increases muskmelon yield and quality (Dufault, 1986).

C. Energy Management

Energy flow within the fruit is the net effect of photosynthate gains from photosynthesis and translocation and losses due to respiration. Light, carbon dioxide concentration, and temperature largely govern the rates of these processes. These environmental constraints are virtually unmanageable in the field, but can be managed in the greenhouse. Carbon dioxide enrichment increases the yield of greenhouse tomatoes (Slack et al., 1988), as well as net photosynthesis and fruit size and weight for several other crops (Mortensen, 1987).

Other cultural practices are used to improve photosynthesis and the partitioning of photosynthates in field production (Pantastico, 1975). Although the supply of sunlight cannot be increased in a field setting, pruning and training increase the number of leaves exposed to sunlight, and can improve quality of grape juice (Morris, 1985; Morris et al., 1985). Thinning the crop load insures that the remaining fruits will be larger. In addition to the direct effects on energy flows in the plant, these practices also alter the relative sink strength of vegetative and reproductive tissue. This alteration, in turn, changes partitioning of water by changing the strength of transpiration sinks, with concomitant increases in calcium supplies to fruit, resulting in improved quality. Row spacing and orientation and plant population within the row also are varied to optimize light interception and crop yield. Fresh-market snap beans grown in a narrow row spacing are lighter in color and higher in ascorbic acid and have a higher seed index than those from wide row spacing (Drake and Silbernagel, 1982; Drake et al., 1984). Leafy vegetable crops grown at close spacings have larger thinner leaves, which are considered to be of lower quality (Pantastico, 1975).

Plant growth regulators (PGRs) are chemicals used in crop production to regulate growth, development, yield, and maturation of plants. The physiological activities of these chemicals and their effects on product quality are complex, but can be understood partially as affecting the partitioning of nutrients and photosynthates within the plant.

The quality effects of preharvest PGR application include improved color, flavor, and uniformity of maturity in peaches and nectarines (Lill et al., 1989; Blanco, 1990) and increased firmness in cherries (Looney and Lidster, 1980). In apples, PGR treatments increase color and firmness, while reducing fruit size and decreasing storage disorders (Looney, 1967; Monselise and Goren, 1987; Wang and Steffens, 1987; Elfving et al., 1990). PGRs increase TSS concentration, acidity, and ascorbic acid over those in untreated fruit in mangoes, primarily by hastening maturity

(Khader, 1990). In grape, preharvest PGR treatments do not affect TSS concentration, but different materials affect storage quality differently (Al-Juboory et al., 1990). In various vegetables, PGRs increase protein, ascorbic acid, and β-carotene content (Graham and Ballesteros, 1980).

Although it is difficult to generalize the effects of PGRs on quality, many of the effects appear to result from changing the rate of maturation or changing the source–sink relationships that affect photosynthate and nutrient partitioning. Earlier and more uniform maturation is reported for several tree fruits in response to PGR application (Bangerth, 1983; Monselise and Goren, 1987; Lill et al., 1989; Blanco, 1990; Khader, 1990). Growers apply PGRs in "chemical thinning" to reduce fruit number, thereby increasing fruit size (Bangerth, 1983; Wang and Gregg, 1990). Whereas larger fruits might be expected to have lower calcium concentrations and subsequent physiological disorders (Bangerth, 1983), blossom end rot in tomatoes is reduced whereas the yield of larger fruits is increased with fruit thinning (Wang and Gregg, 1990). In apples (Monselise and Goren, 1987; Wang and Steffens, 1987), PGRs reduce fruit size and increase fruit calcium concentration, thereby improving postharvest storage quality. In contrast, daminozide application increased core breakdown in pears 2 years out of 6 studied (Meheriuk, 1990). Some PGR treatments also reduce vegetative growth (Wang and Steffens, 1987; Wang and Gregg, 1990), enhancing partitioning of water and calcium to fruits. Increased light penetration in response to reduced vegetative growth improves color in nectarines (Blanco, 1990). The effect of daminozide on apple color (Monselise and Goren, 1987; Elfving et al., 1990) appears to result directly from increasing the accumulation of anthocyanin. Finally, PGR use reduces fruit respiration (Looney, 1967), which improves postharvest quality.

Despite their value in improving fruit and vegetable quality, PGRs are subject to increasing public scrutiny and a certain degree of mistrust. The most striking example of this public concern resulted in the withdrawal of daminozide by the manufacturer in 1989 (Elfving et al., 1990). Growers must be responsive to these concerns to maintain consumer confidence, and must be prepared to adapt their production systems if other currently approved technologies are withdrawn.

D.   Water Management

Since relative humidity cannot be controlled in the field, the most direct method to address water flow in plants is to increase the availability of water in the root zone by irrigation. Drought stress can limit crop yields severely, especially of fruits and vegetables. On the other hand, drought stress may either decrease or increase produce quality, so irrigation response serves as a good example of the potential contradiction between yield and quality. Snap bean yield and quality varies with irrigation treatments, depending on cultivar (Varseveld et al., 1985). In particular, optimal irrigation produces darker snap beans with lower seed index and shear values and higher moisture and ascorbic acid contents than deficit irrigation (Drake et al., 1984). Further, rill irrigation produces higher quality snap beans than sprinkler irrigation (Drake and Silbernagel, 1982).

Increasing the water supplied through drip irrigation increases yield, color, fruit size, and acidity but decreases TSS concentration in tomatoes (Sanders *et al.*, 1989). Similarly, irrigation during fruit maturation increases tomato yield but decreases quality measures such as TSS concentration, acidity, viscosity, and vitamin C content (Rudich *et al.*, 1977). Other reports concerning drought stress (Stevens, 1985) or salinity effects (Mizrahi and Pasternak, 1985; Hobson, 1988; Mizrahi *et al.*, 1988) on tomatoes show that TSS concentration, sugar, acidity, and flavor all increase with stress. The effects appear to be caused by a combination of osmo-regulation and decreased yield, which concentrates the photosynthates into fewer and smaller fruits. De Koning and Hurd (1983) suggest that the improved quality of drought-stressed tomatoes results from an increase in carbohydrate storage in leaves and stems. Saline irrigation improves the taste of melons, but not the taste of lettuce or peanuts (Mizrahi and Pasternak, 1985).

Reduced storage quality in spinach is attributed to high rainfall (Johnson *et al.*, 1989). Irrigation increases the yield of blackberries, but reduces quality as measured by soluble solids, pH, and color (Sims and Morris, 1982; Morris and Sims, 1985). Similarly, irrigation or rainfall during strawberry harvest reduces the sugar content of the fruit (Sistrunk and Morris, 1985). Deficit irrigation reduces grapefruit yield but improves juice quality by increasing TSS concentration and acidity (Metochis, 1989). Summer drought stress (Levy *et al.*, 1978) increases acidity, but does not affect TSS concentration in grapefruit; however, this effect reduces palatability.

In general, drought stress during fruit maturation increases the concentration of most fruit constituents (Stevens, 1985). Whether this response increases or reduces quality depends on the crop and the constituents involved. Note, however, that water stress nearly always reduces yield. The farmer must decide on the acceptable combination of yield and quality as they affect crop value, until irrigation management can be refined to enhance quality without sacrificing yield.

E. Nutrient Management

Fertilizer application, either directly onto the plant or to the soil, is the most direct cultural intervention when nutrient deficiencies threaten to limit crop yield or quality. However, as for irrigation, yield and quality effects are sometimes contradictory. Moreover, climatic effects, nutrient interactions, and physical and chemical processes in the soil complicate the issue of nutrient availability and uptake, making fertilization an indirect strategy for increasing nutrient flows within the plant.

Since nitrogen affects vegetative growth so strongly, nitrogen fertilization to alleviate deficiency can increase yield, as well as quality, greatly by providing adequate photosynthetic surface area. Nitrogen application improves head quality in broccoli (Dufault, 1988) and increases yield, protein, and β-carotene in spinach (Kansal *et al.*, 1981). In cabbage, nitrogen increases dietary fiber and ascorbic acid (Sørensen, 1984), but also increases the incidence of storage disorders (Berard, 1990a). High nitrogen applications are blamed for increased hollow stem damage in broccoli (Vigier and Cutcliffe, 1984). Increased weight loss in sweet potato during storage is related to nitrogen application (Hammett and Miller, 1982). A negative interaction between nitrogen and the accumulation of carbohydrates leads to a decline in flavor of celery (Van Wassenhove *et al.*, 1990) and soluble solids in pears (Quamme and Gray, 1985) and strawberries (Sistrunk and Morris, 1985). Nitrogen application also decreases color, firmness, and edible quality of peaches (Sistrunk, 1985) and fruit color and soluble solids in grapes (Kliewer, 1977).

Nitrogen application also can lead to potentially harmful accumulations of nitrate nitrogen, especially in leafy greens (Eppendorfer, 1978; Blom-Zandstra, 1989) and potatoes (Munshi and Mondy, 1988). Nitrate nitrogen concentration in tissue is higher when low light intensity limits carbohydrate production (Blom-Zandstra, 1989). Foliar application of indoleacetic acid reduces nitrate nitrogen content in potato tubers (Ponnampalam and Mondy, 1986), as does fertilization with either magnesium (Mondy and Ponnampalam, 1985) or molybdenum (Munshi and Mondy, 1988).

Calcium deficiency causes a large number of the most devastating physiological disorders affecting the quality of many crops, including blossom end rot in tomatoes, peppers, and watermelons; tipburn in lettuce and cabbage; blackheart in celery; and corkspot and bitter pit in apples (Shear, 1975). Unfortunately, calcium nutrition is very complex and is affected by many environmental and physiological factors, especially water movement in the plant (Shear, 1975; Bangerth, 1979; Collier and Tibbitts, 1982; Ho, 1988). Interactions with other nutrients affect calcium uptake and use. Molybdenum increases calcium concentration in tomato fruits (Kheshem *et al.*, 1988), whereas ammonium nitrogen increases the incidence of blossom end rot in tomatoes (Pill *et al.*, 1978). Research on the storage quality of apples showed that fruits with higher calcium concentrations stored better, with less senescent breakdown, rot, and scald (Bramlage *et al.*, 1985; Marmo *et al.*, 1985).

Calcium disorders cannot be prevented by a straightforward method such as applying a calcium source to the soil, and foliar application has led to mixed results. Watkins *et al.* (1989) reported that foliar application of soluble calcium salts increases firmness and reduces bitter pit and senescent breakdown in apples. In contrast, Davenport and Peryea (1990) observed no effect on foliar calcium application on fruit mineral element composition or postharvest quality of apples. Moreover, increasing fruit calcium decreases firmness in strawberries (Chéour *et al.*, 1990), and fails to reduce fruit rot in peaches (Conway *et al.*, 1987). Foliar calcium sprays reduce soluble solids but do not affect firmness in blackberries (Morris *et al.*, 1980). Shear (1975) concluded that calcium sprays are effective in supplying calcium to leafy vegetables. However, to increase calcium supply to developing fruits, topical sprays must be applied to the fruit surface and absorbed directly. The amount of calcium that can be supplied using this approach is limited in most crops, and the calcium required by fruits must be absorbed through the roots. Partitioning in the transpiration stream is generally unmanageable or, at best, influenced indirectly through irrigation and growth management.

All plant nutrients affect quality, either directly or indirectly. For the sake of brevity, the reader is referred to other publications for details on the quality effects of phosphorus (Letham, 1969), potassium (Winsor, 1979; Morris *et al.*, 1983; Sørensen, 1984; Stevens, 1985; Usherwood, 1985; Dick and Shattuck, 1987) and micronutrients (Hårdh and Takala, 1979; Locascio and Fiskell, 1987; Smith and Clark, 1989; Berard, 1990b; Berard *et al.*, 1990).

F. Pest Management

Pest management consists of limiting pest damage to plant tissues or the harvested plant product to acceptable levels. Public concerns about the safety of pesticide and other chemical residues on fresh and processed horticultural products are forcing the production industry to develop or use pest management practices that require fewer chemical inputs. However, preliminary information suggests that consumers are not willing to sacrifice cosmetic quality for lower pesticide residues (Ott and Maligaya, 1989). Because of concerns for public safety and awareness of the risks of environmental contamination, efforts are mounting to increase pesticide regulation. Therefore, there is increased interest in integrated pest management (IPM) systems that are designed to reduce pesticide applications. Specific programs have been developed for fruit crops such as apples (Whalon and Croft, 1984) and vegetable crops such as greenhouse-grown tomatoes (Berlinger *et al.*, 1988) and field-grown celery (Trumble, 1990).

IPM systems are intended to integrate the use of preventive measures to attain effective protection under biological, technological, and socioeconomic limitations (Geier, 1982). In other words, IPM seeks to balance the benefits and costs of various pest control measures with the benefits or costs in terms of crop quality and yield. This approach uses biological and economic models to assess the economic threshold of pest damage, for instance, to identify when to control a specific pest on a specific crop chemically to optimize profitability. These models, however, are not appropriate when a producer cannot monitor pests adequately or when the damage caused by the pest is difficult to measure, for example, product quality loss or disease transmission (Mumford and Norton, 1984). Under these conditions, producers are unwilling to assume the level of risk the models based on ecnomic threshold entail and usually adopt a regular chemical control schedule. Additional research in the area of pest effects on product quality would improve the reliability of the models and perhaps improve farmer confidence in the use of the models, thereby further reducing dependence on pesticides.

Fruit and vegetable producers use many management practices to reduce pest damage. These practices include use of resistant cultivars; selection of planting times to avoid unacceptable pest pressures (Ezueh, 1982); selection of various crop management practices that reduce pest pressure, including tillage practices and maintenance of sanitary growing conditions designed to reduce host plants or materials harboring insects or diseases or, conversely, the management of complementary weed and crop interactions (William, 1981); use of reflective mulches or floating row covers to repel pests or reduce pest contact with crops (Trumble, 1990); and increased use of biological control agents (Stephens, 1990; Trumble, 1990).

G. Harvest Management

Harvest represents the culmination of cultural management. The processes of plant growth, reproduction, and fruit maturation occur over an extended period of time. Consequently, individual fruits on a given plant will vary in maturity and quality at a given time, and entire populations of plants will represent an even greater range of maturity, quality, and productivity. Thus, harvest timing can have a profound effect on the quality of produce offered to consumers (Pantastico, 1975; Smittle and Maw, 1988; Lill *et al.*, 1989; Shewfelt, 1990). Further, the farmer must decide when to harvest based on considerations other than quality. Grower prices for fresh fruits and vegetables are very volatile, so farmers often harvest prior to optimum maturity to sell under favorable market conditions. However, this practice may actually limit returns by limiting repeat sales later in the season. In addition, harvesting before the product attains maximum size limits yield. Multiple hand harvests can increase the proportion of a crop harvested near optimal maturity; also hand picking generally causes less bruising than mechanical harvesting. Nevertheless, the rising cost and decreasing availability of harvest labor increasingly necessitate once-over mechanical harvesting. This strategy can cause bruising, which leads to subsequent decay, and requires that plants of all maturity levels in a field be harvested in bulk. The choice of harvest methods and timing often is constrained by economic, logistical, and weather considerations. The grower must balance the potential for gain in yield or quality in delaying with the risk of loss to excess maturity, pest or weather damage, or unfavorable price movements. Thus, as in other production practices, quality will be a consideration in harvest management, but only to the extent that it will help improve the potential for profitability for the farmer.

IV. Coordinating Production and Marketing to Enhance Quality

Whereas the postharvest handler and produce marketer strive for uniformity and predictability in product volume and quality, the nature of fresh-market fruit and vegetable production is variable and volatile. Certainly, growers would prefer to have stability of production and quality, which can be achieved to a degree in relatively uniform climates, such as arid zones. However, such events as rain storms and freezes keep field production out of reach of total predictability and management. Although the postharvest segment can emphasize quality control, the production industry is limited largely to constraint reduction or damage control.

What, then, is the potential for communication and coordination between growers and handlers to enhance the quality of fresh produce? Can the marketplace inform and motivate growers to change their production systems to improve quality for the consumer? Such a potential does exist, but requires mutual understanding by growers and postharvest workers of the needs and capabilities of the entire marketing system.

First, to achieve a given level of quality, the objective must be defined clearly. Presumably, consumer demands at retail and consumption will dictate optimal quality at each stage of the marketing system. Still, growers respond directly to the demands of first receivers. Thus, the entire marketing channel must agree on the ultimate quality goal, and on the product characteristics at each stage of the marketing system that are required to achieve that goal so farmers can deliver the necessary product at harvest.

Second, farmers must be rewarded adequately for achieving quality objectives. The highest prices received by farmers are often for the earliest fruits and vegetables, which are often of very poor quality. Delaying harvest would improve quality greatly, but almost certainly would decrease the profitability of the crop. Similarly, multiple hand harvests, rather than once-over harvesting, would improve the quality at each harvest, but would increase the total cost of production. Enhancing quality at harvest no doubt will require more intensive management, and will expose the grower to greater risk of physical loss or price volatility, with little additional cost to the rest of the marketing channel. Growers currently receive about one-fourth of the retail value of fresh fruits and vegetables (Anonymous, 1991). Their management cost and risk of loss must be compensated to change production systems for the sake of quality. For example, to gain greater control over pest management or harvest practices, the postharvest segment could assume more of the risk of the producer, for example, by forward contracting. Until economic or other incentives are available, however, growers ultimately must make production decisions to optimize their own objectives.

Third, postharvest handlers should understand the potential trade off between desired quality attributes. For instance, flavor typically increases, but the ability to withstand the rigors of transport (i.e., firmness) decreases, with increasing maturity (Stevens, 1985). If firmness and flavor cannot be maximized simultaneously, which combination of these factors at harvest will best satisfy consumer desires? The postharvest system must identify and communicate these conditions to growers, and be willing to accept unavoidable trade offs.

Fourth, the postharvest system must be willing to adapt to changes itself, whether initiated by consumers or growers, to enhance quality. The wholesale marketing channel, in particular, continues to demand traditional packaging methods and package sizes, which often do not meet the needs of either growers or retailers. The virtual failure of the effort in the mid-1980s to standardize package size and convert to metric measures is one example of reluctance to change in the wholesale industry. If growers are to change production systems to enhance quality, then marketers must be willing to change handling systems to accommodate growers. This cooperation may be more feasible with greater vertical integration, in which produce would not change possession as many times between grower and consumer.

Finally, the inherent variability of produce may prove to be an advantage rather than a disadvantage. There has been some successful product differentiation in fresh produce: Washington State apples, Idaho potatoes, Vidalia onions, and Sunkist citrus are among the more notable examples. Although these products are in some ways superior to the alternatives, their particular value is that they are distinguishable, and therefore perceived to be superior, by consumers. Branding has been suggested as a means of improving profitability of fresh produce, but branding will require adequate volumes of consistent quality through a long marketing season. This differentiation, even more than superiority to alternatives, elevates some produce from the status of a commodity to that of an identifiable product. From a production perspective, it appears that, rather than striving for standardization and bland consistency, fresh produce offers the opportunity for boundless variety in sizes, shapes, and colors, even within the same species. Perhaps the postharvest system should use consumer education and promotion to capitalize on the excitement and vitality that the inherent variability of biological and environmental conditions ascribe to fresh produce.

V.   Future Directions in Production Research and Management

Viewing plant growth and quality development as the results of integrated effects of energy, water, and nutrient flows suggests a number of avenues of research and management. A greater understanding of the quality implications of processes affecting photosynthesis, respiration, nutrient uptake, water uptake, transport, and partitioning is needed. In particular, the complex interactions between factors within the plant need to be understood, and the current fragmentary knowledge on these subjects needs to be integrated into a more holistic understanding of quality production and maintenance. Such an agenda demands an interdisciplinary basic research program under controlled conditions dealing primarily with individual whole plants.

Whole-field production management also presents research needs. Cultivar development to satisfy consumer demand and decrease production inputs will help alleviate environmental and health concerns. New crops and cultivars with greater pest resistance and drought tolerance or decreased fertilizer requirements will reduce dependence on purchased inputs. However, the cultivars also must satisfy consumer desires for sensory quality and nutrition, and must be able to withstand the rigors of the marketing system. Unfortunately, development of such a "super plant" is unlikely, since attributes that confer pest resistance often lower product quality, and may lead to accumulation in the crop of substances that are toxic to humans. Therefore, the need for pesticides and fertilizers will remain. Research on management systems to improve the ability of these materials to accomplish production goals without harming the environment or consumers must continue. Finally, the increasing cost and decreasing availability of harvest labor favor mechanical harvesting. Research on harvest aids to facilitate hand harvest and on methods to optimize mechanical harvest yield and quality remains relevant.

In contrast to these research issues that focus on plant physiology and crop ecology, production management will continue to respond largely to social conditions communicated by economic, legal, and regulatory messages. How can growers best adapt production systems that integrate these principles of energy, water, and nutrient flows into the production of high-quality fruits and vegetables in the face of growing economic and environmental constraints? First, growers can adopt appropriate crops for their regions. Lima beans, okra, and sweet potatoes are well adapted to production in the southeastern United States. However, demand for these crops is limited, so appropriate marketing programs to promote consumption will be required. Still, developing consumer acceptance of appropriate crops is a better long-term solution than developing production and pest management strategies for poorly adapted crops. Similarly, developing better-adapted cultivars of all crops will decrease dependence on production inputs.

Finally, growers must become increasingly sophisticated in their production systems and adopt the most appropriate technology available to insure compliance with environmental and other regulations. Crop production and protection chemicals probably will not be removed from use completely, but growers must manage these chemicals very carefully to avoid environmental or consumer problems. Soil testing, plant tissue analysis, crop modeling, disease and insect forecasting, and field scouting all will be required to insure timely and efficient application of fertilizers and pesticides. Ultimately, irrigation water may become the limiting factor in production, and every effort must be made to insure the quality and availability of this resource.

In summary, the fruit and vegetable production industry faces fantastic potential in the future because of increasing demands among an aging and increasingly health-conscious population. The successful growers will be those who best manage their production systems to optimize the flows of energy, water, and nutrients into the harvested plant product, then harvest and market those products with proper attention to the economic and environmental consequences of their production decisions.

Bibliography

Al-Juboory, K. H., Jumma'a, F., Shaban, A., and Skirvin, R. M. (1990). Preharvest treatment with growth regulators improves quality of 'Thompson Seedless' grapes (*Vitis vinifera* L.) during cold storage. *Fruit Var. J.* **44,** 124–127.

Anonymous (1991). Growers receive small share of dollar. *The Packer* **19 Jan.** 71B.

Bangerth, F. (1979). Calcium-related physiological disorders of plants. *Annu. Rev. Phytopathol.* **17,** 97–122.

Bangerth, F. (1983). Hormonal and chemical preharvest treatments which influence postharvest quality, maturity, and storeability of fruit. *NATO Adv. Study Inst. Ser. A. Life Sci.* **46,** 331–354.

Bartels, D., Gebhardt, C., Knapp, S., Rohde, W., Thompson, R., Uhrig, H., and Salamini, F. (1989). Combining conventional plant breeding procedures with molecular based approaches. *Genome* **31,** 1014–1026.

Beevers, H. (1985). Regulation of carbon partitioning in photosynthetic tissue. *Proc. 8th Symp. Plant Physiol., Univ. Calif., Riverside,* 367–369.

Berard, L. S. (1990a). Effects of nitrogen fertilization on stored cabbage. I. Development of physiological disorders on tolerant and susceptible cultivars. *J. Hort. Sci.* **65,** 289–296.

Berard, L. S. (1990b). Effects of nitrogen fertilization on stored cabbage. III. Changes with time and distribution in outer-head leaves of the mineral contents. *J. Hort. Sci.* **65,** 417–422.

Berard, L. S., Senecal, M., and Vigier, B. (1990). Effects of nitrogen fertilization on stored cabbage. II. Mineral composition in midrib and head tissues of two cultivars. *J. Hort. Sci.* **65,** 409–416.

Berlinger, M. J., Dahan, R., and Mordechi, S. (1988). Integrated pest management of organically grown greenhouse tomatoes in Israel. *Appl. Agric. Res.* **3,** 233–238.

Björkman, O., Badger, M. R., and Armond, P. A. (1978). Response and adaptation of photosynthesis to high temperatures. *In* "Adaptations of Plants to Water and High Temperature Stress" (N. C. Turner and P. J. Kramer, eds.), pp. 233–249. John Wiley & Sons, New York.

Blanco, A. (1990). Effects of paclobutrazol and of ethephon on cropping and vegetative growth of 'Crimson Gold' nectarine trees. *Sci. Hortic.* **42,** 65–73.

Blom-Zandstra, M. (1989). Nitrate accumulation in vegetables and its relationship to quality. *Ann. Appl. Biol.* **115,** 533–561.

Boyer, C. D., and Shannon, J. C. (1983). The use of endosperm genes for sweet corn improvement. *Plant Breeding Rev.* **1,** 139–161.

Bramlage, W. J., Weis, S. A., and Drake, M. (1985). Predicting the occurrence of poststorage disorders of 'McIntosh' apples from preharvest mineral analyses. *J. Amer. Soc. Hort. Sci.* **110,** 493–498.

Chéour, F., Willemot, C., Arul, J., Desjardins, Y., Makholouf, J., Charest, P. M., and Gosselin, A.

(1990). Foliar application of calcium chloride delays postharvest ripening of strawberry. *J. Amer. Soc. Hort. Sci.* **115**, 789–792.

Collier, G. F., and Tibbits, T. W. (1982). Tipburn of lettuce. *Hort. Rev.* **4**, 49–65.

Conway, W. S., Greene, G. M., and Hickey, K. D. (1987). Effects of preharvest and postharvest calcium treatments of peaches on decay caused by *Monilinia fructicola. Plant Dis.* **71**, 1084–1086.

Daie, J. (1985). Carbohydrate partitioning and metabolism in crops. *Hort. Rev.* **7**, 69–108.

Daie, J. (1988). Mechanism of drought-induced alterations in assimilate partitioning and transport in crops. *CRC Crit. Rev. Plant Sci.* **7**, 117–137.

Davenport, J. R., and Peryea, F. J. (1990). Whole fruit mineral element composition and quality of harvested 'Delicious' apples. *J. Plant Nutr.* **13**, 701–712.

De Koning, A., and Hurd, R. G. (1983). A comparison of winter-grown tomato plants grown with restricted and unlimited water supply. *J. Hort. Sci.* **58**, 575–581.

Dick, J., and Shattuck, V. (1987). Influence of potassium fertilization on blotchy ripening in processing tomatoes. *Can. J. Plant Sci.* **67**, 359–363.

Drake, S. R., and Silbernagel, M. J. (1982). The influence of irrigation and row spacing on the quality of processed snap beans. *J. Amer. Soc. Hort. Sci.* **107**, 239–242.

Drake, S. R., Silbernagel, M. J., and Dyck, R. L. (1984). The influence of irrigation, soil preparation and row spacing on the quality of snap beans, *Phaseolus vulgaris. J. Food Qual.* **7**, 59–66.

Dufault, R. J. (1986). Influence of nutritional conditioning on muskmelon transplant quality and early yield. *J. Hort. Sci.* **111**, 698–703.

Dufault, R. J. (1988). Nitrogen and phosphorus requirements for greenhouse broccoli production. *HortScience* **23**, 576–578.

Elfving, D. C., Lougheed, E. C., Chu, C. L., and Cline, R. A. (1990). Effects of daminozide, paclobutrazol, and uniconazole treatments on 'McIntosh' apples at harvest and following storage. *J. Amer. Soc. Hort. Sci.* **115**, 750–756.

Eppendorfer, W. H. (1978). Effects of N-fertilisation on amino acid composition and nutritive value of spinach, kale, cauliflower, and potatoes. *J. Sci. Food Agric.* **29**, 305–311.

Ezueh, M. I. (1982). Effects of planting dates on pest infestation, yield, and harvest quality of cowpea *(Vigna unguiculata). Exp. Agric.* **18**, 311–318.

Fallahi, E., and Larsen, F. E. (1981). Rootstock influences on 'Bartlett' and 'd'Anjou' pear fruit quality at harvest and after storage. *HortScience* **16**, 650–651.

Farquhar, G. D., and Sharkey, T. D. (1982). Stomatal conductance and photosynthesis. *Annu. Rev. Plant Physiol.* **33**, 317–345.

Fellers, P. J. (1985). Citrus: Sensory quality as related to rootstock, cultivar, maturity, and season. *In* "Evaluation of Quality of Fruits and Vegetables" (H. E. Pattee, ed.), pp. 83–128. AVI, Westport.

Finch-Savage, W. E., and McKee, J. M. T. (1990). The influence of seed quality and pregermination treatment on cauliflower and cabbage transplant production and field growth. *Ann. Appl. Biol.* **116**, 365–369.

Forshey, C. G., and Elfving, D. C. (1989). The relationship between vegetative growth and fruiting in apple trees. *Hort. Rev.* **11**, 229–287.

Gasser, C. S., and Fraley, R. T. (1989). Genetically engineering plants for crop improvement. *Science* **244**, 1293–1299.

Geier, P. W. (1982). The concept of pest management—Integrated approaches to pest management. *Prot. Ecol.* **4**, 247–250.

Gent, M. P. N., and Enoch, H. Z. (1983). Temperature dependence of vegetative growth and dark respiration: A mathematical model. *Plant Physiol.* **71**, 562–567.

Graham, H. D., and Ballesteros, M. (1980). Effect of plant growth regulators on plant nutrients. *J. Food Sci.* **45**, 503–508.

Grange, R. I., and Hand, D. W. (1987). A review of the effects of atmospheric humidity on the growth of horticultural crops. *J. Hort. Sci.* **62**, 125–134.

Gull, D. D., Stoffella, P. J., Locascio, S. J., Olson, S. M., Bryan, H. H., Everett, P. H., Howe, T. K., and Scott, J. W. (1989). Stability differences among fresh-market tomato genotypes, II. Fruit quality. *J. Amer. Soc. Hort. Sci.* **114**, 950–954.

Guzman, V. L., Wolf, E. A., and Martin, F. G. (1983). Effect of compensated-rate seeding and seed protectants on yield and quality of a *Shrunken-2* sweet corn hybrid. *HortScience* **18**, 338–340.

Hammett, L. K., and Miller, C. H. (1982). Influence of mineral nutrition and storage on quality factors of 'Jewel' sweet potatoes. *J. Amer. Soc. Hort. Sci.* **107**, 972–975.

Hårdh, J. E., and Takala, E. (1979). Micronutrients and the quality of tomato. *Acta Hort.* **93**, 361–365.

Hayman, G. (1988). Tomatoes—A review of current research and development on glasshouse environment and fruit quality in the UK. *Appl. Agric. Res.* **3**, 269–274.

Ho, J. C. (1988). The physiological basis for improving dry matter content and calcium status in tomato fruit. *Appl. Agric. Res.* **3**, 275–281.

Hobson, G. E. (1988). Pre- and post-harvest strategies in the production of high quality tomato fruit. *Appl. Agric. Res.* **3**, 282–287.

Johnson, J. R., McGuinn, J. R., and Rushing, J. W. (1989). Influence of preharvest factors on postharvest quality of prepackaged fresh market spinach. *Appl. Agric. Res.* **4**, 141–143.

Jones, H. G. (1981). Carbon dioxide exchange of developing apple (*Malus pumila* Mill.) fruits. *J. Exp. Bot.* **32**, 1203–1210.

Jones, H. G., Lakso, A. N., and Syvertsen, J. P. (1985). Physiological control of water status in temperate and subtropical fruit trees. *Hort. Rev.* **7**, 301–344.

Jones, R. A. (1986). Breeding for improved post-harvest tomato quality: Genetical aspects. *Acta Hort.* **190**, 77–87.

Kader, A. A., Heintz, C. M., and Chordas, A. (1982). Postharvest quality of fresh and canned clingstone peaches as influenced by genotypes and maturity at harvest. *J. Amer. Soc. Hort. Sci.* **107**, 947–951.

Kansal, B. D., Singh, B., Bajaj, K. L., and Kaur, G. (1981). Effect of different levels of nitrogen and farmyard manure on yield and quality of spinach (*Spinacea oleracea* L.) . *Qual. Plant. Plant foods Hum. Nutr.* **31**, 163–170.

Khader, S. E. S. A. (1990). Orchard application of paclobutrazol on ripening, quality, and storage of mango fruits. *Sci. Hort* **41**, 329–335.

Kheshem, S. A., Kochan, W. J., Boe, A. A., and Everson, D. O. (1988). Calcium translocation and tomato plant and fruit responses to molybdenum and daminozide. *HortScience* **23**, 582–584.

Kinet, J. M., Sachs, R. M., and Bernier, G. (1985). "The Physiology of Flowering," Vol. III. CRC Press, Boca Raton, Florida.

Kliewer, W. M. (1977). Influence of temperature, solar radiation, and nitrogen on coloration and composition of 'Emperor' grapes. *Am. J. Enol. Vitic.* **28**, 96–103.

Laughnan, J. R. (1953). The effect of the sh_2 factor on carbohydrate reserves in the mature endosperm of maize. *Genetics* **38**, 485–499.

Lee, D. R. (1990). A unidirectional water flux model of fruit growth. *Can. J. Bot.* **68**, 1286–1290.

Letham, D. S. (1969). Influence of fertilizer treatment on apple fruit composition and physiology. II. Influence on respiration rate and contents of nitrogen, phosphorus, and titratable acidity. *Aust. J. Agric. Res.* **20**, 1073–1085.

Levy, Y., Bar-Akiva, A., and Vaadia, Y. (1978). Influence of irrigation and environmental factors on grapefruit acidity. *J. Amer. Soc. Hort. Sci.* **103**, 73–76.

Lill, R. E., O'Donoghue, E. M., and King, G. A. (1989). Postharvest physiology of peaches and nectarines. *Hort. Rev.* **11**, 413–452.

Lipton, W. J. (1977). Ultraviolet radiation as a factor in solar injury and vein tract browning of cantaloupes. *J. Amer. Soc. Hort. Sci.* **102**, 32–36.

Locascio, S. J., and Fiskell, J. G. A. (1987). Vegetable needs for micronutrients in perspective. *Proc. Soil Crop Sci. Soc. Florida* **47**, 12–18.

Looney, N. E. (1967). Effect of *N*-dimethylaminosuccinamic acid on ripening and respiration of apple fruits. *Can. J. Plant Sci.* **47**, 549–553.

Looney, N. E., and Lidster, P. D. (1980). Some growth regulator effects on fruit quality, mesocarp composition, and susceptibility to postharvest surface marking of sweet cherries. *J. Amer. Soc. Hort. Sci.* **105**, 130–134.

McGrady, J. (1990). Transplant nutrient conditioning improves cauliflower early yield. *Proc. Natl. Symp. Stand Estab. Hort. Crops, Minnesota,* 147–150.

Marmo, C. A., Bramlage, W. J., and Weis, S. A. (1985). Effects of fruit maturity, size, and mineral concentrations on predicting the storage life of 'McIntosh' apples. *J. Amer. Soc. Hort. Sci.* **110**, 499–502.

Melton, R. R., and Dufault, R. J. (1991). Nitrogen, phosphorus, and potassium fertility regimes affect tomato transplant growth. *HortScience* **26,** 141–142.

Meheriuk, M. (1990). Effects of diphenylamine, gibberellic acid, daminozide, calcium, high CO_2, and elevated temperatures on quality of stored 'Bartlett' pears. *Can. J. Plant Sci.* **70,** 887–892.

Mengel, K. (1979). Influence of exogenous factors on the quality and chemical composition of vegetables. *Acta Hort.* **93,** 133–151.

Meredith, C. P. (1982). Genetic engineering. The new techniques and their potential. *Calif. Agric.* **36(8),** 5.

Metochis, C. (1989). Water requirement, yield, and fruit quality of grapefruit irrigated with high-sulphate water. *J. Hort. Sci.* **64,** 733–737.

Mizrahi, Y., and Pasternak, D. (1985). Effect of salinity on quality of various agricultural crops. *Plant Soil* **89,** 301–307.

Mizrahi, Y., Taleisnik, E., Kagan-Zur, V., Zohar, Y., Offenback, R., Matan, E., and Golan, R. (1988). A saline irrigation regime for improving tomato fruit quality without reducing yield. *J. Amer. Soc. Hort. Sci.* **113,** 202–205.

Mondy, N. I., and Ponnampalam, R. (1985). Effect of magnesium fertilizers on total glycoalkaloids and nitrate-N in Katahdin tubers. *J. Food Sci.* **50,** 535–536.

Monselise, S. P., and Goren, R. (1987). Preharvest growing conditions and postharvest behavior of subtropical and temperate-zone fruits. *HortScience* **22,** 1185–1189.

Morris, J. R. (1985). Grape juice: Influences of preharvest, harvest, and postharvest practices on quality. *In* "Evaluation of Quality of Fruits and Vegetables" (H. E. Pattee, ed.), pp. 129–176. AVI, Westport.

Morris, J. R., and Sims, C. A. (1985). Effects of cultivar, soil moisture, and hedge height on yield and quality of machine-harvested erect blackberries. *J. Amer. Soc. Hort. Sci.* **110,** 722–725.

Morris, J. R., Cawthon, D. L., Nelson, G. S., and Cooper, P. E. (1980). Effects of preharvest calcium sprays and postharvest holding on firmness and quality of machine-harvested blackberries. *HortScience* **15,** 33–34.

Morris, J. R., Sims, C. A., and Cawthon, D. L. (1983). Effects of excessive potassium levels on pH, acidity, and color of fresh and stored grape juice. *Am. J. Enol. Vitic.* **34,** 35–39.

Morris, J. R., Sims, C. A., and Cawthon, D. L. (1985). Yield and quality of 'Niagara' grapes as affected by pruning severity, nodes per bearing unit, training system, and shoot positioning. *J. Amer. Soc. Hort. Sci.* **110,** 186–191.

Mortensen, L. M. (1987). Review: CO_2 enrichment in greenhouses. Crop responses. *Sci. Hort.* **33,** 1–25.

Mumford, J. D., and Norton, G. A. (1984). Economics of decision making in pest management. *Annu. Rev. Entomol.* **29,** 157–174.

Munger, H. M. (1979). The potential of breeding fruits and vegetables for human nutrition. *HortScience* **14,** 247–250.

Munshi, C. B., and Mondy, N. I. (1988). Effect of soil applications of molybdenum on the biochemical composition of Katahdin potatoes: Nitrate nitrogen and total glycoalkaloids. *J. Agric. Food Chem.* **36,** 688–690.

Ott, S. L., and Maligaya, A. (1989). "An Analysis of Consumer Attitudes toward Pesticide Use and the Potential Market for Pesticide Residue-Free Fresh Produce." Division of Agricultural Economics, University of Georgia, Faculty Series. 89–09. Griffin, Georgia.

Pantastico, E. B. (1975). Preharvest factors affecting quality and physiology after harvest. *In* "Postharvest Physiology, Handling, and Utilization of Tropical and Subtropical Fruits and Vegetables" (E. B. Pantastico, ed.), pp. 25–40. AVI, Westport.

Patrick, J. W. (1988). Assimilate partitioning in relation to crop productivity. *HortScience* **23,** 33–40.

Picken, A. J. F. (1984). A review of pollination and fruit set in the tomato (*Lycopersicon esculentum* Mill.) *J. Hort. Sci.* **59,** 1–13.

Pill, W. G., Lambeth, V. N., and Hinckley, T. M. (1978). Effects of nitrogen form and level on ion concentrations, water stress, and blossom-end rot incidence in tomato. *J. Amer. Soc. Hort. Sci.* **103,** 265–268.

Ponnampalam, R., and Mondy, N. I. (1986). Effect of foliar application of indoleacetic acid on the total glycoalkaloids and nitrate nitrogen content of potatoes. *J. Agric. Food Chem.* **34,** 686–688.

Qualset, C. O. (1982). Agricultural applications. Integrating conventional and molecular genetics. *Calif. Agric.* **36(8),** 29–30.

Quamme, H. A., and Gray, J. I. (1985). Pear fruit quality and factors that condition it. *In* "Evaluation of Quality of Fruits and Vegetables" (H. E. Pattee, ed.), pp. 47–61. AVI, Westport.

Risse, L. A., and McDonald, R. E. (1990). Quality of supersweet corn film-overwrapped in trays. *HortScience* **25**, 322–324.

Rudich, J., Kalmar, D., Geizenberg, C., and Harel, S. (1977). Low water tensions in defined growth stages of processing tomato plants and their effects on yield and quality. *J. Hort. Sci.* **52**, 391–399.

Sanders, D. C., Howell, T. A., Hile, M. M. S., Hodges, L., Meek, D., and Phene, C. J. (1989). Yield and quality of processing tomatoes in response to irrigation rate and schedule. *J. Amer. Soc. Hort. Sci.* **114**, 904–908.

Shaw, D. V. (1990). Response to selection and associated changes in genetic variance for soluble solids and titratable acids contents in strawberries. *J. Amer. Soc. Hort. Sci.* **115**, 839–843.

Shear, C. B. (1975). Calcium-related disorders of fruits and vegetables. *HortScience* **10**, 361–365.

Shewfelt, R. L. (1990). Sources of variation in the nutrient content of agricultural commodities from the farm to the consumer. *J. Food Qual.* **13**, 37–54.

Shewfelt, R. L., Prussia, S. E., Jordan, J. L., Hurst, W. C., and Resurreccion, A. V. A. (1986). A systems analysis of postharvest handling of fresh snap beans. *HortScience* **21**, 470–472.

Sims, C. A., and Morris, J. R. (1982). Effects of cultivar, irrigation, and ethephon on the yield, harvest distribution, and quality of machine-harvested blackberries. *J. Amer. Soc. Hort. Sci.* **107**, 542–547.

Sinclair, T. R., and deWit, C. T. (1975). Photosynthate and nitrogen requirements for seed production by various crops. *Science* **189**, 565–567.

Sistrunk, W. A. (1985). Peach quality assessment: Fresh and processed. *In* "Evaluation of Quality of Fruits and Vegetables" (H. E. Pattee, ed.), pp. 1–42. AVI, Westport.

Sistrunk, W. A., and Morris, J. R. (1985). Strawberry quality: Influence of cultural and environmental factors. *In* "Evaluation of Quality of Fruits and Vegetables" (H. E. Pattee, ed.), pp. 217–256. AVI, Westport.

Skapski, H., and Oyer, E. B. (1964). The influence of pre-transplanting variables on the growth and development of cauliflower plants. *Proc. Amer. Soc. Hort. Sci.* **85**, 374–385.

Slack, G., Fenlon, J. S., and Hand, D. W. (1988). The effects of summer CO_2 enrichment and ventilation temperatures on the yield, quality, and value of glasshouse tomatoes. *J. Hort. Sci.* **63**, 119–129.

Smith, G. S., and Clark, C. J. (1989). Effect of excess boron on yield and post-harvest storage of kiwifruit. *Sci. Hort.* **38**, 105–115.

Smittle, D. A. (1986). Influence of cultivar and temperature on lima bean yield and quality. *J. Amer. Soc. Hort. Sci.* **111**, 655–659.

Smittle, D. A. and Maw, B. W. (1988). Effects of maturity and harvest method on storage and quality of onions. *HortScience* **23**, 141–143.

Sørensen, J. N. (1984). Dietary fiber and ascorbic acid in white cabbage as affected by fertilization. *Acta Hort.* **163**, 221–230.

Stephens, C. T. (1990). Minimizing pesticide use in a vegetable management system. *HortScience* **25**, 164–168.

Stevens, M. A. (1985). Tomato flavor: Effects of genotype, cultural practices, and maturity at picking. *In* "Evaluation of Quality of Fruits and Vegetables" (H. E. Pattee, ed.), pp. 367–386. AVI, Westport.

Stevens, M. A., and Rudich, J. (1978). Genetic potential for overcoming physiological limitations on adaptability, yield, and quality in the tomato. *HortScience* **13**, 673–678.

Tibbitts, T. W. (1979). Humidity and plants. *BioScience* **29**, 358–363.

Tigchelaar, E. C., McGlasson, W. B., and Beuscher, R. W. (1978). Genetic regulation of tomato fruit ripening. *HortScience* **13**, 508–513.

Trumble, J. T. (1990). Vegetable insect control with minimal use of insecticides. *HortScience* **25**, 159–163.

Usherwood, N. R. (1985). The role of potassium in crop quality. *In* "Potassium in Agriculture" (R. D. Munson, ed.), pp. 489–513. American Society of Agronomy/Crop Science Society of American/Soil Science Society of America. Madison, Wisconsin.

Van Wassenhove, F. A., Dirinck, P. J., Schamp, N. M., and Vulsteke, G. A. (1990). Effect of nitrogen fertilizers on celery volatiles. *J. Agric. Food Chem.* **38**, 220–226.

Varseveld, G. W., Mack, H. J., and Baggett, J. R. (1985). Green beans: Effects of modified cultural practices and varietal improvement on sensory quality. *In* "Evaluation of Quality of Fruits and Vegetables" (H. E. Pattee, ed.), pp. 329–347. AVI, Westport.

Vigier, B., and Cutcliffe, J. A. (1984). Effect of boron and nitrogen on the incidence of hollow stem in broccoli. *Acta Hort.* **157,** 303–309.

Wadge, A., and Hutton, M. (1986). The uptake of cadmium, lead, and selenium by barley and cabbage grown on soils amended with refuse incinerator fly ash. *Plant Soil* **96,** 407–412.

Wang, C. Y., and Steffens, G. L. (1987). Postharvest responses of 'Spartan' apples to preharvest paclobutrazol treatment. *HortScience* **22,** 276–278.

Wang, Y. T., and Gregg, L. L. (1990). Uniconazole controls growth and yield of greenhouse tomato. *Sci. Hort.* **43,** 55–62.

Wasserman, B. P. (1990). Expectations and role of biotechnology in improving fruit and vegetable quality. *Food Tech.* **44,** 68–71.

Watkins, C. R., Hewett, E. W., Bateup, C., Gunson, A., and Triggs, C. M. (1989). Relationship between maturity and storage disorders in 'Cox's Orange Pippin' apples as influenced by preharvest calcium or ethephon sprays. *N.Z. J. Crop Hort. Sci.* **17,** 283–292.

Whalon, M. E., and Croft, B. A. (1984). Apple IPM implementation in North America. *Annu. Rev. Entomol.* **29,** 435–470.

White, P. L. (1979). Challenge for the future: Nutritional quality of fruits and vegetables. *HortScience* **14,** 257–258.

William, R. D. (1981). Complementary interactions between weeds, weed control practices, and pests in horticultural cropping systems. *HortScience* **16,** 508–513.

Winsor, G. W. (1979). Some factors affecting the quality and composition of tomatoes. *Acta Hort.* **93,** 335–341.

Wolf, E. A., and Showalter, R. K. (1974). "Florida-Sweet, A High Quality sh_2 Sweet Corn Hybrid for Fresh Market." Florida Agricultural Experiment Station Circular S-226. Gainesville, Florida.

CHAPTER 5

MEASURING QUALITY AND MATURITY

Robert L. Shewfelt

99

Consumer perception of quality and value is the driving force in a systems approach to postharvest handling. Although quality is defined in general terms in Chapter 3, the quality attributes of importance vary from one fruit or vegetable to the next. Attributes of importance also tend to vary with the end use of the product (fresh or processed, raw or cooked, and so forth).

I. Quality and Acceptability

Since quality is defined by the buyer (Kramer and Twigg, 1970; Chapter 3), perception of the quality of a product changes as it travels through the handling system. The grower buys seeds or plants of a selected cultivar as well as a series of inputs (water, fertilizer, or pest protection) that will help provide a good yield at a level of quality acceptable to the first buyer. Quality of a fresh product early in the postharvest system (at packinghouses or warehouses) usually is evaluated against grades and standards. Such grades and standards tend to be based on attributes that can be determined readily visually—color, size, shape, and absence of defects. Visual sorting and grading operations use these attributes to determine the acceptance or rejection of shipments of a fresh product.

For any given lot of a fresh crop, a grade can be established at harvest, usually at the packing facility. Theoretically, the grade of that lot will not change, but the condition of the commodity will change during handling and storage as the product senesces. Perishability of a commodity is a function of how rapidly the condition of the product deteriorates under a given commercial storage regime.

Maturity of a crop is an assessment of physiological development. Physiological maturity is described as "the stage of development when a plant or plant part will continue ontogeny even if detached" whereas commercial maturity is defined as "the stage of development when a plant or plant part possesses the prerequisites for utilization by consumers for a particular purpose" (Watada et al., 1984). Maturity of a crop at harvest directly affects color and size of an item and, thus, its grade. Other important quality characteristics such as texture, flavor, and nutrient content as well as perishability and susceptibility to adverse handling and storage conditions are a function of harvest maturity.

Although grade and condition are the primary factors influencing buying decisions for fresh items from the farmer to the consumer, the consumer uses different criteria to judge quality. Quality attributes can be divided into purchase quality and consumption quality. Purchase quality is composed of those characteristics that are important to the consumer when deciding whether to buy a particular commodity and which item(s) to select. Purchase attributes may include color, size, shape, absence of defects, firmness to the touch, and aroma. Consumption quality consists of those characteristics assessed by the consumer to determine how much that item is liked during eating. Consumption attributes include flavor (taste and aroma) and mouthfeel. In addition to purchase and consumption quality other attributes are hidden such as wholesomeness, nutritional value, and safety. These attributes are considered hidden, because they cannot be detected readily either by visual inspection or by consumption but require sophisticated analysis. Perception of these hidden attributes plays an important role in the consumer purchase decision.

Quality characteristics constitute part of a wider range of factors leading to food acceptability that is defined as "the level of continued purchase or consumption by a specified population" (Land, 1988). Extrinsic attributes or other factors that affect acceptability, for example, packaging, price, marketing practices, and merchandising techniques, are discussed in Chapters 7 and 14. More detailed descriptions of food acceptability (Thomson, 1988), quality measurement (Kapsalis, 1987), and quality attributes of specific commodities (Salunkhe and Desai, 1984 a,b,c,d; Pattee, 1985; Eskin, 1989) are the subjects of other books. This chapter focuses on the intrinsic attributes of a fruit or vegetable that affect its acceptability, and places measurement of maturity and quality in a systems context.

II.   Commodity-Specific Quality Attributes

A set of characteristics important to consumer acceptance is associated with each fresh fruit or vegetable. Broccoli should be green but green peaches are rejected. Celery should be crisp and crunchy, but strawberries are expected to be soft and succulent. Bland flavors are associated with lettuce and potatoes but are not desirable in tomatoes and blueberries.

Determining characteristics that are important in consumer acceptance is not as easy as it might seem. Few investigators determine consumer acceptability of specific fruits and vegetables in their research. More consumer acceptance studies have been performed on the tomato than on any other fresh commodity. Research conducted in four separate countries establishes that external factors (particularly firm to the touch with uniform, but not fully ripe color) are of primary importance in tomato purchase (Gormley and Egan, 1978; Beattie et al., 1983; Puig and Casado, 1983; Alvensleben, 1986). A list of characteristics identified as important in consumer acceptance of specific fruits and vegetables is provided by Kader (1985). Unfortunately, a single test does not establish acceptance once and for all, since consumer tastes change with time and are influenced by cultural factors. Carefully planned studies identify specific target markets. For example, German consumers prefer a fuller-flavored grapefruit whereas French consumers want one that is sweeter (Goldman and Givon, 1986). Regional preferences exist for apples in the United States; New Englanders prefer a tart taste and Southerners a sweeter fruit (Moyer and Aitkin, 1980).

Despite good intentions to serve the consumer, the grower must satisfy the immediate buyer (packer or distributor) to stay in business. Most packers, wholesale distributors, and retail sales operators buy fruits and vegetables on the basis of grades and standards, as mentioned earlier. Many postharvest systems thus are biased toward purchase quality attributes that are closer to grades and standards. Since these standards relate primarily to appearance of a product, they are criticized for keeping "organic" products out of United States markets for cosmetic reasons (National Research Council, 1989). Blaming the lack of acceptance of organic products on grades and standards ignores the importance consumers place on appearance when buying fresh produce (Zind, 1990). In the United States, organic products from commercial operations have been removed from shelves of major grocery chains because they do not have the purchase appeal of conventional fruits and vegetables (Prevor, 1990; Brumback, 1990).

When designing specifications for quality management programs of a specific commodity, attributes must be selected that can be used to predict both purchase and consumption quality as perceived by the consumer, as well as being readily quantifiable throughout the handling system. Techniques should be identified or developed for each attribute that would provide a single number on a linear scale to distinguish clearly between products of acceptable and unacceptable quality. Instrumental techniques usually are preferred to any other methods if they are rapid and provide reproducible results. Nondestructive instrumental methods are preferred to destructive ones since they decrease waste and permit repeated measures on the same items over time. Chemical methods usually are preferred to sensory techniques, primarily for reproducibility. For some products, no instrumental or chemical analyses are available that adequately predict consumer response. In these cases, objective scales are developed for a commodity and items are evaluated by an expert judge. Wang and Hruschka (1977) describe such a series of scales used to evaluate broccoli quality. At a minimum, the use of any technique must be validated by its relationship to sensory perception of small, experienced, or trained panels. When possible, these attributes should be tested to determine their ability to predict consumer acceptability in large, untrained panels.

When evaluating a system, quantification of key attributes should be made at major points in the handling system. A technique that is dependent on a single expensive instrument bound to a particular location is not useful. Quick, reliable, reproducible methods that can be performed by available personnel at each critical step are ideal. Clearly written quality specifications coupled with defined actions for specific circumstances are beneficial. Monitoring quality attributes should start as close to the field as practicable. The earlier in the handling system a problem can be detected, the greater are the chances of taking corrective action to minimize economic losses. For example, if a quality check reveals that a harvested crop is deteriorating more rapidly than normal, a decision can be made to (1) expedite shipping and handling to market and distribute it directly to consumers while the quality is still acceptable, (2) grade and sort items to save those that will be able to withstand normal handling and discard those that will not, or (3) stop shipment immediately and discard the lot before any additional input costs are incurred. If appropriate databases and models are available, mathematical prediction of quality attributes as described in Chapter 8 can be beneficial when making these decisions.

Some postharvest operations collect additional product at each sampling step and partition the sample into subsamples that will be analyzed immediately and those that will be stored under anticipated handling conditions. This practice helps increase the chances of detecting potential problems while they are still manageable and can provide insight into whether a problem is the result of inferior product or abusive handling conditions. Use of temperature recorders during transport or in storage rooms and of time–temperature indicators on the boxes of fresh product also can provide information about temperature abuse.

III. Sample Collection and Preparation

Part of any quality specification includes the number of samples and the frequency of collection. Requirements for individual commodities vary widely; specific recommendations are beyond the scope of this book. Factors that must be considered are normal fruit-to-fruit variation within a lot, seasonal and regional variability, degree of precision needed to predict acceptability, and the capabilities of the analytical facilities. A compromise must be reached between collecting so few samples that resultant information is meaningless and collecting more samples than can be analyzed accurately. Sampling plans should enhance the chances of detecting fruit-to-fruit variation in the lot at the expense of detecting variation in the methodology, that is, increase the number of fruit analyzed but not duplicate measures on the same fruit. Two-tiered sampling plans are also useful, in which a certain result triggers more detailed sampling.

All quality management programs must be well grounded in statistics, from the development of sampling schedules to the interpretation of results. Statistical methods cannot merely be added on to a fully developed management program, but must be integrated thoroughly into the entire process. Pitfalls to be avoided include (1) undercollection of data so no valid conclusions can be drawn, (2) overcollection of data to answer questions that are not relevant to management problems, (3) subtle changes in collection techniques that invalidate the results, and (4) failure to appreciate and account for the dynamic changes that occur in senescing plant tissue.

Once specifications, including sampling schedules, have been established, every effort must be made to provide the necessary equipment, supplies, and personnel at each sampling location. A full commitment to the quality program is needed to reap any benefits. Any scaling back of monitoring efforts must be done only after a careful assessment of the implications of the changes. For more details on this subject, see Chapter 13.

IV. Maturity Indices

Maturity at harvest is an important factor affecting quality perception and the rate of change of quality during postharvest handling. Thus, it is critical that measures of maturity be obtained. An ideal maturity index can be measured nondestructively and is "different at distinct levels of maturity and [does] not change with time of storage" (Shewfelt, 1984). Unfortunately, few such ideal measures exist.

Maturity indices can be determined in many ways, including estimation of the duration of development; measurement of size, weight, or density; physical attributes such as color, firmness, and moisture or solids content; other chemical attributes such as starch, sugar, or acid content; or morphological evaluation. Development of such indices can help separate maturity effects from storage and handling effects, thus permitting more effective predictive modeling. A list of maturity indices developed for different commodities is provided in Table I.

Table I

Indices Developed to Assess Maturity at Harvest of Selected Horticultural Crops

Crop	Assessment technique	Reference
Apple	Firmness	Magness and Taylor (1925); Lidster and Porritt (1978)
	Iodine-starch test	Bigelow et al. (1905); Reid et al. (1982)
Avocado	Nondestructive firmness	Peleg et al. (1990)
Bean, snap	Seed weight	Farkas (1967)
	Weight-to-length ratio	Ramaswamy et al. (1980)
Bean, lima	Alcohol-insoluble solids	Kramer and Smith (1947)
Cabbage	Density and weight	Isenberg et al. (1975)
Cantaloupe	Delayed light emission	Forbus and Senter (1989)
	Soluble solids	USDA (1961)
Cherry, sweet	Soluble solids	Drake and Fellman (1987)
Corn, sweet	Alcohol-insoluble solids	Twigg et al. (1956)
	Centrifugation layering	Arnold (1974)
	Density	Miller and Hughes (1969)
	Soluble solids	Drake and Nelson (1979)
Cucumber	Days after anthesis	Kanellis et al. (1986)
Grape	Degree days	Morris et al. (1980)
	Light sorting	Ballinger and McClure (1983)
Kiwifruit	Firmness and soluble solids	Crisosto et al. (1984)
Okra	Pod size	Gupta and Mukherjee (1982)
	Nonmucilaginous alcohol-insoluble solids	Ramaswamy and Ranganna (1982)
Orange	Brix-to-acid ratio	Ben-Arie and Lurie (1986)
Papaya	Delayed light emission	Forbus et al. (1987)
Peach	Color chips	Delwiche and Baumgardner (1985)
	Colorimetry	Delwiche et al. (1987)
	Delayed light emission	Forbus and Dull (1990)
Pear	Firmness	Leonard et al. (1954)
Pea, green	Tenderometer	Voisey and Nonnecte (1973)
Pea, southern	Alcohol-insoluble solids	Hoover and Dennison (1953)
	Corrected total solids	Shewfelt et al. (1985)
Pineapple	Specific gravity	Smith (1988)
Raspberry	Total anthocyanins	Sjulin and Robbins (1987)
Strawberry	Size	Morris et al. (1978)
Tomato	Color charts	USDA (1975); McGlasson et al. (1986)
	Colorimeter	Yang et al. (1987)
	Evaluation of internal gel formation	Brecht (1987)
	X-Ray tomography	Brecht et al. (1991)

A. Measuring Maturity

An estimation of the duration of development is a useful tool to determine the maturity of an item for research purposes. Tagging flowers and calculating the number of days from date of anthesis to harvest provides perhaps the best index of maturity for many crops. Some variation in development will be noted among cultivars and with variation in field conditions, but for many crops such as cucumbers

(Kanellis *et al.*, 1986) and onions (Steiner and Akintobi, 1986), this is the most reliable method. With large samples in commercial settings, the technique of tagging flowers becomes too burdensome. Degree-day accumulations have been modeled to predict optimal harvest dates for juice grapes (Morris *et al.*, 1980).

Size, weight, and density also are employed as maturity indices. Each individual fruit becomes larger as it matures on the plant, but other production factors such as amount of moisture, soil fertility, and source–sink relationships also affect the size of a fruit or vegetable. Pod size is used to describe maturity in okra (Gupta and Mukerjee, 1982) and fruit size is considered the most effective means of separating strawberries of different maturity classes (Sistrunk and Morris, 1985). Circumference measurements can be determined using calipers at the largest diameter of spherical products such as apples and oranges or at the largest or smallest diameter of odd-shaped items such as summer squash. Since seeds tend to change little after harvest, seed weight has been used as an index for snap bean maturity (Farkas, 1967), but weight-to-length ratio provides a better maturity index (Ramaswamy *et al.*, 1980). Specific gravity is used to index pineapple maturity (Smith, 1988).

Color is a useful measure for maturity of fruits. Chlorophyll tends to disappear as the fruit matures on the plant. Unfortunately, at least for the sake of maturity measurements, chlorophyll also tends to degrade during handling and storage. Anthocyanins are synthesized in light when attached to the plant, but are not affected greatly during postharvest handling. Total anthocyanin content shows the greatest potential as a maturity index for raspberries (Sjulin and Robbins, 1987). Standard color chips are available for assessment of maturity of many commodities including peaches (Delwiche and Baumgardner, 1985) and color charts are used to assess tomatoes (U.S. Department of Agriculture, 1975; McGlasson *et al.*, 1986). Colorimeters show potential for maturity assessment of peaches (Delwiche *et al.*, 1987) and tomatoes (Yang *et al.*, 1987). Fiber-optic light sorting can partition grapes into four maturity classes (Ballinger and McClure, 1983). Delayed light emission has the advantage of not being affected by orientation of multicolored samples in papayas (Forbus *et al.*, 1987), cantaloupes (Forbus and Senter, 1989), and peaches (Forbus and Dull, 1990).

Firmness of fruits is affected by maturity and storage. One of the first methods used to assess maturity (Magness and Taylor, 1925), firmness is a simple, rapid technique for measuring harvest maturity of apples (Lidster and Porritt, 1978) and pears (Quamme and Gray, 1985). A nondestructive firmness tester is available for assessing maturity of avocado fruit (Peleg *et al.*, 1990). Firmness is also an important quality attribute and usually is combined with another attribute, such as color, for use as an index of both maturity and quality. For example, Crisosto *et al.* (1984) recommend the use of firmness and soluble solids as an index for kiwifruit maturity.

Moisture or solids concentrations are used effectively as maturity indices. A simple moisture or total solids/100 g fresh weight measurement is not usually sufficient , since these characteristics vary with water conditions during crop production operations and postharvest handling. Mathematical modeling techniques are used to correct for rapid changes in southern peas during postharvest handling, given a time–temperature history of a specific lot (Shewfelt *et al.*, 1985). Alcohol-insoluble solids tend to be less variable, however, and can be used as maturity

Table II
Scoring of "Mature Green" Tomatoes Based on Visual Evaluation of Sliced Fruit

Score	Gel formation	Color change
M1	None	None
M2	Observed in at least one but not all locules	None
M3	Observed in all locules	None
M4	Observed in all locules	Red coloration in one or more locules but none on fruit surface

indices for southern peas (Hoover and Dennison, 1953; Sistrunk *et al.*, 1965), sweet corn (Twigg *et al.*, 1956), and lima beans (Kramer and Smith, 1947). Nonmucilaginous alcohol-insoluble solids are proposed as a maturity index for okra (Ramaswamy and Ranganna, 1982).

Measurement of starch and sugars is an additional means of assessing maturity. Iodine staining of starch content is used frequently in crops such as apples to provide a qualitative measure (Bigelow *et al.*, 1905; Reid *et al.*, 1982) by comparison of stained samples with pictures on a chart. Sugars are the industry standard to assess the maturity of cantaloupes (U.S. Department of Agriculture, 1961) and other melons. Soluble solids are suggested as a maturity index for sweet cherries (Drake and Fellman, 1987). Maturity standards for citrus fruits are based on Brix-to-acid ratios (Ben-Arie and Lurie, 1986). Both soluble solids and acidity are used to assess grape maturity (Morris, 1985).

Maturity can be assessed in some crops by morphological examination. Most notably, mature green tomato fruit can be subdivided into four distinct physiological stages that are not distinguishable by external visual evaluation. By slicing the fruit, however, and looking at internal morphology, the stages can be distinguished by observing locular gel formation (Table II). The difference in ripening to breaker stage in fruit not treated with exogenous ethylene is only 2–3 days for stage M4 and more than 16 days for stage M1 (Brecht, 1987). Potential use of X-ray tomography for nondestructive evaluation of tomato maturity had been demonstrated (Brecht *et al.*, 1991; Chapter 11).

B. Evaluating the Effects of Maturity on Quality

Although the effects of maturity at harvest on quality and storage stability of numerous commodities are accepted widely, the use of maturity indices to separate "maturity effects" from "handling effects" has not been exploited sufficiently. Two techniques are available to quantify maturity effects:

- Separate maturity into discrete classes and plot the change in a particular quality attribute of each class as a function of storage time.
- Treat maturity as a continuous variable and plot change in a particular quality attribute at distinct steps in the handling process as a function of the maturity index.

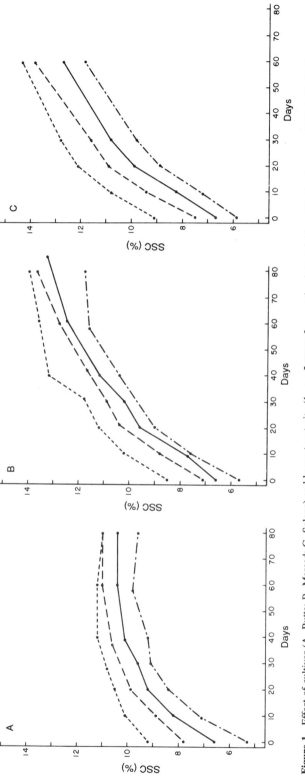

Figure 1 Effect of cultivar (A, Butte; B, Merced; C, Solano) and harvest maturity (1, ‒‒‒; 2, ‒‒‒; 3, ——; 4, —·—) on soluble solids development (SSC) in kiwifruit stored for 80 days at 0°C (Reprinted with permission from Crisosto *et al.*, (1984). *Am. Soc. Hort. Sci.* **109(4)**, 584–587.

An example of maturity class plots, shown in Fig. 1, is the interaction of maturity and cultivar on soluble solids development in kiwifruit during postharvest storage. In this specific case, harvesting 'Merced' and 'Solano' at a more advanced stage of maturity is more critical than harvesting 'Butte' at a particular stage. Other examples of these plots are published for muscadine grapes (Ballinger and McClure, 1983) and tomatoes (Shewfelt *et al.*, 1987b). Such information, when combined with the limits of acceptability for critical quality attributes and expected postharvest lives, can be used to determine mathematically the limits of optimal maturity at harvest for a commodity. These maturity specifications must be evaluated carefully under commercial conditions and might require some adjustment, but the process puts maturity evaluation on a much more solid scientific basis than does the generally accepted earlier-is-better approach (see Chapter 4).

An example of a maturity index as a continuous, dependent variable for summer squash is shown in Fig. 2. In this specific case, a hue angle of 92° is the optimal hue for sensory color acceptance, indicating that a squash with an external length of 15 cm from proximal to distal end will have the best color acceptance within this defined system, which consists of 10 days from harvest to sale. Such plots provide a clearer picture of the optimal maturity range than the maturity class plots, but they must be viewed with some caution. Use of continuous indices is preferable when the index is highly accurate and can be measured precisely, and the relationship of maturity and quality is correlated highly. Both techniques are dependent on how closely the defined test system mirrors actual handling conditions. It can be seen that changes in storage duration within the system could have profound effects on recommended maturity levels. For example, shorter periods between harvest and

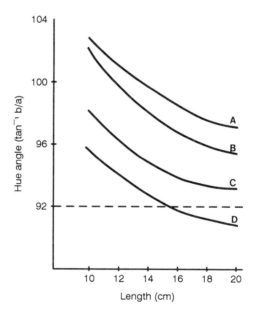

Figure 2 Effect of maturity (external length from proximal to distal end) and simulated handling step (A, packing-house departure 1 day postharvest; B, wholesale warehouse arrival after 2 days at 5°C; C, wholesale warehouse departure after 4 days at 10°C; D, retail sale after 3 days at 10°C) on hue angle of summer squash. Hue angle of 92° was considered optimal by a sensory panel (——).

sale would permit the sale of more mature (> 15 cm) squash at optimal maturity but might limit the sale of smaller squash that could be perceived as too green.

V. Measuring Quality

A. Visual Evaluation

Visual evaluation of quality characteristics by an expert judge, despite limitations, is still a widely used and accepted technique. Numerical scales for specific attributes are available for commodities when no chemical or physical measure is available that relates to a specific purchase characteristic. A listing of many of these scales is provided in Table III. Such scales are treated as objective measures, but they suffer from many of the problems of sensory analysis without having many of the safeguards of those techniques:

- Scoring is subject to variability by expert judge.
- It is almost impossible to "blind" the judge to treatments, particularly when the same samples are evaluated over time in a storage study.
- The full range of the scales rarely is used, since studies usually are stopped when the sample drifts into the lower (poor quality) end of the scale.
- Results tend to be analyzed, assuming linearity of the scale, although no clear evidence exists in many cases that points on the scale are at equal intervals.

On the other hand, an experienced judge can detect subtle changes well before any differences can be detected by instruments or sensory panels. Sample variability

Table III
Rating Scales Developed for Specific Commodities

Commodity	Attributes	Range	Reference
Broccoli	Color, compactness, decay, flavor, odor, opening of florets, salability, turgor	0–10	Wang (1979)
Cauliflower	Incidence and extent of curd spots, low oxygen injury, soft rot	1–9	Lipton and Harris (1976)
Cucumbers	Severity of chilling injury	1–5	Wang and Baker (1979)
Lettuce	Firmness, butt discoloration, decay, visual quality, wilting, other defects, overall quality	1–5	Kader et al. (1973)
Bell peppers	Decay and shrivel	1–5	Sherman et al. (1982)
	Visual quality	1–9	
Tomatoes	Finger firmness	1–9	Kader et al. (1978)
Zucchini squash	Firmness	1–5	Mencarelli et al. (1983)
	Color, overall quality, severity of low oxygen and chilling injury	1–9	

within a treatment, the short length of most storage studies (which frequently span weekends), and large sample size requirements prevent the use of sensory panels in experimental studies or for routine quality control checks. Examples of attributes for which no adequate physical or chemical measure is available include wilting of leafy vegetables (Kader *et al.*, 1973; Yano *et al.*, 1981), visual color of broccoli (Brennan and Shewfelt, 1989) and other green vegetables (Gnanasekharan *et al.*, 1992), and visual evidence of physiological disorders.

When faced with the need to use a visual evaluation technique or another sense (such as smell) to evaluate quality without consumption, these guidelines are recommended:

1. When possible, use a previously published scale so results can be related to previous studies.
2. Evaluate only those characteristics that relate directly to purchase quality attributes of the intended end use of the product.
3. Select as the expert judge someone who has little or no detailed knowledge of the design of the study and no stake in the results (i.e., not the graduate student who designed the study).
4. Use the same judge throughout each study or, in a quality control environment, minimize the number of judges. When possible, have two or more judges independently evaluate each item.
5. Periodic discussions should be held to refresh the judge(s) on definitions of key terms. (These discussions, however, should never be conducted in the middle of an on-going experiment in which it is essential to maintain consistency of interpretation.)
6. The scale should be evaluated for its ability to predict likability by an untrained consumer panel (30 or more panelists) for both the specific attribute(s) and overall acceptability.

B. Color

Measurement of color is an important means of quality assessment of food products. Although color of fruits and vegetables is an external manifestation of composition and form of plant pigments, a simple compositional analysis of extracted pigments does not necessarily predict visual impact. Fruit ripening and vegetable yellowing frequently involve the unmasking of yellow-to-orange xanthophylls and carotenes by the disappearance of chlorophyll (Gross, 1987). A direct measure of chlorophyll concentration, however, is a poor predictor of the visual impact of broccoli color (Shewfelt *et al.*, 1984). Anthocyanins are the primary pigments in blueberries, present in the fruit in metal-ion complexes. When extracted, however, the pigment is red with little resemblance to the blue coloration of whole fruit. Coloration of anthocyanins is highly dependent on the intracellular environment, particularly pH (Franics, 1989). Traditional spectrophotometric methods for total anthocyanins (Fuleki and Francis, 1968), betalains (von Elbe *et al.*, 1972), chlorophyll (Lebermann *et al.*, 1968), and carotenoids are being replaced by HPLC methods that separate individual pigments. Published HPLC separation methods exist for the anthocyanins of several items, including grapes (Williams *et al.*, 1978) and cranberries (Camire and Clydesdale, 1979), as well as for betalains of beets (Schwartz and von Elbe, 1980), carotenoids (Bushway, 1986), and chlorophylls and their degradative products of green vegetables (reviewed by Schwartz and Lorenzo, 1990).

Measurement of changes in pigments is important in understanding the physiology of ripening and senescence. In measuring changes in visual impact, however, it is more important to detect physical changes in the appearance. Although appearance is a function of more than just color, and there are instruments available to detect these other factors (Hunter, 1987), this discussion will focus on color measurement. Many color scales have been developed but the predominant scale used for fruits and vegetables is the Hunter "Lab" or its variant CIE L*a*b*. For most applications, either scale provides meaningful information. Since most investigators are switching to the CIE L*a*b* system, it is the scale of choice. For the sake of simplicity, the following discussion refers to Lab.

In selecting color-measuring equipment, careful attention must be paid to the specific applications desired and the range of commodities or products to be tested. In the experimental stage, sample orientation and light aperture are critical. These factors are described in detail elsewhere (Clydesdale, 1991) and are not covered in this chapter.

The most frequent error in color measurement is the use of Lab results directly without conversion to hue, value, and chroma. The primary reason food scientists use food colorimeters is that the readings are related to human color perception; this perception influences consumer acceptance of the product. Humans and colorimeters "see" color differently. Humans see the color of a product in terms of its lightness, hue (color name such as red, blue, or green), and chroma (brightness or saturation) by integrating some very complex signals into these three components. Colorimeters do not have the capacity to integrate directly and, thus, must break the signal down into a simpler construct. Instruments "see" in terms of lightness (L), red–green character in the absence of yellow or blue components (a), and yellow–blue character in the absence of red or green components (b). L, a, and b measures are machine language, whereas hue, chroma, and lightness are terms that relate to human perception. Fortunately, we can convert the machine language, through some rather simple mathematical calculations, to numbers that have relevance to humans.

As soon as the specific terms of hue (for example, red or yellow) are used, different things are being said in machine and human terms. To the machine, an increase in yellowness is signaled by an increase in the magnitude of $+b$, whereas in human terms an increase in yellowness is signaled by the closeness of the hue angle ($\tan^{-1} b/a$) to 90°. Thus, in human terms, the yellowness of a sample can increase even if the $+b$ reading decreases if the $+a$ reading exhibits a greater decrease. Likewise, the yellowness of a sample can decrease even if the $+b$ reading increases if the $+a$ reading exhibits a greater increase.

In the example shown in Table IV and Fig. 3, apple S is more yellow than apple R, which is in turn more yellow than apple T to the instrument. In terms of human perception, however, the ranking of yellowness of the samples is just the opposite. Differences in chroma also may affect human perception in this case, but hue usually is more important in perception of fruit and vegetable quality.

Although color is related primarily to maturity or purchase quality, it also may

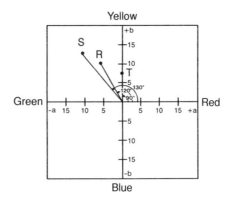

Figure 3 Illustration of the misleading conclusions drawn on use of "+b" readings to determine yellowness of a sample. (See Table IV.)

contribute to consumption quality. Johnson and Clydesdale (1982) show that dark-colored beverages are perceived to be sweeter than lighter-colored counterparts presented with the same flavorings and sugar concentration. Flesh color of many fruits and vegetables may not be observed until the time of consumption and may provide a different quality perception than external color. When measuring flesh color, the sample should be measured as soon after cutting the fruit as possible to avoid changes due to browning or desiccation. In addition, it is a good practice to clean any juice from the sample port between measurements.

Other noninvasive techniques employing irradiation with visible or ultraviolet light have been used to develop nondestructive quality measurements. Most notable of these techniques are those that relate reflected and transmitted radiation to soluble solids concentrations in many fruits (see reviews by Deshpande *et al.*, 1984; Dull, 1986). A detailed description of nondestructive evaluation techniques is provided in Chapter 11.

C. Texture

Firmness is the primary textural attribute measured in fruits and vegetables. Firmness usually is measured by destructive puncture tests including hand-held Effegi (Abbott *et al.*, 1976), foot pedal modifications of hand-held instruments (Shewfelt *et al.*,

Table IV
Hypothetical Example to Demonstrate Differences in Yellowness of 'Golden Delicious' Apples as Perceived by Humans and Instruments[a]

Apple	L	a	b	b/a	Hue angle (°)
R	35	−5.0	+10.0	−2.00	120
S	35	−10.0	+12.5	−1.25	130
T	35	0	+7.5	∞	90

[a]See text for explanation of abbreviations used.

1987a,b) and mechanized Instron tests. An indication of firmness is obtained by the force necessary to cause penetration of a standard probe a specified distance into the product. These tests are being replaced by more nondestructive deformation tests (Bourne, 1973, 1982a; Watada *et al.*, 1976). Although deformation tests are thought to be more accurate, Kader *et al.* (1978) showed similar correlation coefficients for the relationships of puncture and deformation tests with finger firmness of tomatoes. Deformation tests do have the advantage over puncture tests of being repeatable on the same fruit throughout a storage study. Whether these deformation tests are truly nondestructive, however, is still a matter of debate, since even small deformations may induce membrane damage or generate wound ethylene (Yang and Pratt, 1978). Such concerns may be minor, since all means of sampling and measurement may introduce similar responses.

As in color measurement, sample presentation for textural analysis is important. The size of the surface area for puncture or deformation, the geometry of the sample, the means of support, and the interaction of the instrument and the sample all affect results (Bourne, 1982a). In puncture tests, a decision must be made about whether the peel should be retained or removed. In analyzing tomatoes, the peel usually is retained, but in analyzing peaches it usually is removed. Temperature of the samples can affect measurements (Bourne, 1982b) and should be standardized. Penetration instruments yield data about firmness as a force. Thus, the SI unit of force, the Newton (N), should be used to report all results; probe diameter also must be reported. For more details on measurement of food texture, see Bourne (1982a).

D. Flavor

Chemical analysis of fruit and vegetable composition is used primarily to estimate consumption quality and hidden attributes. Sweetness is a function of sugar concentration and sourness a function of acidity. Consumer perception of sweetness or sourness is related to the ratio of sugars and acids. Sugar composition usually is estimated by measuring the percentage of soluble solids (Brix) using a refractometer (Ruck, 1963). Acidity is determined by titration with standard base and expressing the results in terms of the predominant acid present. The Brix-to-acid ratio (BAR) is used to assess relative sweetness or sourness for most fruits. In tomatoes, however, flavor impact tends to be related to the total sugars *and* acids present (Jones and Scott, 1983). More detailed analysis and separation of individual sugars and acids can be determined using HPLC (Spanos and Wrolstad, 1987).

Volatile compounds are responsible for the distinctive aromas associated with fruits and vegetables. These compounds, in combination with taste sensations (sweet, sour, and bitter), form characteristic flavors. The volatile constituents of numerous fruits have been isolated and characterized. As many as 225 volatile compounds are found in grapes (Schreier *et al.*, 1976) using gas chromatography. Many fruits contain a single volatile compound, known as a character impact compound, that conveys the flavor message by itself. Examples of character impact compounds are ethyl 2-methylbutyrate in apples, γ-decalactone in peaches, and isoamylacetate in bananas (Shewfelt, 1986). Full aroma of any fruit, however, is a subtle combination of many compounds, which is why duplicating fruit flavors in artificial beverages is so difficult. Determining the volatile compounds responsible for aroma and flavor is a difficult task. In one of the most comprehensive studies of a fruit flavor to date, Buttery *et al.* (1989) note that 30 volatiles contribute to tomato flavor. The complexity of flavor and our inability to relate peaks of a few compounds to consumer perception of flavor adequately greatly limits our ability to incorporate flavor into quality evaluation programs. Presence of bitter compounds such as limonene and naringen in citrus fruits (Shewfelt, 1986) or absence of a critical flavor component such as (Z)-3-dexenal in chilled tomato fruits (Buttery *et al.*, 1987) are examples of flavor problems identified using gas chromatography.

E.   Nutrients

Vitamins and minerals are hidden attributes that affect consumer perception. Nutrient composition varies widely in raw commodities because of genetics, preharvest factors (soil fertility, moisture content of the soil, growth temperature, growth regulators, and cultural practices), maturity at harvest, and postharvest handling conditions (mechanical damage, storage times, temperatures, relative humidity, gaseous atmosphere, and use of additives). Despite the importance of these compounds, little is known about the rates of degradation of nutrients during postharvest handling (Shewfelt, 1990). Consumers buy certain items as good sources of specific nutrients, for example, leafy green vegetables for vitamin A, oranges for vitamin C, and bananas for magnesium and potassium. Without sophisticated analytical equipment, however, the consumer cannot detect differences in individual products at the point of purchase (Institute of Food Technologists, 1990). Thus, there is little incentive to measure nutrient content in a quality control program unless specific nutritional claims can be made. The two most commonly measured nutrients in fruits and vegetables are ascorbic acid (vitamin C) and β-carotene (pro-vitamin A). Ascorbic acid is measured by iodometric titration (Jacobs, 1958) or HPLC (Bushway *et al.*, 1989). Spectrophotometric and HPLC (Rodriquez-Amaya, 1989) methods are available for β-carotene. For a recent review, see Eitenmiller (1990). Mineral analysis usually is performed by ashing and atomic absorption (Pomeranz and Meloan, 1987).

VI.   Sensory Evaluation Techniques

Known widely as "taste" testing, sensory evaluation incorporates a much wider range of senses than merely taste. Taste is the sense that detects chemical properties of foods in the mouth in the absence of aroma. Usually considered to be limited to sweet, sour, salty, and bitter sensations, taste is more complex (O'Mahony, 1991). The senses of smell (Lawless, 1991), to detect aroma, which combines with taste to form flavor, sight (Clydesdale, 1991), which detects color and other appearance characteristics, kinesthetics, which detect textural attributes by hands and mouth (mouthfeel) (Szczeniak, 1991), and even sound (Vickers, 1991), which is an indication of crispness and crunchiness, all can play a role in an understanding of sensory perception of food quality.

A. Types of Sensory Tests

Sensory tests are divided into affective and analytical tests (Institute of Food Technologists, 1981). Affective tests provide information on the preference (liking one sample better than other) or acceptance (how much is a sample liked or disliked) of products. Analytical tests seek to determine the level of specific attributes or the sensitivities of panelists. Most postharvest studies and quality control tests are designed to answer questions that require affective tests, whereas most tests conducted tend to be analytical. Examples of questions requiring affective tests include

- Which treatment results in the preferred product?
- Is this product acceptable and will it remain acceptable long enough to satisfy consumer needs?

Unfortunately, analytical sensory tests are not designed to provide meaningful answers to such questions.

1. Affective tests

A minimum of 24 untrained panelists is essential to place any confidence in preference test results; usually 50–100 panelists are needed to provide adequate information. A demographic profile of the panelists is important to provide insight into wider applicability of the results (e.g., 24 white Anglo-Saxon men might not provide an accurate projection of consumers in New York City). Score sheets typically ask panelists to rank the samples in order of preference or rate each product from 9 (like extremely) to 1 (dislike extremely), known as hedonic scaling. Ranking tests give more direct information about which sample is preferred, but give no information about how much a sample is preferred and why. Hedonic scales are treated statistically as linear equal-interval scales, although panelists tend to ignore both extremes. These scales are more readily adaptable to obtaining information about some specific attributes. More details on consumer acceptance studies are available in Chapter 7.

2. Analytical tests

Analytical tests can be subdivided into descriptive and discriminative tests. Descriptive tests measure and quantify specific attributes of a product, for example, sweetness, juiciness, or flesh color, whereas discriminative tests determine differences in samples and products.

In descriptive tests, the panelist is asked to rate the intensity of a particular attribute on a scale. Two such scaling techniques are quantitative descriptive analysis (QDA) (Stone et al., 1974) and magnitude estimation (Moskowitz and Sidell, 1971). An example of a QDA score sheet for peaches is provided in Fig. 4. Note that the panelist is not asked to indicate which sample is preferred. For example, some panelists may prefer sweet apples whereas others prefer tart ones, but such opinions are not relevant to these tests. Descriptive tests provide important information about specific attributes and should be incorporated into any study in which appropriate chemical or physical tests cannot be developed for critical consumption attributes. Normally, descriptive panels consist of 5–15 trained panelists, usually

SENSORY EVALUATION OF PEACHES

Name _____ Date _____ A.M./P.M.   Set _____

Please evaluate these samples of peaches using the rating scales below. Place vertical marks on each of the scales to indicate your rating of each sample. Label each mark with the code number of the sample it represents.

THE SAMPLE CODE NOS. ARE: _____   _____   _____   _____   _____   _____

YOU SHOULD HAVE SIX MARKS ON EACH SCALE WHEN YOU COMPLETE THIS.

FLESH COLOR
      green   yellow   red-orange

FLAVOR
Sweetness
      too bland   about right   too sweet

Sourness
      too bland   about right   too sour

Peach flavor
intensity
      weak   moderate   strong

Overall flavor
intensity
      weak   moderate   strong

Off flavor
(IF ANY)
      slight   moderate   strong

DESCRIBE OFF-FLAVOR _____

PLEASE TAKE YOUR SCORE CARD TO THE KITCHEN AREA TO EVALUATE SAMPLES FOR COLOR, FIRMNESS TO THE TOUCH, AND OVERALL PREFERENCE.

FLESH COLOR
      green   yellow   red-orange

OVERALL
EXTERNAL COLOR
      fair   good   excellent

FIRMNESS TO
THE TOUCH
      not firm   moderately firm   very firm

Figure 4   Sensory panel sheet for evaluation of peaches.

permanent support staff personnel. Experienced panels contain panelists with a familiarity with the terminology and quality characteristics of the product. Panelists have normal sensory acuity, in contrast with members of highly trained panels, who can detect subtle changes in a product. It is critical that, within a given study, changes in the composition of a panel are minimized. A panelist whose scores differ from the norm still can be useful if judgments are consistent, but can skew results greatly if present for only part of the study. Managers, who rarely can be tied down to a specific location at a specific time on a predictable schedule, and students, who tend to graduate, make poor panelists.

Discriminative tests can be subdivided further into difference and sensitivity tests. Difference tests such as paired-comparison, duo-trio, or triangle tests can be used to determine if two products differ from each other. Postharvest scientists largely have ignored this useful technique. An excellent example, however, is provided for walnuts by Ingels *et al.* (1990). A difference test can be used to determine if a new handling technique results in a detectable difference in overall quality or in a specific attribute. For example, if a new handling technique is introduced to improve the firmness of fresh avocadoes during shipment, demonstration of detectable differences in firmness would provide strong support to adopt the technique. However, finding no *significant* difference is not equivalent to finding no difference! Statistical tests are designed to minimize the risk of a Type I error (stating there is a difference when none exists) at the expense of making a Type II error (stating there is no difference when one exists) (Steel and Torrie, 1980). Unfortunately, there are no simple tests to determine if a modification of a system (for example, changing precooling temperature requirements) will result in a product of comparable quality (O'Mahony, 1982). Standard tests can detect only significant differences in quality, if they exist.

B. Sample Preparation and Presentation

Sensory evaluation tests usually are performed in special facilities housing a number of individual booths. These booths should provide an atmosphere conducive to making sound judgements—clean, adequately lighted and ventilated, free from audio and visual distractions, equipped with a sink for rinsing and expectoration, with ready access to the food preparation area. Samples should be presented to panelists in a form in which and at a temperature at which the item is consumed normally. Samples should be coded in a fashion that will not bias a panelist (three-digit codes extracted from a random number table are sufficient) and should be presented in a random fashion to avoid first or last sample biases. Consider the number of samples provided at a sitting to avoid panelist fatigue. Consuming an unsalted cracker between strong samples, such as raw onions, helps prevent carryover. Otherwise, a water rinse is usually sufficient. Distilled water is preferable, particularly if a pronounced flavor is present in tap water. Tests should be conducted at the same time each day, preferably not to interfere with normal break times or close to a normal mealtime. For more details on panel environment and sample preparation, see Institute of Food Technologists (1981) or Larmond (1987).

Design of a proper questionnaire for a sensory test is critical. Does the questionnaire adequately address the test objectives? Is the questionnaire readily understandable using unequivocal language? Is it too long, so it taxes the panelist?

Does it present the samples to the panelist in the same order in which they are presented physically? As in any other analytical test, attention to detail is essential to generate valid, accurate data from sensory tests.

C. Evaluating Purchase and Consumption Attributes

Most sensory tests are associated with consumption attributes such as flavor and mouthfeel. As described earlier, sensory evaluation may be the only valid measure for consumption attributes for certain crops. Purchase attributes also can be evaluated by sensory techniques. Color, other appearance attributes, firmness to the touch, and aroma are purchase attributes that can be assessed. Usually, purchase attributes are not evaluated in booths, but are measured on a well-lighted counter top. No communication is allowed between panelists. Unless individual items are small (peas or blueberries, for example), they should be evaluated individually and not in clusters of 2 or 3. More samples can be evaluated for purchase attributes by a panelist, since fatigue is usually not as much of a factor as it is for consumption attributes. When measuring consumption and purchase attributes as part of the same test, it is usually preferable to perform consumption tests first, followed by purchase tests, since the latter are more likely to bias the former than vice versa. When filtered light is being used to screen out color differences, however, some time must be permitted between consumption and purchase evaluation to adapt to normal lighting, or the purchase attributes should be evaluated first. In any case, coding of samples for purchase and consumption samples should be different. It is always tempting to compare purchase and consumption attributes, although experience has suggested consistently that purchase attributes are not reliable predictors of consumption attributes. One goal of a systems approach is to improve consumption quality while maintaining acceptable purchase quality.

D. Correlating Sensory and Physicochemical Results

Quality tests are only meaningful if they relate to consumer acceptance. In the absence of consumer acceptance data for many commodities, most chemical and physical tests are evaluated for their ability to correlate with sensory results. In many cases, simple correlation coefficients are used; a coefficient of 0.9 ($R^2 = 0.81$) is preferable and 0.8 ($R^2 = 0.64$) is considered acceptable. More sophisticated techniques have been developed using cluster analysis and factorial analysis (Powers et al., 1977; Resurreccion et al., 1987), which help reduce the data required to discriminate between samples from multiple attributes to a few critical ones. In any decision-making process, it is critical not to let the level of statistical significance obscure the practical implications of the results.

VII. Quality in a Systems Context

Quality is only one of the factors that influence consumer acceptability of a fruit or vegetable, but it is the only factor that is intrinsic to the item and the factor most directly affected by handling and storage conditions. Quality can be divided into purchase, consumption, and hidden attributes. Each commodity has a unique set of quality attributes desired by the consumer. Maturity and quality indices have been developed for many commodities to permit quality evaluation of an item through the system and to separate "maturity effects" from "handling effects." Sensory evaluation represents an important means of assessing quality, but frequently is misapplied in postharvest experiments. An understanding of the interaction of production systems and subsequent handling steps to affect quality represents the greatest potential application of a systems approach to postharvest handling.

Bibliography

Abbott, J. A., Watada, A. E., and Massie, D. R. (1976). Effe-gi, Magness–Taylor, and Instron fruit pressure testing devices for apples, peaches, and nectarines. *J. Amer. Soc. Hort. Sci.* **101**, 698–700.

Alvensleben, R. V. (1986). Consumer and wholesale preferences for fresh vegetables, especially tomatoes in Federal Republic of Germany. Paper presented at Optimal Yield Management Symposium-AgriTech 86, Tel Aviv, Israel. Reprint available from author at Institute for Horticultural Economics, University of Hanover.

Arnold, C. Y. (1974). A preliminary evaluation of a centrifuge technique for measuring maturity in sweet corn. *HortScience* **9**, 78–79.

Ballinger, W. E., and McClure, W. F. (1983). The effect of ripeness on storage quality of 'Carlos' muscadine grapes. *Sci. Hort.* **18**, 241–245.

Beattie, B. B., Kavanagh, E. E., McGlasson, W. B., Smith, E. F., and Best, D. J. (1983). Fresh market tomatoes: A study of consumer attitudes and quality of fruit offered for sale in Sydney 1981–82. *Food Technol. Austral.* **35**, 450–454.

Ben-Arie, R., and Lurie, S. (1986). Prolongation of fruit life after harvest. *In* "CRC Handbook of Fruit Set and Developement" (S. P. Monseliese, ed.), pp. 493–520. CRC Press, Boca Raton, Florida.

Bigelow, W. D., Gore, H. C., and Howard, D. J. (1905). "Studies on Apples." USDA Bureau of Chemistry Bulletin No. 94. U.S. Govt. Printing Office, Washington D.C.

Bourne, M. C. (1973). Use of the penetrometer for deformation testing of foods. *J. Food Sci.* **38**, 720–721.

Bourne, M. C. (1982a). "Food Texture and Viscosity: Concept and Measurement." Academic Press, Orlando, Florida.

Bourne, M. C. (1982b). Effect of temperature on firmness of raw fruits and vegetables. *J. Food Sci.* **47**, 440–444.

Brennan, P. S., and Shewfelt, R. L. (1989). Effect of cooling delay at harvest on broccoli quality during postharvest storage. *J. Food Qual.* **12**, 13–22.

Brecht, J. K. (1987). Locular gel formation in developing tomato fruit and the initiation of ethylene production. *HortScience* **22**, 476–479.

Brecht, J. K., Shewfelt, R. L., Garner, J. C., and Tollner, E. W. (1991). Using X-ray-computed tomography to nondestructively determine maturity of green tomatoes. *HortScience* **26**, 45–47.

Brumback, N. (1990). Organically grown produce: How to tell if it's the real thing. *Produce Business* **6(5)**, 38–43.

Bushway, R. J. (1986). Determination of α and β-carotene in some raw fruits and vegetables by high-performance liquid chromatography. *J. Agric. Food Chem.* **34**, 409–412.

Bushway, R. J., Helper, P. R., King, J., Perkins, B., and Krishman, M. (1989). Comparison of ascorbic acid content of supermarket versus roadside stand produce. *J. Food Qual.* **12**, 99–105.

Buttery, R. G., Teranishi, R., and Ling, L. C. (1987). Fresh tomato aroma volatiles: A quantitative study. *J. Agric. Food Chem.* **35**, 540–544.

Buttery, R. G., Teranishi, R., Flath, R. A., and Ling, L. C. (1989). Fresh tomato volatiles: Composition and sensory studies. *In* "Flavor Chemistry Trends and Developments" (R. Teranishi, R. G. Buttery, and F. Shahidi, eds.). pp. 213–222. ACS Symposium Series 388. American Chemical Society, Washington, D.C.

Camire, A. L., and Clydesdale, F. M. (1979). High-pressure liquid chromatography of cranberry anthocyanins. *J. Food Sci.* **44**, 926–927.

Clydesdale, F. M. (1991). Color perception and food quality. *J. Food Qual.* **14**, 61–73.

Crisosto, G. U., Mitchell, F. G., Arpaia, M. L., and Mayer, G. (1984). The effect of growing location and harvest maturity on the storage performance and quality of 'Hayward' kiwifruit. *J. Amer. Soc. Hort. Sci.* **109**, 584–587.

Delwiche, M. J., and Baumgardner, R. A. (1985). Ground color as a peach maturity index. *J. Amer. Soc. Hort. Sci.* **110**, 53–57.

Delwiche, M. J., Tang, S., and Rumsey, J. W. (1987). Color and optical properties of clingstone peaches related to maturity. *Trans. ASAE* **30**, 1873–1879.

Deshpande, S. S., Cheryan, M., Gunasekaran, S., Paulsen, M. R., and Salunkhe, D. K. (1984). Nondestructive optical methods of food quality evaluation. *Crit. Rev. Food Sci. Nutr.* **21**, 323–379.

Drake, S. R., and Fellman, J. S. (1987). Indicators of maturity and storage quality of 'Ranier' sweet cherry. *HortScience* **22**, 283–285.

Drake, S. R., and Nelson, J. W. (1979). A comparison of three methods of maturity determination in sweet corn. *HortScience* **14**, 546–548.

Dull, G. G. (1986). Nondestructive evaluation of quality of stored fruits and vegetables. *Food Technol.* **40(5)**, 106–110.

Eitenmiller, R. R. (1990). Strengths and weaknesses of assessing vitamin content of foods. *J. Food Qual.* **13**, 7–20.

Eskin, N. A. M. (1989). "Quality and Preservation of Vegetables." CRC Press, Boca Raton, Florida.

Farkas, D. F. (1967). Use of seed size for controlling snap bean quality for processing. *Food Technol.* **21**, 789–791.

Forbus, W. R., and Dull, G. G. (1990). Delayed light emission as an indicator of peach maturity. *J. Food Sci.* **55**, 1581–1584.

Forbus, W. R., and Senter, S. D. (1989). Delayed light emission as an indicator of cantaloupe maturity. *J. Food Sci.* **54**, 1094–1095.

Forbus, W. R., Senter, S. D., and Chan, H. T. (1987). Measurement of papaya maturity by delayed light emission. *J. Food Sci.* **52**, 356–360.

Francis, F. J. (1989). Food colorants anthocyanins. *Crit. Rev. Food Sci. Nutr.* **28**, 273–314.

Fuleki, T., and Francis, F. J. (1968). Quantitative methods for anthocyanins. 2. Determination of total anthocyanin and degradation index for cranberry juice. *J. Food Sci.* **33**, 78–83.

Gnanasekharan, V., Shewfelt, R. L., and Chinnan, M. S. (1992). Detection of color changes in green vegetables. *J. Food Sci.* **57**, 149–154.

Goldman, A., and Givon, M. (1986). "Taste Preferences for Grapefruit in Germany and France." Research Report. Citrus Marketing Board of Israel.

Gormley, R., and Egan, S. (1978). Firmness and colour of the fruit of some tomato cultivars from various sources during storage. *J. Sci. Fd. Agric.* **29**, 534–538.

Gross, J. (1987). "Pigments in Fruits." Academic Press, Orlando, Florida.

Gupta, V. K., and Mukherjee, D. (1982). Effect of wax emulsion with and without morphactin on the shelf-life of okra pods of different maturities. *Sci. Hort.* **16**, 201–207.

Hoover, M. W., and Dennison, R. A. (1953). A study of certain biochemical changes occurring in the southern pea, *Vigna sinensis,* at six stages of maturity. *Proc. Amer. Soc. Hort. Sci.* **63**, 402–408.

Hunter, R. S. (1987). "The Measurement of Appearance." John Wiley & Sons, New York.

Institute of Food Technologists (1981). Sensory evaluation guide for testing food and beverage products. *Food Technol.* **35(11)**, 50–59.

Institute of Food Technologists (1990). Quality of fruits and vegetables. A scientific status summary by the Institute of Food Technologists' Expert Panel on Food Safety and Nutrition. *Food Technol.* **44(6)**, 99–106.

Ingels, C. A., McGranahan, G. H., and Noble, A. C. (1990). Sensory evaluation of selected Persian walnut cultivars. *HortScience* **25**, 1446–1447.

Isenberg, F. M. R., Pendergress, A., Carroll, J. E., Howell, L., and Oyer, E. B. (1975). The use of weight, density, heat units, and solar radiation to predict the maturity of cabbage for storage. *J. Amer. Soc. Hort. Sci.* **100**, 313–316.

Jacobs, M. B. (1958). "The Chemical Analysis of Foods and Food Products," 3d Ed. Van Nostrand, New York.

Johnson, J., and Clydesdale, F. M. (1982). Perceived sweetness and redness in colored sucrose solutions. *J. Food Sci.* **47,** 747–752.

Jones, R. A., and Scott, S. J. (1983). Improvement of tomato flavor by genetically increasing sugar and acid contents. *Euphytica* **32,** 845–855.

Kader, A. A. (1985). Quality factors: Definition and evaluation for fresh horticultural crops. *In* "Postharvest Technology of Horticultural Crops" (A. A. Kader, ed.), pp. 118–121. Agriculture and Natural Resources Publications, University of California, Berkeley.

Kader, A. A., Lipton, W. J., and Morris, L. L. (1973). Systems for scoring quality of harvested lettuce. *HortScience* **8,** 408–409.

Kader, A. A., Morris, L. L., and Chen, P. (1978). Evaluation of two objective methods and a subjective rating scale for measuring tomato fruit firmness. *J. Amer. Soc. Hort. Sci.* **103,** 70–73.

Kanellis, A. K., Morris, L. L., and Saltveit, M. E. (1986). Effect of stage of development on the postharvest behavior of cucumber fruit. *HortScience* **21,** 1165–1167.

Kapsalis, J. G. (1987). "Objective Methods in Food Quality Assessment." CRC Press, Boca Raton, Florida.

Kramer, A., and Smith, H. R. (1947). "Effect of Variety, Maturity, and Canning Procedures on Quality and Nutritive Value of Lima Beans." University of Maryland Agricultural Experimental Station Bulletin No. A47. College Park, Maryland.

Kramer, A., and Twigg, B. A. (1970). "Quality Control for the Food Industry," 3d Ed. Vol. 3 AVI/van Nostrand, New York.

Land, D. G. (1988). Negative influences on acceptability and their control. *In* "Food Acceptability" (D.M.H. Thomson, ed.) pp. 475–483. Elsevier, New York.

Larmond, E. (1987). Sensory evaluation can be objective. *In* "Objective Methods in Food Quality Assessment" (J. G. Kapsalis, ed.), pp. 3–14. CRC Press, Boca Raton, Florida.

Lawless, H. (1991). The sense of smell in food quality and sensory evaluation. *J. Food Qual.* **14,** 33–60.

Lebermann, K. W., Nelson, A. I., and Steinberg, N. P. (1968). Postharvest change of broccoli stored in modified atmosphere. 1. Respiration of shoots an color of flower heads. *Food Technol.* **22,** 487–490.

Leonard, S., Luh, B. S., Hinreiner, E., and Simone, M. (1954). Maturity of Bartlett pears for canning. *Food Technol.* **8,** 478–482.

Lidster, P. D., and Porritt, S. W. (1978). Influence of maturity and delay of storage on fruit firmness and disorders in 'Spartan' apple. *HortScience* **13,** 253–254.

Lipton, W. J., and Harris, C. M. (1976). Response of stored cauliflower (*Brassica oleracea* L., Botrytis group) to low-O_2 atmospheres. *J. Amer. Soc. Hort. Sci.* **101,** 208–211.

McGlasson, W. B., Beattie, B. B., and Kavanagh, E. E. (1986). "Tomato Ripening Guide." Department of Agriculture, New South Wales, Ag. Fact. H8.4.5.

Magness, J. R., and Taylor, G. F. (1925). An improved type of pressure tester for the determination of fruit maturity. USDA Circular No. 350. U.S. Govt. Printing Office, Washington D.C.

Mencarelli, F., Lipton, W. J., and Peterson, S. J. (1983). Responses of 'zucchini' squash to storage in low-O_2 atmospheres at chilling and nonchilling temperatures. *J. Amer. Soc. Hort. Sci.* **108,** 884–890.

Miller, C. H., and Hughes, G. R. (1969). Harvest indices for pickling cucumbers in once-over harvested systems. *J. Amer. Soc. Hort. Sci.* **94,** 485–487.

Morris, J. R. (1985). Grape juice quality. *In* "Evaluation of Quality of Fruits and Vegetables" (H. E. Pattee, ed.), pp. 129–176. AVI/Van Nostrand, New York.

Morris, J. R., Kattan, A. A., Nelson, G. S., and Cawthon, D. L. (1978). Developing a mechanized system for production, harvesting and handling of strawberries. *HortScience* **13,** 413–422.

Morris, J. R., Cawthon, D. L., Spayd, S. E., May, R. D., and Bryan, D. R. (1980). Prediction of 'Concord' grape maturation and sources of error. *J. Amer. Soc. Hort. Sci.* **105,** 313–318.

Moskowitz, H. R., and Sidel, J. L. (1971). Magnitude and Hedonic scales of food acceptability. *J. Food Sci.* **36,** 677–680.

Moyer, J. C., and Aitken, H. C. (1980). Apple juice. *In* "Fruit and Vegetable Juice Processing Technology," 3rd ed. (P. E. Nelson and D. K. Tressler, eds, pp. 212–267. AVI/Van Nostrand, New York.

National Research Council (1989). "Alternative Agriculture." National Academy Press, Washington, D.C.

O'Mahony, M. (1982). Some assumptions and difficulties with common statistics for sensory analysis. *Food Technol.* **36(11)**, 75–82.

O'Mahony, M. (1991). Taste perception, food quality, and consumer acceptance. *J. Food Qual.* **14**, 9–31.

Pattee, H. E. (1985). "Evaluation of Quality in Fruits and Vegetables." AVI/Van Nostrand, New York.

Peleg, K., Ben-Hanan, U., and Hinga, S. (1990). Classification of avocado by firmness and maturity. *J. Text Stud.* **21**, 123–139.

Pomeranz, Y., and Meloan, C. E. (1987). "Food Analysis Theory and Practice," 2d Ed. AVI/Van Nostrand Reinhold, New York.

Powers, J. J., Godwin, D. R., and Bargmann, R. E. (1977). Relations between sensory and objective measurements for quality evaluation of green beans. *In* "Flavor Quality: Objective Measurement" (R. A. Scanlan, ed.), pp 51–70. American Chemical Society, Washington, D.C.

Prevor, J. (1990). A boost for organics. *Produce Business* **6(5)**, 14.

Puig, E., and Casado, C. (1983). Consumer attitudes toward fresh tomatoes. *Acta Hort.* **135**, 357–363.

Quamme, H. A., and Gray, J. I. (1985). Pear fruit quality and factors that condition it. *In* "Evaluation of Quality of Fruits and Vegetables" (H. E. Pattee, ed.), pp. 47–61. AVI/Van Nostrand, New York.

Ramaswamy, H. S., and Ranganna, S. (1982). Maturity parameters for okra (*Hibiscus esculentus* L. Moench var. *Pusa Sawani*). *Can. Inst. Food Sci. Technol. J.* **15**, 140–143.

Ramaswamy, H. S., Ranganna, S., and Govindarajan, V. A. (1980). A nondestructive test for determination of optimum maturity of French (green) beans *(Phaseolus vulgaris).* *J. Food Qual.* **3**, 11–23.

Reid, M. S., Padfield, C. A. S., Watkins, C. B., and Harman, J. E. (1982). Starch iodine pattern as a maturity index for Granny Smith apples. 1. Comparison with flesh firmness and soluble solids content. *N.Z.J. Agric. Res.* **25**, 239–243.

Resurreccion, A. V. A., Shewfelt, R. L., Prussia, S. E., and Hurst, W. C. (1987). Relationships between sensory and objective measures of postharvest quality of snap beans as determined by cluster analysis. *J. Food Sci.* **52**, 113–123.

Rodriguez-Amaya, D. (1989). Critical review of provitamin A determination in plant foods. *J. Micronutr. Anal.* **5**, 191–225.

Ruck, J. A. (1963). "Chemical Methods for Analysis of Fruit and Vegetable Products." Canadian Dept. Agric. Publ. 1154.

Salunkhe, D. K., and Desai, B. B. (1984a). "Postharvest Biotechnology of Fruits," Vol. I. CRC Press, Boca Raton, Florida.

Salunkhe, D. K., and Desai, B. B. (1984b). "Postharvest Biotechnology of Fruits" Vol. II. CRC Press, Boca Raton, Florida.

Salunkhe, D. K., and Desai, B. B. (1984c). "Postharvest Biotechnology of Vegetables," Vol. I. CRC Press, Boca Raton, Florida.

Salunkhe, D, K., and Desai, B. B. (1984d). "Postharvest Biotechnology of Vegetables," Vol. II. CRC Press, Boca Raton, Florida.

Schwartz, S. J., and Lorenzo, T. V. (1990). Chlorophylls in foods. *Crit. Rev. Food Sci. Nutr.* **29**, 1–17.

Schwartz. S. J., and von Elbe, J. H. (1980). Quantitative determination of individual betacyanin pigments by high performance liquid chromaography, *J. Agric. Food Chem.* **28**, 540–543.

Schreier, P., Drawert, F., and Junker, A. (1976). Identification of volatile constituents from grapes. *J. Agric. Food Chem.* **24**, 331–336.

Sherman, M., Kasmire, R. F., Shuler, K. D., and Botts, D. A. (1982). Effect of precooling methods and peduncle lengths on soft rot decay of bell peppers. *HortScience* **17**, 251–252.

Shewfelt, R. L. (1984). Quality and maturity indices for postharvest handling of southern peas. *J. Food Sci.* **49**, 389–392.

Shewfelt, R. L. (1986). Flavor and color of fruits as affected by processing. *In* "Commerical Fruit Processing," 2d Ed. (J. G. Woodroof and B. S. Luh, eds.), pp. 479–529. AVI/Van Nostrand, New York.

Shewfelt, R. L. (1990). Sources of variation in the nutrient content of agricultural commodities from the farm to the consumer. *J. Food Qual.* **13**, 37–54.

Shewfelt, R. L., Heaton, E. K., and Batal, K. M. (1984). Nondestructive color measurement of fresh broccoli. *J. Food Sci.* **49**, 1612–1613.

Shewfelt, R. L., Prussia, S. E., Hurst, W. C., and Jordan, J. L. (1985). A systems approach to the evaluation of changes in quality during postharvest handling of southern peas. *J. Food Sci.* **50,** 769–772.

Shewfelt, R. L., Myers, S. C., Prussia, S. E., and Jordan, J. L. (1987a). Quality of fresh-market peaches within the postharvest handling system. *J. Food Sci.* **52,** 361–364.

Shewfelt, R. L., Prussia, S. E., Resurreccion, A. V. A., Hurst, W. C., and Campbell, D. T. (1987b). Quality of vine-ripened tomatoes within the postharvest handling system. *J. Food Sci.* **52,** 661–664, 672.

Sistrunk, W. A., and Morris, J. R. (1985). Strawberry quality influence of cultural and environmental factors. *In* "Evaluation of Quality of Fruits and Vegetables" (H. E. Pattee, ed.), pp. 217–256. AVI/Van Nostrand, New York.

Sistrunk, W. A., Bailey, F. L., and Kattan, A. A. (1965). Influence of maturity and yield on quality of fresh and canned southern peas. *Proc. Amer. Soc. Hort. Sci.* **86,** 491–497.

Sjulin, T. M., and Robbins, J. (1987). Effects of maturity, harvest date, and storage time on postharvest quality of red raspberry fruit. *J. Amer. Soc. Hort. Sci.* **112,** 481–487.

Smith, L. G. (1988). Indices of physiological maturity and eating quality in Smooth Cayenne Pineapples. I. Indices of physioloical maturity. *Queensland J. Agric. Anim. Sci.* **45,** 213–218.

Spanos, G. A., and Wrolstad, R. E. (1987). Anthocyanin pigment, nonvolatile acid, and sugar composition of red raspberry juice. *J. Assoc. Off. Anal. Chem.* **70,** 1036–1046.

Steel, R. G. D., and Torrie, J. H. (1980). "Principles and Procedures of Statistics: A Biometrical Approach," 2d Ed. McGraw-Hill, New York.

Steiner, J. J., and Akintobi, D. C. (1986). Effect of harvest maturity on viability of onion seed. *HortScience* **21,** 1220–1221.

Stone, H., Sidel, J., Oliver, S., Woolsey, A., and Singleton, R. C. (1974). Sensory evaluation by Quantitative Descriptive Analysis. *Food Technol.* **28(11),** 24–34.

Szczesniak, A. S. (1991). Textural perception and food quality. *J. Food Qual.* **14,** 75–85.

Thomson, D. M. H. (1988). "Food Acceptability." Elsevier Applied Science, New York.

Twigg, B. A., Kramer, A., Falen, H. N., and Southerland, F. L. (1956). Objective evaluation of the maturity factor in processed sweet corn. *Food Technol.* **10,** 171–174.

U.S. Dept. of Agriculture (1961). "United States Standards for Cantaloupes." F. R. Doc. #61-2272.

U.S. Dept. of Agriculture (1975). Color classification requirements in tomatoes. USDA Visual Aid TM-L-1. The John Henry Company. P. O. Box 17099, Lansing Michigan 48901.

Vickers, Z. (1991). Sound perception and food quality. *J. Food Qual.* **14,** 87–96.

Voisey, P. W., and Nonnecke, I. L. (1973). Measurement of pea tenderness. V. The Ottawa pea tenderometer in relation to the pea tenderometer and FTC texture test system. *J. Text. Stud.* **4,** 323–343.

von Elbe, J. H., Sy, S. H., Maing, I. Y., and Gabelman, W. H. (1972). Quantitative analysis for betacyanins in red table beets *(Beta vulgaris). J. Food Sci.* **37,** 932–934.

Wang, C. Y. (1979). Effect of short-term high CO_2 treatment on the market quality of stored broccoli. *J. Food Sci.* **44,** 1478–1482.

Wang, C. Y., and Baker, J. E. (1979). Effects of two free radical scavengers and intermittent warming on chilling injury and polar lipid composition of cucumber and sweet pepper fruits. *Plant Cell Physiol.* **20,** 243–251.

Wang, C. Y., and Hruschka, H. W. (1977). Quality maintenance in polyethylene-packaged broccoli. USDA Marketing Research Report No. 1085. U.S. Government Printing Office, Washington, D.C.

Watada, A. E., Abbott, J. A., and Finney, E. E. (1976). Firmness of peaches measured nondestructively. *J. Amer. Soc. Hort. Sci.* **101,** 404–406.

Watada, A. E., Herner, R. C., Kader, A. A., Romani, R. J., and Staby, G. L. (1984). Terminology for the description of developmental stages of horticultural crops. *HortScience* **19,** 20–21.

Williams, M., Hrazdina, G., Wilkinson, M. M., Sweeney, J. G., and Iacobucci, G. A. (1978). High-pressure liquid chromatographic separation of 3-glucosides, 3,5-diglucosides, 3-(6-O-*p*-coumaryl)-glucosides, and 3-(6-O-*p*-coumaryl-glucoside)-5-glucosides of anthocyanidins. *J. Chromatogr.* **155,** 389–398.

Yang, C. C., Brennan, P., Chinnan, M. S., and Shewfelt, R. L. (1987). Characterization of tomato ripening process as influenced by individual seal-packaging and temperature. *J. Food Qual.* **10,** 21–33.

Yang, S. F., and Pratt, H. K. (1978). The physiology of ethylene in wounded plant tissue. *In* "Biochemistry of Wounded Plant Tissues" (G. Kahl, ed.). pp. 595–622. deGruyter, Berlin.

Yano, T., Kojima, I., and Torikata, Y. (1981). Role of water in withering of leafy vegetables. *In* "Water Activity: Influences on Food Quality" (L. B. Rockland and G. F. Stewart, eds.), pp. 765–780. Academic Press, New York.

Zind, T. (1990). Fresh Trends '90—A profile of fresh produce consumers. *The Packer Focus* **96(54),** 37–68.

CHAPTER 6

MICROBIAL QUALITY

Robert E. Brackett

Postharvest Handling: A Systems Approach
Copyright © 1993 by Academic Press, Inc. All rights of reproduction in any form reserved.

The produce industry always has been aware of the importance of microorganisms in maintaining produce quality. However, most of this awareness was, until relatively recently, concerned primarily with the so-called field or market diseases responsible for decay. Consequently, most of the research in fruit and vegetable microbiology fell into the domain of the plant pathologist. Microorganisms do more than cause spoilage and affect the profits of farmers. Food scientists now recognize that microorganisms and their by-products are also important quality attributes of fresh fruits and vegetables that affect the product long after it leaves the farm. Thus, all handlers of fresh produce need to be aware of how various microbes grow and how they affect foods at various stages of production and marketing. This chapter is designed to provide the reader with an integrated approach to fruit and vegetable microbiology that includes aspects of both traditional plant pathology and food microbiology.

I.   Microorganisms of Concern

Microbiology is the study of living things that are difficult or impossible to view with the unaided eye. Although seemingly specialized, microbiology actually is quite varied and involves the study of many different creatures. Of the many types of microorganisms, only relatively few affect the overall quality of fresh produce. Before one can discuss specific organisms, however, it is important to understand how major groups of microorganisms differ.

A.   Viruses

Viruses are among the smallest and simplest of microorganisms. They are incapable of the metabolic activity characteristic of other living things, yet they are still capable of efficiently reproducing themselves and can be disseminated over large areas.

Viruses contain no metabolic system of their own. They rely on both the energy and the biosynthetic capabilities of their hosts for their existence. However, many viruses contain active enzymes that enable them to gain access to the inside of host cells. The normal life cycle of a virus involves several steps. First, the virus particle attaches to the outer surface of the host cell. Next, the virus injects its genetic material (either DNA or RNA) into the host cell. The virus genes then direct the host cell to produce new viruses. The growth cycle usually culminates when the host cell ruptures and releases the new viruses into the environment to repeat the cycle.

The specific host that a given virus attacks depends on the virus in question. Most viruses are relatively host specific. The most well studied viruses, bacteriophages or phages, attack only bacteria. Likewise, some viruses can attack various fruit or vegetable plants and cause diseases that negatively affect the quality of the product. Although some viruses can attack several types of similar hosts, they cannot attack dissimilar hosts, that is, plant viruses will not attack animals nor will animal viruses attack bacteria. Only those viruses that cause field disease or are pathogenic to humans are of concern to the produce industry. The latter group of viruses is discussed in greater detail in Chapter 15.

B. Bacteria

Bacteria are the most simple single-celled organisms capable of independent existence. Unlike viruses, bacteria contain all or most of the cellular components for growth and reproduction. They are able to extract energy from organic (and sometimes inorganic) materials and produce by-products from their growth. Bacteria are prokaryotic organisms, meaning they lack the discrete well-defined nucleus that characterizes the higher organisms, eukaryotes.

Bacteria reproduce in a manner quite different from viruses. In essence, the bacterial cell simply splits in two whenever it becomes too large. This process, known as binary fission, causes populations of bacteria to increase exponentially rather than additively. The length of time between each new splitting is known as the generation time. The generation time can range from minutes to weeks. The exact time is characteristic of specific bacteria but is influenced also by environmental conditions such as temperature. Most bacteria of concern in foods can increase a million-fold easily in one day if growth conditions are suitable.

The manner in which a bacterial culture grows is quite predictable. In general, bacterial cultures pass through four stages of growth when they are introduced into a new environment. When a bacterial cell first arrives in a suitable growth environment, there is a period of acclimation during which no growth takes place. This period, known as the lag phase of growth, is dependent on both the bacterium and the growth conditions. Lag periods will usually be longer in less favorable environments. Following the lag phase, a period of exponential growth begins. This period, appropriately called the log or exponential phase, continues until the population of cells reaches its maximum. At that point the cells enter the stationary phase of growth in which the number of new cells equals the number of dying cells. The stationary phase usually accounts for at least half the total growth cycle. Eventually, accumulation of waste products and depletion of nutrients causes the population to decrease in the death phase. After the death phase, the culture either dies out completely or some cells become dormant.

Bacteria are classified in several different ways. The most common way to differentiate bacteria is with the Gram stain. In this color reaction, bacterial cells are subjected to a series of stains. Gram-positive cells appear blue or violet whereas gram-negative cells appear red. The particular color that a cell becomes after staining is an indication of the basic chemical composition of the bacterial cell wall. Bacteria also are described by their shapes and growth characteristics, and by normal taxonomic convention.

C. Fungi

Fungi is the collective term for a rather large group of related eukaryotic organisms. The yeasts and molds are probably the most well known and have the greatest impact on postharvest quality of fruits and vegetables. The taxonomy of fungi is rather complex; a detailed discussion is beyond the scope of this chapter but can be found in several discussions on the topic (Pitt and Hocking, 1985b; Beneke and Stevenson, 1987). Some of the main characteristics that are used in classifying fungi are morphology, form of reproduction, and whether they are multi- or single celled. Molds are primarily multicelled and filamentous whereas yeasts are usually single celled. However, the morphology of these organisms is often dependent on growth conditions. Fungi may be capable of asexual or sexual reproduction.

Like bacteria, fungi are free living and capable of both breakdown of nutrients and biosynthesis of needed biochemicals. However, fungi are able to carry out even more complex processes than bacteria, allowing fungi to produce a wider range of enzymes and other metabolites than bacteria. In addition, most fungi often can grow in a wider range of environmental conditions than bacteria, particularly at acidic pH or low water activity.

D. Parasites

Parasites are the fourth major group of microorganisms of importance in fruits and vegetables. These organisms are of less concern for their influence on quality of the product than for their effect on safety. Parasites are the most complex of the microorganisms and actually are related more closely to animals than to plants. They can be either single or multicelled and are represented by both protozoa and helminthes (worms). Parasites differ from bacteria and fungi in that they do not grow on the fruit or vegetable. Usually, the eggs or cysts of parasites are deposited on the food when contaminated water is used for irrigation or washing. The parasite eggs or cysts then remain dormant until eaten by a suitable host, such as a human. At that time, the life cycle of the parasite continues in its normal way. Illness often results when vital organs or other body parts are infected by the live parasites. A more detailed discussion of specific parasites and associated illness is presented in Chapter 15.

II. Factors Affecting Microbial Growth

Many factors influence the growth and survival of microorganisms in foods. A basic understanding of these factors is essential to an understanding of how microorganisms behave in foods and how they are controlled.

A. Temperature

Temperature is one of the most important factors influencing the growth of microorganisms. As do all other organisms, microorganisms only grow within a limited range of temperatures. They grow optimally within an even narrower range. Cellular reproduction either ceases or is imperceptibly slow below the minimum growth temperature. However, limited metabolism often continues at temperatures above freezing. In contrast, temperatures above the maximum growth temperature not only stop reproduction, but are often lethal. Cellular reproduction and metabolism is fastest at the optimum growth temperature. Consequently, the generation time is shortest at the optimum temperature.

The specific minimum, maximum, and optimum growth temperatures differ greatly for various microorganisms. Most microorganisms grow only or primarily at ambient temperatures (10–40°C). These organisms, known as mesophiles, include those associated with warm-blooded animals as well as many important spoilage organisms. The plant pathogen *Erwinia caratovora* and the human pathogen *Salmonella typhimurium* are examples of mesophiles.

Some microorganisms, known as psychrotrophs, grow well at refrigeration temperatures. However, warmer temperatures (20–30°C) are optimum for almost all psychrotrophs. The ability to grow at refrigeration temperatures makes these organisms among the most important in fruits and vegetables. Many important spoilage organisms are psychrotrophic (Table I).

The types of microorganisms found on freshly harvested produce normally will include psychrotrophs and mesophiles. However, the warm growing conditions usually will lead to a predominance of the latter group. Refrigerated storage after harvest can lead to changes in the microflora. Often the cold temperatures will select for psychrotrophs at the expense of mesophiles. For example, Brackett (1989) found that the percentage of psychrotrophs in fresh broccoli increased from 0.3 to 20% of the total aerobic microflora during refrigerated storage.

Refrigerated storage can also change the type of spoilage encountered. Mold spoilage is more likely to occur at warmer temperatures, particularly in acidic products such as fruits. At refrigeration temperatures, psychrotrophs more successfully compete with the slower growing molds and mesophilic bacteria. For example, psychrotrophic *Pseudomonas* species are common spoilage agents in refrigerated vegetables whereas the mesophilic *Erwinia* seldom are found. In contrast, the latter genus is quite common in the postharvest spoilage of vegetables or in vegetables held at relatively warm temperatures (Lund, 1982).

Table I
Psychrotrophic Spoilage Organisms

Bacteria		
Acinetobacter	*Aeromonas*	*Alcaligenes*
Arthrobacter	*Bacillus*	*Chromobacterium*
Citrobacter	*Clostridium*	*Corynebacterium*
Enterobacter	*Erwinia*	*Escherichia*
Flavobacterium	*Klebsiella*	*Lactobacillus*
Leuconostoc	*Listeria*	*Microbacterium*
Micrococcus	*Moraxella*	*Proteus*
Pseudomonoas	*Serratia*	*Streptococcus*
Streptomyces	*Vibrio*	*Yersinia*
Molds		
Aspergillus	*Penicillium*	
Yeasts		
Candida	*Cryptococcus*	*Torulopsis*

Source: Brackett (1993).

Table II
Limits of pH Allowing Initiation of Growth of
Various Microorganisms

Organism	pH range
Gram-negative bacteria	
Escherichia coli	4.4–9.0
Erwinia caratovora	5.6–9.3
Pseudomonas fluorescens	6.0–8.5
Salmonella typhimurium	5.6–8.0
Gram-positive bacteria	
Bacillus subtilis	4.5–8.5
Clostridium botulinum	4.7–8.5
Lactobacillus spp.	3.8–7.2
Staphylococcus aureus	4.0–9.8
Streptococcus lactis	4.3–9.2
Molds	
Aspergillus oryzae	1.6–9.3
Fusarium oxysporum	1.8–11.1
Penicillium italicum	1.9–9.3
Yeasts	
Candida pseudotropicalis	2.3–8.8
Saccharomyces cerevisiae	2.3–8.6
S. pastori	2.1–8.8

Sources: Corlett and Brown (1980); Brackett (1993).

B. pH

Acidity is another important factor influencing microorganisms. The pH can have a direct effect on both the microbiological safety and the quality of fruits and vegetables. Most microorganisms can grow over a range of pH values. The pH for growth of most microorganisms ranges from 4.5 to 8.0. The optimum pH for most microorganisms is about 7. Few fruits or vegetables are alkaline enough to affect microbial growth. However, many produce items are acidic, and microorganisms differ substantially in their tolerance to acids (Table II). In fact, one can usually predict both predominant microflora and potential spoilage organisms based on pH. Fungi are quite tolerant of acidic pH and are therefore common spoilage organisms in fruits, which are normally quite acidic. Bacteria, on the other hand, vary widely in their tolerance to acidity. Certain genera of bacteria can survive and grow at pH values of 2 or less (Brock et al., 1984) but rarely, if ever, are found in foods. The lactic acid bacteria are the most common acid-tolerant bacteria found in foods. Their ability to produce lactic acid allows them to cause unintentional souring of foods. Other microorganisms, such as molds, can neutralize mildly acidic foods, (Mundt and Norman, 1982; Draughon et al., 1988), which can lead to safety problems because acidity is often relied on to prevent the growth of pathogenic bacteria.

C. Water Activity

The presence of adequate water is important to all living things. Microorganisms are no exception. However, the mere presence of water is not enough to support growth. The water must be in a form that the organisms can use. The presence of solutes such as salts or sugars can reduce the amount of water available to microorganisms. The amount of water available for use by microorganisms or in biochemical reactions is indicated by a measurement known as water activity (A_w). A_w is defined by the formula

$$A_w = \frac{\text{vapor pressure of solution (or food)}}{\text{vapor pressure of pure water}}$$

Thus, the A_w ranges from 0 (total dryness or complete unavailability) to 1.0 (pure water). The A_w is lowered either by adding solutes such as NaCl or by dehydrating to concentrate the solutes already present.

The range of water activities over which microorganisms grow is relatively narrow, although various groups of microorganisms differ (Table III). Fungi, as a group, are able to tolerate and survive at lower water activities. This ability to withstand low water activities is sometimes referred to as osmotolerance. Some relatively uncommon osmotolerant yeasts can, and even must, grow at water activities as low as 0.70 (Corry, 1987). However, such water activities are found only in foods, such as jelly, that contain high concentrations of sugar or salt. In contrast, bacteria are more sensitive to A_w than are fungi. Although some bacteria can withstand A_w values as low as 0.85, most require water activities above 0.97. It is important to note that almost all microorganisms grow better at high water activities.

Most foods have an A_w between 0.8 and 0.99. However, fresh fruits and

Table III
Approximate Minimum Water Activities for Growth of Some Organisms

Organisms	Water activity (A_w)
Bacteria	
Bacillus subtilis	0.90
Clostridium botulinum type b	0.94
Escherichia coli	0.95
Halobacterium halobium	0.75
Pseudomonas fluorescens	0.97
Salmonella sp.	0.95
Molds	
Alternaria citri	0.84
Aspergillus flavus	0.78
Botrytis cinerea	0.93
Xeromyces bisporus	0.61
Penicillium patulum	0.81
Rhizopus nigricans	0.93
Yeasts	
Debaryomyces hansenii	0.83
Saccharomyces bailii	0.80
S. cerevisiae	0.90
S. rouxii	0.75

Source: Christian (1980).

vegetables have an A_w at the upper end of this range. Thus, there is ample water available for the growth of virtually all microorganisms. This high A_w makes most fruits and vegetables very susceptible to microbial infection and decay. Indeed, the processing of fruits such as grapes into jelly or the dehydration of grapes into raisins is aimed at inhibiting the growth of most spoilage bacteria. Dehydration of various vegetables is done for a similar purpose. For more information on water activities, see Christian (1980) and Gould (1985).

Another important issue to consider with respect to water is liquid-phase water that is in contact with fruits or vegetables as a result of condensation or washing. Such water affects the microflora in several ways. First, it dissolves usable carbohydrates present or exuding from the product and serves as a growth medium. Consequently, any microorganisms present are more likely to grow to higher populations on the products. Second, free water tends to raise humidity in enclosed environments such as packages, which increases the likelihood that osmotolerant organisms, most notably molds, will grow.

D. Atmosphere

The atmosphere to which microorganisms are exposed can have a profound effect on their growth and survival. Some organisms (aerobes) requires oxygen for growth and survival whereas others (anaerobes) will grow only in the absence of oxygen (O_2). Many bacteria (facultative anaerobes), however, can grow either with or without O_2. However, facultative organisms almost always will grow best when O_2 is present.

Gases other than O_2 also affect microorganisms. Carbon dioxide (CO_2) is inhibitory to many microorganisms. The gram-positive bacteria are among the most sensitive, whereas gram-negative bacteria are less so (Clark and Takács, 1980; Jay, 1986). Molds are also relatively sensitive to CO_2, but yeasts tend to be more resistant (Daniels *et al.*, 1985). In general, at least 5% CO_2 is needed to repress microbial growth (Daniels *et al.*, 1985). Moreover, the antimicrobial effect of CO_2 is influenced by factors such as temperature. The gas is usually most effective at colder temperatures but becomes less effective as the temperature increases (Jay, 1986). The exact mechanisms by which CO_2 inhibits microorganisms are not understood completely yet. Some investigators have suggested that the gas acidifies the cytoplasm of microorganisms or causes inhibition of essential enzymes (Clark and Takács, 1980).

E. Availability of Nutrients

Microorganisms must have an energy source and other biochemical precursors to grow. The specific nutrients required by an organism vary widely. Moreover, the ability of a specific organism to obtain the necessary nutrients can vary also. Some bacteria can use only simple sugars and require preformed vitamins. Others are able to use many forms of organic matter efficiently for energy and biosynthetic precursors, and can synthesize the nutrients they need. The most resourceful organisms, such as the pseudomonads and the molds, are often the ones that are most likely to cause spoilage. They can produce various degradative enzymes to help them obtain the necessary raw materials for growth and survival. The action of these enzymes produces the symptoms that we associate with spoilage. However, most microorganisms require a minimum amounts of nutrients to enable them to produce enzymes. Additionally, all organisms grow best if easily used nutrients are present. Thus, foods high in sugars and amino acids serve as superior growth media. Many fresh fruits and vegetables contain at least enough sugar to enable microorganisms to start growing. Once active, these organisms may produce pectinases, cellulases, or amylases that provide more sugars.

F. Competitive Microorganisms

Microorganisms rarely exist as pure cultures in nature. Foods, like the rest of the natural environment, are dynamic habitats that support the growth and reproduction of many different types of microorganisms (Skovgaard, 1984; Hobbs, 1986). This diversity increases the chances that a spoilage organism is present on a food. As one would expect, these various populations of microorganisms are not without effect on one another. Depletion of or limited access to available nutrients and accumulation of waste products limit the growth of microbes in foods.

Bacteria, yeasts, and molds compete for both physical space and nutrients. Organisms that can colonize most desirable locations will fare better than those in less desirable locations. Consequently, having flagellae to move about or aerial hyphae to extend is a competitive advantage. Once located, the ability to stick to the food source and produce biofilms not only helps prevent competitive organisms from colonizing the same spot but prevents the original microorganisms from being washed away.

Microorganisms have several strategies for competing for available nutrients. Some organisms rely on rapid growth to outgrow competitors, giving them an initial advantage in colonizing or using food sources. However, once the nutrient source is depleted, their advantage is diminished. Other microorganisms may grow more slowly but are able to use a wide rather than a limited range of nutrients. The ability to adapt more easily to changing nutrient supplies serves as a competitive advantage.

Finally, resistance to and creation of adverse environmental stresses also aid an organism in competing successfully. Organisms that can resist heat, desiccation, or harmful chemicals will outlast those that are less resistant. Some organisms produce metabolites that are antagonistic to other microorganisms. For example, various lactic acid bacteria and *Pseudomonas* species can produce antibiotics (Sinell, 1980). Likewise, many species of molds produce toxic secondary metabolites (Davis and Diener, 1987).

G. Antimicrobial Factors

Antimicrobial factors in fruits and vegetables can affect the growth and survival of microorganisms in these foods. The antimicrobial factor can be either intrinsic or foreign, that is, the compound can be produced by the fruit or vegetable itself or arise from other sources. Among the common intrinsic factors are organic acids found in many fruits (Doores, 1983). Benzoic acid, for instance, is a common antimicrobial compound found in several fruits, most notably cranberries (Chipley, 1983). Many fresh vegetables produce another common group of antimicrobial factors, phytoalexins. Often, these compounds are produced in response to invasion by or presence of microorganisms (Bulgarelli and Brackett, 1991).

Organic acids, antibiotics, and toxic secondary metabolites produced by bacteria and fungi are examples of foreign antimicrobial factors. Regardless of source, most antimicrobials affect various types of microorganisms differently. Some are general antimicrobials that affect a wide range of microorganisms whereas others are quite specific and affect some groups of microorganisms more than others. Thus, the presence of an antimicrobial can affect the general microflora by allowing some microorganisms to grow while inhibiting others.

Although not antimicrobial factors in the strictest sense, physical barriers also influence the microflora of fresh fruits and vegetables. Most fruits and vegetables possess at least some type of skin or peel that serve as the primary barrier to microbial infection. Some vegetables, such as potatoes, have specialized wound healing mechanisms that allow them to repair damage that otherwise might allow microorganisms access to inner tissues (Friend *et al.*, 1973).

III.   Factors Affecting Microbial Quality

Because various environmental and biological factors affect microbes in different ways, it is not surprising that these factors also would affect the ultimate quality of fruits and vegetables. Thus, the factors discussed in the following section deserve consideration with respect to produce quality.

A.   Microorganisms Present

As mentioned previously, microorganisms exist in a dynamic state and can have either minor or major effects on each other. The specific microorganisms present, as well as the number of individual organisms present, affect the microbiological quality of fruits and vegetables.

1.   Types

The type of microorganisms present on fresh produce can vary widely. Table IV lists some of the many microorganisms reported to have been isolated from various products. These organisms gain access to the products from dust or soil, air, irrigation waters, insects, or animals. Thus, the specific microorganisms present depend on the source from which they came. For example, soil is more likely to contaminate the product with molds and gram-positive bacteria whereas irrigation water would be a better source of gram-negative bacteria (Brackett, 1993). Once present, microorganisms subsist on traces of carbohydrates, proteins, and minerals dissolved in plant exudate (Brackett and Splittstoesser, 1992). However, growth is limited by intrinsic properties of the particular fruit or vegetable as well as by storage conditions. In general, fungi, because of their resistance to acidity, are

Table IV
Microorganisms Reported to Have Been Isolated from Fresh Vegetables

Vegetable	Organisms isolated
Asparagus	Bacteria: *Aeromonas hydrophila*
Bell peppers	Molds: *Aspergillus* sp., *Fusarium* sp.
Broccoli	Bacteria: *A. hydrophila*
Cabbage	Bacteria: *Pseudomonas* sp.
	Molds: *Alternaria* sp., *Aureobasicium pullulans*, *Botrytis cinerea*, *Cladosporium* sp., *Penicillium* sp.
Carrots	Bacteria: *Bacillus* sp., *Erwinia* sp., *Pseudomonas* sp.
Cauliflower	Bacteria: *A. hydrophila*
Collards	Bacteria: *Citrobacter freundii*, *Enterobacter agglomerans*, *E. cloacae*, *Escherichia coli*, *Hafnia alvei*, *Klebsiella oxytoca*, *Serratia rubidea*
Corn	Bacteria: *Enterobacter* sp., *E. agglomerans*, *E. cloacae*, *Enterococcus faecalis*, *E. faecium*, *Flavobacterium* sp., *Pseudomonas* sp., *Serratia* sp., *Xanthomonas* sp.
	Molds: *Aspergillus niger*, *Cladosporium cladosporioides*, *Penicillium oxalicum*, *P. expansum*, *P. funciulosum*
Cucumbers	Bacteria: *Citrobacter* sp., *Enterobacter cloacae*, *Erwinia* sp.
Green beans	Molds: *Aspergillus pullans*, *A. tenuis*, *Cladosporium fimeti*, *E. nigrum*, *Fusarium* sp., *Mucor* sp., *Phoma* sp., *Rhizopus nigricans*
Lettuce	Bacteria: *Aeromonas hydrophila*, *Citrobacter amalonticus*, *C. freundii*, *Enterobacter aerogenes*, *E. agglomerans*, *E. cloacae*, *Proteus morganii*, *P. rettgeri*, *P. stuartii*, *P. vulgaris*
Peas	Bacteria: *Enterobacter agglomerans*, *E. cloacae*, *Serratia marcescens*
Potatoes	Bacteria: *Bacillus cereus*, *B. lichenformis*, *Enterobacter caratovora*
Southern peas	Molds: *Alternaria* sp., *Aspergillus* sp., *Cladosporium* sp., *Fusarium* sp., *Phoma* sp.
Tomato	Bacteria: *Acinetobacter* sp., *Corynebacteria*, *Enterobacter cloacae*, *Escherichia intermedia*, *Flavobacterium*, *Klebsiella* sp., *Lactobacillus* sp., *Micrococcus luteus*, *Pseudomonas* sp., *Xanthomonas* sp.
	Molds: *Alternaria* sp., *Cladosporium* sp., *Penicillium* sp.

Adapted from Brackett, 1993.

predominant organisms in fruits (Brackett, 1987a; Splittstoesser 1987; Bulgarelli and Brackett, 1991). In contrast, aerobic bacteria predominate on fresh vegetables (Brackett, 1987a, 1993; Brackett and Splitstoesser, 1992).

Although the specific organisms present are quite varied, not all microorganisms are equally important. Most organisms isolated from fresh produce are innocuous and have little or no effect on sensory qualities. Some microorganisms, however, can spoil the product and are a major economic problem for growers and shippers. Organisms that cause spoilage can be divided into two main groups; true plant pathogens and opportunists. The difference between the two is that plant pathogens are able to overcome the natural defenses of the fruit or vegetable and invade the tissue. Opportunists include organisms that normally do not invade the tissue but can cause spoilage if damage or disease compromises the defenses of the product.

Many different organisms can spoil fresh fruits and vegetables. However, relatively few are responsible for the majority of spoilage. *Erwinia caratovora* frequently is cited as the most important spoilage bacterium of vegetables (Lund, 1971, 1981, 1982, 1983). *Pseudomonas fluorescens* and *Xanthomonas campestris* also commonly are implicated in spoilage. Several species of *Bacillus* and *Clostridium* also are found to cause spoilage occasionally (Lund *et al.*, 1981; Brackett, 1993). In addition, many genera of fungi will initiate spoilage.

The type of organism likely to cause a specific incidence of spoilage depends on the product and the storage conditions. In general, fruits usually are spoiled only by fungi because their acidity inhibits most bacteria (Miller, 1979; Pitt and Hocking, 1985a,d). Vegetables, because of their more neutral pH, can be spoiled by both molds and bacteria (Webb and Mundt, 1978; Pitt and Hocking, 1985a,d; Brackett, 1987b; Bulgarelli and Brackett, 1991). The predominant spoilage organisms also will differ between refrigerated and nonrefrigerated products. However, faster growth rates allow bacteria to compete more successfully with fungi in many situations. For example, *Erwinia* only grows at temperatures above about 3°C, whereas the psychotrophic pseudomonads can grow at even colder temperatures (Brocklehurst and Lund, 1981; Lund, 1983). Thus, the latter group of bacteria is more likely to spoil refrigerated vegetables (Lund, 1983). More details regarding spoilage organisms and their effects on specific products are discussed in review on the subject (Lund, 1971, 1981, 1982, 1983; Splittstoesser, 1987; Brackett, 1993).

2. Populations

The number of microbial cells present on a food frequently is considered a measure of food quality. Although this relationship is true in some cases, such numbers do

Table V
Populations of Microorganisms Found on Fresh Produce

Organism	Product	Population (per gm)
Bacteria	Asparagus	31,600
	Beet	3,200
	Bell peppers	132,000
	Broccoli	10,000–2,500,000
	Cabbage	4–100,000
	Collards	3,200,000–6,300,000
	Lettuce	100,000–1,000,000
	Lima beans	1–150
	Potatoes	75–28,000
	Southern peas	25,100,000
Mold and yeasts	Bell pepper	2,900
	Blackberries	1,400
	Carrots	44,000
	Celery	50,000
	Grapes	150–45,000
	Green peas	150
	Lima beans	34,000
	Okra	1,700,000
	Peaches	50
	Raspberries	50–3,000
	Southern peas	10,000
	Strawberries	50–200

Adapted from Koburger and Farhat (1975); King (1986); Brackett (1993).

not always reflect quality accurately. Fruits and vegetables can be contaminated heavily with soil and plant debris. Consequently, the populations of microorganisms can be quite high, even on freshly harvested products (Table V). For example, total counts of bacteria in fresh vegetables range from ten to millions of cells per gram (Brackett, 1993). Moreover, the populations on a given product are influenced by extrinsic factors such as weather and, thus, change from day to day (Goepfert, 1980; Brackett, 1987a, 1993).

In many cases, the population of microorganisms found on a product is mean-ingless in and of itself. This information is only useful in comparisons or when meeting some standard. For example, counts of microorganisms might be compared at various steps during processing. A dramatic increase in populations immediately after a particular processing procedure might indicate a sanitation problem with that procedure. This information is useful because it indicates to the processor that perhaps equipment is not being cleaned adequately or that a source of contamination is present at that point. An example of how counts might be used to meet a standard is determining the number of fecal coliforms, an indicator of fecal contamination, in wash water.

B. Physiology

Fresh fruits and vegetables are affected by microorganisms but the reverse is also true. Because fresh produce is metabolically active, it has the ability to change itself and its surroundings. Unfortunately, little has been published about the effects of most of these changes on the microflora. Most of the research that has been published had addressed changes in atmospheric gases brought about by respiration.

As produce respires, O_2 is consumed and CO_2 is produced (Brecht, 1980). These changes in O_2 and CO_2 can be dramatic, especially when products are packaged or stored in enclosed areas. In general, lower O_2 and higher CO_2 concentrations will repress gram-negative bacteria and molds, but select for lactic acid bacteria and yeasts (Goepfert, 1980). If O_2 concentrations become sufficiently low, anaerobic bacteria grow, which was demonstrated by Sugiyama and Yang (1975) to occur in fresh mushrooms. They demonstrated that the respiration in packaged mushrooms was able to reduce the O_2 concentration to the point where the obligately anaerobic *Clostridium botulinum* was able to grow.

The production of phytoalexins is another example of how a plant product can affect microorganisms. Phytoalexins are metabolites produced by some plants in response to invasion by microorganisms or other stresses (Bulgarelli and Brackett, 1991). These compounds are toxic to fungi but also may affect humans.

C. Processing

Processing and handling are among the biggest influences on the microbiology of fruits and vegetables. However, they differ from other environmental factors because the processor can control them, at least to some degree. To achieve control, however, the handler must know how the various facets of the system affect the microflora.

1. History of land

An appreciation for how land has been used prior to production of food crops is a much overlooked but important detail. Repeated planting of the same product over several seasons encourages plant diseases in that product (Lund, 1983). Land in which decaying vegetation is allowed to accumulate also can harbor higher populations of spoilage organisms, such as *Erwinia carotovora,* that actively degrade plant matter. Higher populations of spoilage organisms therefore could increase the incidence of spoilage in vegetables grown in those fields.

The history of the land on which produce is grown also affects the microbial safety of the product. Soils that have been used recently to graze animals are more likely to be contaminated with human pathogens. Similarly, the routine use of animal waste or even treated sewage as fertilizer also increases the risk. Allowing time for any pathogens to die off will help decrease the risk that fruits or vegetables would become contaminated. However, some pathogenic microorganisms can persist in soils for extended times. For instance, Watkins and Sleath (1981) found that *Listeria monocytogenes* survived the sewage treatment process and survived for over 8 wk in soil on which the sewage sludge was deposited. This observation perhaps should not be surprising, because *L. monocytogenes* can survive for years in decaying vegetation and inhabits even fallow fields (Brackett, 1988). Thus, it is prudent to grow vegetables destined for the fresh market on fields with a known history that have not been contaminated by animal or human waste.

2. Handling

The manner in which a fruit or vegetable is handled during harvest not only will affect the sensory quality of the product directly but also will influence microbiological quality. Virtually every procedure used in handling fresh produce can affect the microbiology of these products. In general, anything that induces damage in fruits or vegetables makes them more susceptible to microbial decay. Delays in removing field heat from the product after harvest can cause undue stress of fruits and vegetables and also can allow microorganisms to grow faster during that time.

Equipment also can influence the microbial quality of fresh fruits and vegetables. Poorly maintained equipment can serve both to contaminate and to damage products. Damaged equipment and containers can puncture or abrade the outer surfaces of fruits and vegetables and expose inner tissues to opportunistic spoilage organisms (Brackett, 1987b). Such damage also allows juices to leak onto the equipment. These juices then serve as growth media and allow biofilms containing high populations of microorganisms to become established on the surface of equipment. For example, the fungus *Geotrichum candidum* is sometimes called machinery mold because it is found often on improperly cleaned equipment (Eisenberg and Cichowicz, 1977). The presence of high populations of microorganisms on equipment or containers increases the potential for equipment to contaminate the products (Goepfert, 1980). The role of workers in cross-contaminating fruits and vegetables also must be remembered. Individuals who handle spoiled or damaged products can serve as vectors and contaminate otherwise sound products.

Most fresh produce items receive some type of wash procedure after harvest. Washing usually is done by dipping the products in tanks of wash water or spraying the products as they pass underneath spray nozzles. Once washed, products either are allowed to air dry or are dried partially in centrifugal spin driers. The main purpose of washing is to remove soil and debris from the product, but it also can wash away some of the microorganisms and decrease the microbial load (Brackett and Splittstoesser, 1992). The degree to which washing reduces microbial load depends on the product in question and the procedure employed. Splittstoesser *et al.* (1961) found that an initial washing of fresh peas reduced microbial populations by about 90%.

The type of wash procedure itself can determine how the microflora is affected. Dipping is more likely to cross-contaminate uncontaminated products with undesirable microorganisms washed from contaminated products. This possibility increases with the number of produce items that pass through a given batch of wash water. Thus, it is possible that dipping can increase the incidence of spoilage, particularly if dirty water is used for washing continually. Spraying, on the other hand, causes minimal cross-contamination if fresh water is used. Obviously, spraying also will lead to cross-contamination if wash water is recycled continually and sprayed on new products. Regardless of how the washing is done, allowing the wash water to remain on the product for extended times will increase the chance for microbial growth.

In most cases, wash water has at least some chlorine added to it to control microbial growth. Usually, the chlorine is added as sodium or potassium hypochlorite, although other forms are used also (Brackett and Splittstoesser, 1992). The concentrations of chlorine normally added to wash waters can vary greatly but generally range from 5 to 250 mg/liter. However, the concentration of chlorine available to kill microorganisms can be much less. Chlorine is very reactive and breaks down quickly in the presence of organic matter (Dychdala, 1983), for example, that present in dirty wash water. Moreover, the effectiveness of chlorine varies with pH; it is more effective at acidic pH but less so at neutral and alkaline pH (Cords, 1983). Thus, chlorine concentrations must be monitored continually to insure that they are at desired levels.

Chlorine is a potent antimicrobial agent and is particularly effective for killing microbes in solutions or on surfaces of clean equipment (Dychdala, 1983). However, several researchers have found chlorine to be relatively ineffective in removing microorganisms from fruits and vegetables. Senter *et al.* (1987) reported that chlorine had little effect on the microflora of tomatoes. Beuchat and Brackett (1991), in contrast, found that dipping tomatoes in 200 to 250 µl/liter chlorine significantly reduced total aerobic microorganisms but not psychotrophs, yeasts, or molds. However, they found no significant difference in any of the microbial groups after four or more days of storage. Beuchat and Brackett (1990b) reported a similar pattern when comparing carrots washed with and without added chlorine. In this case, populations of all groups of microorganisms studied decreased by as much as 90% immediately after being treated with chlorine. Throughout storage, however, chlorine-treated carrots developed significantly higher populations of mesophiles, psychotrophs, and yeasts and molds than did untreated carrots. Thus, the effectiveness of using sanitizers on fresh produce will differ depending on the concentration of sanitizer used, the product, and the amount of organic matter present (Brackett, 1993). As a rule, chlorine should never be used as a substitute for proper sanitation.

Finally, the role of transportation in microbial quality also must be considered. Transporting fresh produce from the point of harvest to market can cause great stress and damage to fruits and vegetables. The agitation and bouncing inherent in transporting products can cause them to be bruised, crushed, and abraded. Any of these processes increases the chance for microorganisms to penetrate the fruit or vegetable and cause spoilage (Brackett, 1987a; Brackett and Splittstoesser, 1992). In addition, trucks and boxcars sometimes are allowed to become hot and humid, thereby encouraging faster microbial growth.

3. Modified-atmosphere storage

The use of modified atmospheres in the storage and packaging of minimally processed fruits and vegetables has become widespread in the fresh produce industry. The main reason for modifying the atmosphere is to extend the shelf life of the product by slowing respiration and, thus, the normal ripening and senescence process (Brecht, 1980). Doing so also delays changes in sensory quality and decay.

Most modified atmosphere techniques involve reducing the concentration of O_2 while increasing the concentration of CO_2 (Brecht, 1980). The concentrations of the various gases used depend on the product under consideration; guidelines for specific products are available (Salunkhe and Desai, 1984). However, most fruits and vegetables are damaged by CO_2 concentrations above about 20% (Brecht, 1980). Often, these concentrations are not optimal for controlling the growth of microorganisms (Brackett, 1993). Effective inhibition of microorganisms usually requires a minimum of 5% CO_2; many microorganisms require higher concentrations than are acceptable for use with fresh produce (Daniels et al., 1985).

As mentioned previously, various gases affect different microorganisms differently. Nitrogen has little direct effect on microorganisms and is used primarily to displace O_2 (Brackett, 1993). CO_2, on the other hand, not only acts by displacing O_2 but also has direct effects on microorganisms. Most of the direct effects of CO_2 are caused by a reduction of pH in the cytoplasm of microbe cells as well as interference with normal cellular metabolism (Daniels et al., 1985). However, not all microorganisms are repressed similarly by CO_2. Aerobic gram-negative bacteria are most sensitive to CO_2, whereas both facultative and obligate anaerobes are quite resistant to the gas (Jay, 1986). Similarly, the aerobic molds are more sensitive to CO_2 whereas fermentative yeasts are affected only minimally (Daniels et al., 1985). However, the degree to which microorganisms are affected by CO_2 is influenced by such factors as temperature and contact time (Jay, 1986).

Although an antimicrobial effect of CO_2 can be demonstrated readily in the laboratory, the same is not always true in practical use. Data regarding the effects of modified atmospheres on microorganisms in fresh produce are conflicting. Berrang et al. (1990) found that a modified atmosphere containing 10% CO_2 and 11% O_2 significantly inhibited the growth of total aerobic microorganisms in fresh broccoli. In contrast, Priepke et al. (1976) found that similar concentrations of CO_2 had only minimal effects on aerobic microorganisms on salad vegetables. Beuchat and Brackett likewise reported that modified atmosphere storage of lettuce (1990a), carrots (1990b), and tomatoes (1991) extended shelf life but had little or no effect on the microflora.

Usually, the oncentrations of CO_2 required for inhibition of microbes exceed those tolerated by fruits and vegetables. For example, Yackel *et al.* (1971) found that atmospheres that appeared to be best for extending the shelf life of several fruits did not always repress mold growth adequately, particularly if molds already had started growing on the fruit. Similarly, Deák (1984) found that gas concentrations that were optimal for inhibiting growth of several bacteria on laboratory media led to increased spoilage when used to store cauliflower. Thus, gas concentrations often must be tailored to specific situations rather than to inhibition of microorganisms.

4.   Minimal processing

As consumers demand more convenience, more food processors are marketing prepared ready-to-eat fruits and vegetables. Cutting, slicing, and sometimes mixing are fundamental steps in preparing such products. These actions often can have dramatic effects on the microflora, particularly on microbial growth. Splittstoesser (1973) demonstrated that cutting, slicing, and chopping of several vegetables increased microbial populations as much as 6- or 7-fold. Likewise, Splittstoesser and Corlett (1980) and Priepke *et al.* (1976) confirmed these effects on broccoli and salad vegetables, respectively.

There are several reasons for the increases in microbial growth. Cutting and slicing allows fruit and vegetable juices to leak onto the surfaces of utensils and equipment, and onto the products themselves. These juices then serve as a readily available growth medium for microorganisms and promote faster growth and larger populations. In addition, cutting and slicing also increases the surface area of fruits and vegetables, further allowing larger populations.

In addition to allowing higher populations, cutting and slicing also can increase the chance for spoilage of fresh produce. The action of cutting exposes internal tissues of the product, thereby allowing spoilage organisms to bypass the protective peels and skins. In addition, the tissue damage caused by cutting can allow normally innocuous microorganisms to serve as opportunistic spoilage organisms (Brackett, 1993).

The science of packaging is important not only because it improves marketing of produce but because it can be used to increase the shelf life of products (Ben-Yehoshua, 1985). Packaging also can influence the microflora of fruits and vegetables, both directly and indirectly.

Not all processing procedures necessarily increase microbial growth. For example, vegetables such as cucumbers or rutabagas often are coated with edible wax. Waxing is designed primarily to minimize dehydration but also can supplement peels in providing barriers to microbial penetration.

IV.   Methods to Evaluate Microbial Quality

The details of the methodology used for determining microbiology of specific fruits and vegetables are beyond the scope of this chapter. The reader is directed to guides describing the techniques that are used best for fruits and vegetables (Brackett and Splittstoesser, 1992).

A. Microbiological

1. Plate counts

The standard plate count, sometimes also referred to as the total plate count, is probably the most widely used technique for evaluating microorganisms in foods. The purpose, as its name implies, is to estimate the number of viable microorganism cells in a given sample of food. Although the standard plate count provides information about the total microbial load in a food, it also has some limitations. First, the standard plate count only tells how many cells but not what kinds of cells are present. Second, only relatively rapidly growing aerobic organisms such as bacteria are enumerated. Anaerobic bacteria and many fungi will not grow under the conditions used with the standard plate count. Thus, many important organisms are missed by this procedure.

In general, it is not necessary to determine the populations of microorganisms of fresh produce to assure quality. Populations of microorganisms in fruits and vegetables often bear little relationship to quality. The main application for the standard plate count in the fruit and vegetable industry is in monitoring sanitation procedures or in tracing microbiological problems. For example, this technique could be used to determine that various pieces of equipment are being cleaned and sanitized properly. Low populations of microorganisms would indicate that equipment is being maintained properly whereas an increase might point out inadequacies in cleaning procedures. However, what actually constitutes "low" or "high" populations depends on the piece of equipment and on what is considered a normal population. Before such data can be used in a meaningful way, baseline populations reflecting adequately cleaned equipment should be determined. Details of the standard plate count are described by Busta et al. (1984).

2. Yeast and mold counts

There are several basic approaches to detecting and enumerating yeasts and molds in foods. The yeast and mold count provides a very rough estimate of the number of molds in air or food. The procedure is much like the standard plate count, except different media and incubation conditions are used. Other techniques such as the Howard mold count, machinery mold count, and indirect quantitative methods are used also to enumerate fungi (Jarvis and Williams, 1987). However, having an estimate of the concentrations of fungal cells is of limited use in most cases. Yeasts and molds are present on almost all fresh produce. Like total plate counts, the numbers may or may not reflect quality. In addition, yeast and mold counts are not always a true reflection of the actual number of fungi present. Most plating methods for enumerating yeasts and molds rely on recovery of the so-called colony forming units (cfu), as do most other plating techniques. The cfu is based on the number of microbial particles that give rise to a colony on the recovery medium. In the case of molds and other multicellular fungi, counts can be overestimated greatly because each cell gives rise to a colony. Thus, it is theoretically possible for one mold, containing 1000 individual cells, to be counted as 1000 separate molds by the traditional yeast and mold count.

Like the standard plate count, methods to isolate and enumerate molds are used best to validate sanitation procedures. These methods are also useful in isolating and identifying molds from produce items that are experiencing an abnormally high rate of spoilage. Knowing the spoilage organism sometimes gives insight into where the problem originates. In addition, knowing the identity of an offending organism may help devise potential solutions, for example, application of a specific anti-mycotic agent. The reader is directed to one of several discussions of food mycology for more details regarding specific methods to isolate and enumerate fungi in foods (King *et al.*, 1984; Pitt and Hocking, 1985c; Jarvis and Williams, 1987; Mislivec *et al.*, 1992).

3. Coliform bacteria

Coliform bacteria often are considered indicators of fecal contamination and, thus, pathogenic enteric bacteria. Although coliform bacteria often are found associated with enteric pathogens, the reverse is not necessarily true. Several coliform bacteria, such as *Erwinia* and *Enterobacter,* are often part of the natural flora of many vegetables and usually do not indicate a potential public health problem (Brackett and Splittstoesser, 1992). A related subset of the coliforms, fecal coliforms, often indicates the presence of *Escherichia coli* in water. The presence of this organism may be an indicator of fecal contamination and be related to use of polluted irrigation water, the presence of feces, or poor sanitation and hygiene (Brackett and Splitts-toesser, 1992). However, vegetables can harbor bacteria other than *E. coli* that give positive fecal coliform tests. For example, Splittstoesser *et al.* (1980) found that only 29% of vegetables that were positive for fecal coliforms harbored recoverable *E. coli*. Other bacteria including *Klebsiella pneumoniae, Enterobacter cloacae, Enterobacter agglomerans,* and *Enterobacter aerogenes* were responsible for giving false positive fecal coliform tests. Thus, monitoring of fecal coliforms may be useful when water quality or proper hygiene is suspect, but positive results do not always indicate a problem. Procedures and equipment needed to determine the presence of fecal coliforms in foods are described by Mehlman (1984).

4. Pathogenic bacteria

It is not necessary for fresh produce to be tested for pathogenic microorganisms routinely. Doing so is expensive, time consuming, and adds little to the safety of the product. However, certain circumstances may dictate that such analyses for pathogens be done. For example, analysis is sometimes necessary to satisfy the requirement of a commercial customer that the product be certified free of a specific pathogen. Often, the customer also will specify the method to be used. Another case in which analysis for pathogens should be conducted is if the product is intended for compromised consumers. Such consumers include the elderly, the infirm, or those with compromised immunity. The methodology for organisms often differs greatly depending on the organism in question. Several books (Doyle, 1990; American Public Health Association, 1992) and government publications (Food and Drug Administration, 1984) describe generally accepted techniques.

B. Quality Attributes

Measuring and quantifying quality attributes related to microorganisms is more difficult than measuring microorganisms themselves. Traditional methods for monitoring quality related to microbiology simply involve categorizing problems by general cause (Harvey, 1978). For example, diminished quality could result from microbial decay, physical damage, or physiological processes. Although it is sometimes difficult to know which of these general categories actually might have caused quality loss in specific instances, experienced graders usually can identify the cause. Once spoilage is categorized, the relative contribution for each possible source can be quantified in a general way. For instance, a packing shed operator may find that about 20% of the products passing through his shed becomes unsalable before shipping. Of that 20%, perhaps half was lost to microbial decay, a third to physical damage, and the rest to physiological injury such as chill injury. From this information, the packing shed operator can prioritize his expense and his efforts to reduce loss by (1) controlling microorganisms, (2) finding ways to reduce damage during harvest or transportation, and (3) determining how the product is being treated that results in physiological injury.

Categorizing the cause of product loss also can help solve microbial spoilage problems. Microbial losses can be broken down further into more detailed groups. Fruit or vegetable spoilage resulting from mold growth usually is quite different from that caused by bacterial rot. Likewise, bacterial and mold decay arise from different environmental factors and can be minimized by different treatments. Knowing which group or microorganism is responsible for a chronic spoilage problem may suggest a solution to the problem. For example, many fungal decays can be controlled by application of antimycotic agents, whereas these treatments would be ineffective against bacteria.

V. Maintaining Optimal Quality

No single practice can maintain optimal microbiological quality in fresh fruits and vegetables. An integrated or systems approach to handling fresh produce offers the greatest promise of maintaining quality of any kind. However, this means that all factors from before planting to after the consumer purchases the product are important.

Prior land use always should be considered when one is concerned with microbiology (Brackett, 1987a). Fields that have been used previously to graze farm animals should be avoided whenever microbial safety is crucial. Likewise, only reliable sources of irrigation water and fertilizer should be used. Adopting microbiologically safe production practices is relatively easy for vertically integrated produce companies. However, guaranteeing these practices becomes more difficult when processors or packers purchase fruits and vegetables from contract producers or independent farmers. In this case, proper production practices should be spelled out explicitly and agreed on by the producer before the product is purchased.

Taking a systems approach to maintaining acceptable microbiological quality requires that a proactive philosophy be adopted at the outset. One should always try to anticipate microbiological problems and take microbial ecology into account. Techniques that extend shelf life may be desirable from a marketing standpoint, but the effect of extended shelf life on microbiology must be considered. In some cases, changing procedures may improve sensory quality but may cause underlying microbiological problems. For example, Berrang *et al.* (1989a,b) found that modified atmosphere storage increased the shelf life and sensory quality of several vegetables. However, the treatment did not inhibit the growth of *Listeria monocytogenes* or *Aeromonas hydrophila*. Thus, these pathogens conceivably could grow to higher than normal concentrations before the vegetables developed signs of spoilage. Presumably, this situation would put consumers eating such produce at greater risk for contracting foodborne illness. Therefore, no change should be made to the system without determining the effect on microbiology.

Most microbiological problems can be avoided by using common sense. Remembering basic principles of microbial growth and survival can do much to protect food handlers and processors. The usual advice of maintaining a high degree of sanitation, properly maintaining equipment, and applying stringent temperature control may seem trite. Nevertheless, heeding such advice is absolutely essential if one is to avoid microbiological problems. Finally, employees are also an essential part of the system. Investing time and money in educating these people in the basics of sanitation also will do much to maintain quality of fresh produce.

Bibliography

American Public Health Association (1992). "Compendium of Methods for the Microbiological Examination of Foods." American Public Health Association, Washington, D.C.

Beneke, E. S., and Stevenson, K. E. (1987). Classification of food and beverage fungi. *In* "Food and Beverage Mycology" (L. R.. Beuchat, ed.), pp. 1–50. AVI/Van Nostrand Reinhold, New York.

Ben-Yehoshua, S. (1985). Individual seal-packaging of fruit and vegetables in plastic film—A new postharvest technique. *HortScience* **20(1),** 1625–1631.

Berrang, M. E., Brackett, R. E., and Beuchat, L. R. (1989a). Growth of *Aeromonas hydrophila* on fresh vegetables stored under a controlled atmosphere. *Appl. Environ. Microbiol.* **55,** 2176–2171.

Berrang, M. E., Brackett, R. E., and Beuchat, L. R. (1989b). Growth of *Listeria monocytogenes* on fresh vegetables stored under a controlled atmosphere. *J. Food Prot.* **52,** 702–705.

Berrang, M. E., Brackett, R. E., and Beuchat, L. R. (1990). Microbial, color and textural qualities of fresh asparagus, broccoli, and cauliflower stored under controlled atmosphere. *J. Food Prot.* **53,** 391–395.

Beuchat, L. R., and Brackett, R. E. (1990a). Survival and growth of *Listeria monocytogenes* on lettuce as influenced by shredding, chlorine treatment, modified atmosphere packaging, and temperature. *J. Food Sci.* **55,** 755–758.

Beuchat, L. R., and Brackett, R. E. (1990b). Inhibitory effects of raw carrots on *Listeria monocytogenes*. *Appl. Environ. Microbiol.* **56,** 1734–1742.

Beuchat, L. R., and Brackett, R. E. (1991). Behavior of *Listeria monocytogenes* inoculated into raw tomatoes and processed tomato products. *Appl. Environ. Microbiol.* **57,** 1367–1371.

Brackett, R. E. (1987a). Microbiological consequences of minimally processed fruits and vegetables. *J. Food Qual.* **10,** 195–206.

Brackett, R. E. (1987b). Vegetables and related products. *In* "Food and Beverage Mycology" (L. R. Beuchat, ed.), pp. 129–154. AVI/Van Nostrand Reinhold, New York.

Brackett, R. E. (1988). Presence and persistence of *Listeria monocytogenes* in food and water. *Food Technol.* **42(4),** 162–164.

Brackett, R. E. (1989). Changes in the microflora of packaged fresh broccoli. *J. Food Qual.* **12,** 169–181.

Brackett, R. E. (1993). Microbiological spoilage and pathogens in minimally processed refrigerated fruits and vegetables. *In* "Minimally Processed Refrigerated Fruits and Vegetables" (R. Wiley, ed.), Van Nostrand Reinhold, New York. (In press).

Brackett, R. E., and Splittstoesser, D. L. (1992). Fruits and vegetables. *In* "Compendium for the Microbiological Examination of Foods" (C. Vanderzant and D. L. Splittstoesser, eds.) pp. 287–293. American Public Health Association, Washington, D.C.

Brecht, P. (1980). Use of controlled atmospheres to retard deterioration of produce. *Food Technol.* **34(3),** 45–50.

Brock, T. D., Smith, D. W., and Madigan, M. T. (1984). The microbe and its environment. *In* "Biology of Microorganisms," pp. 239–270. Prentice-Hall, Englewood Cliffs, New Jersey.

Brocklehurst, T. F., and Lund, B. M. (1981). Properties of pseudomonads causing spoilage of vegetables stored at low temperatures. *J. Appl. Bacteriol.* **50,** 259–266.

Bulgarelli, M. A., and Brackett, R. E. (1991). The importance of fungi in vegetables. *In* "Handbook of Applied Mycology" (D. K. Arora, K. G. Mukerji, and E. H. Marth, eds.), Vol. 3, pp. 179–199. Marcell Dekker, New York.

Busta, F. F., Peterson, E. H., Adams, D. M., and Johnson, M. G. (1984). Colony count methods. *In* "Compendium of Methods for the Microbiological Examination of Foods" (M. L. Speck, ed.), pp. 62–83. American Public Health Association, Washington, D.C.

Chipley, J. R. (1983). Sodium benzoate and benzoic acid. *In* "Antimicrobials in Foods" (A. L. Branen and P. M. Davidson, eds.), pp. 11–35. Marcel Dekker, New York.

Christian, J. H. B. (1980). Reduced water activity. *In* "Microbial Ecology of Foods" (J. H. Silliker, R. P. Elliott, A. C. Baird-Parker, F. L. Bryan, J. H. B. Christian, D. S. Clark, J. C. Olson, Jr., and T. A. Roberts, eds.), Vol. I, pp. 70–91. Academic Press, New York.

Clark, D. S., and Takács, J. (1980). Gases as preservatives. *In* "Microbial Ecology of Foods" (J. H. Silliker, R. P. Elliott, A. C., Baird-Parker, F. L. Bryan, J. H. B. Christian, D. S. Clark, J. C. Olson, Jr., and T. A. Roberts, eds.), Vol. I. pp. 170–192. American Press, New York.

Cords, B. R. (1983). Sanitizers: Halogens and surface-active agents. *In* "Antimicrobials in Foods" (A. L. Branen and P. M. Davidson, eds.), pp. 257–298. Marcel Dekker, New York.

Corlett, D. A., Jr., and Brown, M. H. (1980). pH and acidity. *In* "Microbial Ecology of Foods" (J. H. Silliker, R. P. Elliott, A. C. Baird-Parker, F. L. Bryan, J. H. B. Christian, D. S. Clark, J. C. Olson, Jr., and T. A. Roberts, eds.), Vol. I. pp. 92–111. Academic Press, New York.

Corry, J. E. L. (1987). Relationship of water activity to fungal growth. *In* "Food and Beverage Mycology" (L. R. Beuchat, ed.), pp. 51–99. AVI/Van Nostrand Reinhold, New York.

Daniels, J., Krishnamurthi, A. R., and Rizvi, S. S. H. (1985). A review of effects of carbon dioxide on microbial growth and food quality. *J. Food Prot.* **48,** 532–537.

Davis, N. D., and Diener, U. L. (1987). Mycotoxins. *In* "Food and Beverage Mycology" (L. R. Beuchat, ed.), pp. 517–570. AVI/Van Nostrand Reinhold, New York.

Deák, T. (1984). Microbial–ecological principles in controlled atmosphere storage of fruits and vegetables. *In* "Microbial Associations and Interactions in Food" (I. Kiss, T. Deák, and K. Incze, eds.), pp. 9–22. Hungarian Academy of Science, Budapest.

Doores, S. (1983). Organic acids. *In* "Antimicrobials in Foods" (A. L. Branen and P. M. Davidson, eds.), pp. 75–108. Marcel Dekker, New York.

Doyle, M. P. (1990). "Foodborne Bacterial Pathogens." Marcel Dekker, New York.

Draughon, F. A., Chen, S., and Mundt, J. O. (1988). Metabiotic association of *Fusarium, Alternaria,* and *Rhizoctonia* with *Clostridium botulinum* in fresh tomatoes. *J. Food Sci.* **53,** 120–123.

Dychdala, G. P. (1983). Chlorine and chlorine compounds. *In* "Disinfection, Sterilization, and Preservation," 3rd Ed. (S. S. Block, ed.), pp. 157–182. Lea & Febiger, Philadelphia.

Eisenberg, W. V., and Cichowicz, S. M. (1977). Machinery mold-indicator organism in food. *Food Technol.* **31(2),** 52–56.

Food and Drug Administration (1984). "Bacteriological Analytical Manual," 6th Ed. Association of Official Analytical Chemists, Arlington, Virginia.

Friend, J., Reynolds, S. B., and Aveyard, M. A. (1973). Phenylalanine ammonia lyases, chlorogenic acid, and lignin in potato tuber tissue inoculated with *Phytophtora infestans. Physiol. Plant Pathol.* **3,** 495–507.

Goepfert, J. M. (1980). Vegetables, fruits, nuts, and their products. *In* "Microbial Ecology of Foods" (J. H. Silliker, R. P. Elliott, A. C. Baird-Parker, F. L. Bryan, J. H. B. Christian, D. S. Clark, J. C. Olson, Jr., and T. A. Roberts, eds.), Vol. II, pp. 606–642. Academic Press, New York.

Gould, G. W. (1985). Present state of knowledge of A_w effects on microorganisms. *In* "Properties of Water in Foods in Relation to Quality and Stability" (D. Simatos and J. L. Multon, eds.), pp. 229–245. Martinus Nijhoff, Boston.

Harvey, J. M. (1978). Reduction of losses in fresh market fruits and vegetables. *Ann. Rev. Phytopathol.* **16**, 321–341.

Hobbs, G. (1986). Ecology of food microorganisms. *Microb. Ecol.* **12**, 15–30.

Jarvis, B., and Williams, A. P. (1987). Methods for detecting fungi in foods and beverages. *In* "Food and Beverage Mycology" (L. R. Beuchat, ed.), pp. 129–154. AVI/Van Nostrand Reinhold, New York.

Jay, J. M. (1986). Intrinsic and extrinsic parameters of foods that affect microbial growth. *In* "Modern Food Microbiology," pp. 33–60. Van Nostrand Reinhold, New York.

King, A. D. (1986). Fungal flora and counts on foods. *In* "Methods for the Mycological Examination of Food" (A. D. King, Jr., J. D. Pitt, L. R. Beuchat, and J. E. L. Corry, eds.), pp. 182–185. Plenum Press, New York.

King, A. D., Jr., Pitt, J. I., Beuchat, L. R., and Corry, J. E. L. (1984). "Methods for the Mycological Examination of Food." Plenum Press, New York.

Koburger, J. A., and Farhat, B. Y. (1975). Fungi in foods. VI. A comparison of media to enumerate yeasts and molds. *J. Milk Food Technol.* **38**, 466–468.

Lund, B. M. (1971). Bacterial spoilage of vegetables and certain fruits. *J. Appl. Bacteriol.* **34**, 9–20.

Lund, B. M. (1981). The effect of bacteria on post-harvest quality of vegetables. *In* "Quality in Stored and Processed Vegetables and Fruit" (P. W. Goodenough and R. K. Arkin, eds.), pp. 599–636. Academic Press, New York.

Lund, B. M. (1982). The effect of bacteria on post-harvest quality of vegetables and fruits, with particular reference to spoilage. *In* "Bacteria and Plants" (M. E. Rhodes-Roberts and F. A. Skinner, eds.), pp. 133–153. Academic Press, New York.

Lund, B. M. (1983). Bacterial spoilage. *In* "Postharvest Pathology of Fruits and Vegetables" (C. Dennis, ed.), pp. 219–257. Academic Press, New York.

Lund, B. M., Brocklehurst, T. F., and Wyatt, G. M. (1981). Characterization of strains of *Clostridium puniceum* sp. nov., a pink-pigmented, pectolytic bacterium. *J. Gen. Microbiol.* **122**, 17–26.

Mehlman, I. J. (1984). Coliforms, fecal coliforms, *Escherichia coli,* and enteropathogenic *E. coli. In* "Compendium of Methods for the Microbiological Examination of Foods" (M. L. Speck, ed.), pp. 265–285. American Public Health Association, Washington, D.C.

Miller, M. W. (1979). Yeasts in food spoilage: An update. *Food Technol.* **33(1),** 76–80.

Mislivec, P. B., Beuchat, L. R., and Cousin, M. A. (1992). Yeasts and molds. *In* "Compendium for the Microbiological Examinations of Foods" (C. Vanderzant and D. L. Splittstoesser, eds.). pp. 120–127. American Public Health Association, Washington, D.C.

Mundt, J. O., and Norman, J. M. (1982). Metabiosis and pH of moldy fresh tomatoes. *J. Food Prot.* **45**, 829–832.

Pitt, J. I., and Hocking, A. D. (1985a). The ecology of fungal food spoilage. *In* "Fungi and Food Spoilage," pp. 5–18, Academic Press, New York.

Pitt, J. I., and Hocking, A. D. (1985b). Naming and classifying fungi. *In* "Fungi and Food Spoilage," pp. 19–28. Academic Press, New York.

Pitt, J. I., and Hocking, A. D. (1985c). Methods for isolation, enumeration, and identification. *In* "Fungi and Food Spoilage," pp. 29–65. Academic Press, New York.

Pitt, J. I., and Hocking, A. D. (1985d). Spoilage of fresh and perishable foods. *In* "Fungi and Food Spoilage," pp. 365–381. Academic Press, New York.

Priepke, P. E., Wei, L. S., and Nelson, A. I. (1976). Refrigerated storage of prepackaged salad vegetables. *J. Food Sci.* **41**, 379–382.

Salunkhe, D. K., and Desai, B. B. (1984). "Postharvest Biotechnology of Vegetables," Vols. I and II. CRC Press, Boca Raton, Florida.

Senter, S. D., Bailey, J. S., and Cox, N. A. (1987). Aerobic microflora of commercially harvested, transported, and cryogenically processed collards (*Brassica olearacea*). *J. Food Sci.* **52**, 1020–1021.

Sinell, H. J. (1980). Interacting factors affecting mixed populations. *In* "Microbial Ecology of Foods" (J. H. Silliker, R. P. Elliott, A. C. Baird-Parker, F. L. Bryan, J. H. B. Christian, D. S. Clark, J. C. Olson, Jr., and T. A. Roberts, eds.), Vol. II, pp. 215–31. Academic Press, New York.

Skovgaard, N. (1984). Vegetables as an ecological environment for microbes. *In* "Microbial Associations and Interactions in Food" (I. Kiss, T. Deák, and K. Incze, eds.), pp. 27–33. Hungarian Academy of Science, Budapest.

Splittstoesser, D. F. (1973). The microbiology of frozen vegetables. *Food Technol.* **27(1),** 54–60.

Splittstoesser, D. F. (1987). Fruits and fruit products. *In* "Food and Beverage Mycology" (L. R. Beuchat, ed.), pp. 101–128. AVI/Van Nostrand Reinhold, New York.

Splittstoesser, D. F., and Corlett, D. A., Jr. (1980). Aerobic plate counts of frozen blanched vegetables processed in the United States. *J. Food Prot.* **43,** 717–719.

Splittstoesser, D. F., Queale, D. T., Bowers, J. L., and Wilkison, M. (1980). Coliform content of frozen blanched vegetables packed in the United States. *J. Food Safety* **2,** 1–11.

Splittstoesser, D. L., Wettergreen, W. P., and Pederson, C. S. (1961). Control of microorganisms during preparation of vegetables for freezing. *Food Technol.* **15(7),** 239–331.

Sugiyama, H., and Yang, K. H. (1975). Growth potential of *Clostridium botulinum* in fresh mushrooms packaged in semi-permeable plastic film. *Appl. Microbiol.* **30,** 964–969.

Watkins, J., and Sleath, K. P. (1981). Isolation and enumeration of *Listeria monocytogenes* from sewage, sewage sludge, and river water. *J. Appl. Bacteriol.* **50,** 1–9.

Webb, T. A., and Mundt, J. O. (1978). Molds on vegetables at the time of harvest. *Appl. Environ. Microbiol.* **35(4),** 655–658.

Yackel, W. C., Nelson, A. I., Wei, L. S., and Steinberg, M. P. (1971). Effect of controlled atmosphere on growth of mold on synthetic media and fruit. *Appl. Microbiol.* **22,** 513–516.

CHAPTER 7

MEASURING AND MODELING CONSUMER ACCEPTANCE

Stanley M. Fletcher, Anna V. A. Resurreccion,

and Sukant K. Misra

I. Introduction

Consumer perceptions and attitudes toward products provide crucial information for the successful introduction of new food products. Without such information, marketing or promotional programs cannot be specific, resulting in costly and unpredictable results. To further illustrate this point, approximately 13,244 new consumer food products were introduced in 1990, a 9.8% increase over 1989. For 1991, the number of new products is projected to exceed 15,000 (*Atlanta Journal,* 1991). Associated with this large number of new products are high rates of failure (*Supermarket News,* 1989). This growth of product development and innovation is being promoted highly in the food industry as a means to increase the demand for food. In response to this growth and failure rate, wholesalers and retailers have been starting programs that charge a failure fee if the product does not meet specified conditions, for example, units sold for a particular time period. In addition to these issues, food retailers face limited floor space to allocate among fresh fruits and vegetables, yet the average produce department in a supermarket handled 210 items in 1988 whereas the same department handled only 65 items in 1975. During the summer months the number of product items can expand up to an average of 300 items. To avoid ineffectiveness of product promotional programs, an investigation of consumer preference for products by means of consumer affective tests, as well as identification of the targeted consumer, is necessary.

Consumer affective tests are used to measure acceptance by current and potential users of a product (see Chapter 5). Acceptance may be defined as (1) an experience or feature of experience characterized by a positive attitude and (2) actual use (purchase or consumption). Acceptance may be measured by preference or liking for a specific food item (Amerine *et al.,* 1965). Generally, a large number of respondents is required for such evaluations. The respondents are not trained, but are selected at random to represent target or potential target populations (Institute of Food Technologists, 1981). Individuals are selected according to several specific criteria for screening subjects and identified according to test objectives, which frequently include characteristics such as previous use of the product, size of the family or age of specific family members, occupation of the head of the household, economic or social level, and geographic area.

This chapter deals with the measuring and modeling of consumer acceptance. The first part of the chapter addresses data collection issues. The second part of the chapter examines the alternative modeling approaches available. The reader is cautioned that the material presented in this chapter has not been used in its entirety to analyze consumer preferences for fresh fruit and vegetables since this area has been the subject of little research. However, based on the authors' research experiences, researchers and industry personnel who want to examine consumer acceptance would benefit greatly in their analysis if they followed the methodologies discussed in this chapter.

II. Data Collection

When measuring consumer acceptance, one is interested in collecting primary data. The types of primary data collected are dependent on the objectives of the test. The data collection methods are of two types: self-reported or observed. Self-reported data can be obtained by several routes, including personally face-to-face, by telephone interviews, or with mailed questionnaires. Data on observed behavior can be collected through personal observation or through mechanical or electronically controlled devices.

Sampling is an important factor in the measurement of consumer acceptance. Appropriate sampling procedures are necessary whenever consumer acceptance tests are conducted. A group of consumers is selected as a sample of a larger target population. The subjects participating in an acceptance tests should be qualified on the basis of typical demographic criteria such as age, sex, income, nationality, geographical region, race, education, and employment. Using company employees or research facility personnel rather than actual consumers may create difficulties in relating the results of the analysis to the targeted consumer.

A. Methods

A variety of methods is employed when collecting data for consumer product testing. These methods have been named after the location in which the test is conducted and include laboratory tests conducted in a sensory laboratory, central location tests (CLT), home use tests (HUT), and scanner-generated information from the food retail store.

1. Sensory laboratory

Laboratory tests are the most frequently used among all types of acceptance tests and are conducted in one location—the sensory laboratory of a research or testing facility (White *et al.*, 1988b). The advantage of the laboratory test is the convenience of the location and testing facility, and the access to needed equipment for product preparation and presentation. In the laboratory test, the researcher is able to control environmental conditions of the test, for example, lighting, sound, odors, and other interferences from outsiders or other subjects. Trained personnel and close supervision can be used to permit better control of the psychological environment, including such considerations as prior instruction to the consumer panelists.

A minimum of 25–50 responses is required; preferably, 100 or more responses should be obtained. There are several means of selecting the consumers for the laboratory tests. One is to prerecruit by working through a consumer database maintained specifically for this purpose (Resurreccion and Heaton, 1987). Prerecruitment offers the opportunity to screen the participants. However, mailing lists need to be maintained. Another means is to work through social groups such as churches, schools, and clubs of various types.

The test subjects are assembled singly or in groups in a reception area where demographic information is collected for the first time or is updated. Instructions may be administered prior to the test. Test subjects are led to the test area by trained staff. Usually, the evaluative information is obtained through self-administered questionnaires that have been designed to address specific objectives of the test and preferably have been pretested.

The disadvantages of this method include familiarity with the product, especially if employees are used. Further, product tolerances in preparation or use may be different from those in actual home use.

2. Central location

The central location approach is the most frequently used test involving consumers (Stone and Sidel, 1985). These tests take place at one or a limited number of locations, usually at a shopping center, mall, or similar location accessible to the public. These tests may be administered in permanent mall test facilities or in a mobile laboratory test facility parked in a shopping area (Castleberry and Resurreccion, 1986,1989; McWatters *et al.*, 1990). They also may be scheduled as needed in a large facility such as a convention hall or a hotel conference room (White *et al.*, 1987, 1988b). Another approach used less frequently is to set up a booth or temporary facility at a convention, fair, industrial show, or similar event where crowds of people are likely to congregate, and invite passers by to participate.

Consumers may be prerecruited or intercepted to participate in the tests. When selecting recruitment procedures, consideration should be given to the time required for prerecruitment compared with the participation rate, the test duration, perishability of the product, and variability of the product with time. Prerecruitment may take a large block of time prior to the test, but the tests usually are administered and completed in one day (Resurreccion *et al.*, 1991a,b). When intercepts are used as the recruitment approach, obtaining the needed number of consumer panelists may take several days (McWatters *et al.*, 1990.

When the researcher does not have access to a currently maintained database of consumer panelists, intercepting prospective panelists is the most sensible and appropriate method of obtaining consumer subjects (Resurreccion, 1986; Santos *et al.*, 1989). The project staff selects likely prospects from the traffic flow and asks the necessary screening questions. When a qualified person is located, the test is explained and the person is invited to the test area; the test is administered immediately. Usually a small reward is offered.

Facilities at a central location test site vary. Nevertheless, they should be adequate for stimulus control, that is, the necessary space and equipment should be available to standardize preparation of samples to be presented to the panel. In addition, the test facilities should permit proper control of the physiological and psychological testing environment. Considerations addressed in laboratory testing should be met also: adequate lighting, general comfort, freedom from distractions such as odors and noise, and elimination of interference from other judges and outsiders.

The number of subjects to be handled at one time depends on a number of factors, including the product type, the size of the testing space, and the number of research staff available. One or two persons typically may be tested at one time. However, testing in larger groups may not allow proper control. Too many panelists in a small area is likely to encourage inattention or mutual interference. In addition, if the panelist-to-researcher ratio is too high, mistakes may occur, questions may go unanswered, and missing responses may result. A general rule is that more than 12–15 panelists at one time requires special justification. The practice of administering a test simultaneously to a roomful of people (perhaps 25–100) is ill-advised because of the high probability of loss of control (American Society for Testing and Materials, 1968).

Almost any of the standard affective tests can be used in the central location test. Information may be collected by a trained interviewer but usually the information is obtained using a self-administered questionnaire. The questionnaires must be easy to understand and should not take too long to complete. A total testing time of 15–20 min is usually permissible since most test subjects will lose interest after that time and cooperation can be maintained only for this period. Longer tests should be avoided, unless there are special circumstances, in which case some consideration should be given to giving the subject a substantial cash award.

The advantage to using the central location test method is the large number of responses usually obtained. Employees generally are not used in these tests, thus minimizing bias. A disadvantage of this approach is the larger volume of product and paperwork due to the large number of respondents used. When the test is conducted in locations other than a test facility or a mobile laboratory, there may be limited resources for sample preparation and testing, and limited control over the testing environment.

3. Home use

The distinguishing feature of the home use test is that the test products are used under conditions that approximate the normal and usual patterns. Consumers or households with certain predesignated characteristics are selected. The test may involve 50–100 households or 75–300 per city in three or more cities nationwide. Panelists are provided with a package containing one or more products, instructions, and a questionnaire. Panelists may pick up the product from a designated site, or the product may be delivered or mailed (Resurreccion and Heaton, 1987). The products are left to be used under normal household conditions, over a specified period of time, except when restrictions are stated in the instructions. Later, the responses of the consumer panelists and possibly those of other household members are obtained. In addition to the acceptance responses, consumer opinions about other sensory characteristics, price, utility, and packaging may be obtained.

Various approaches to locating and recruiting respondents for this test include using employees or consumers recruited from an existing consumer database (Resurreccion and Heaton, 1987) or surveying door-to-door in selected areas. An interviewer asks questions both to establish qualifications and to obtain background information. When a qualified household is found, the placement is made on the spot, instructions are given, and arrangements are made for the return interview or return of the questionnaire. An advantage of this approach is that it permits distributing the sample as desired within a given territory. Also, it eliminates the problem of scheduling product delivery, and people are contacted more easily for the final interview. One major drawback is the time and expense associated with contacting and interviewing nonqualifying consumer households. Another approach is to use the telephone for preliminary screening. Calling may be random within a given area, or selected systematically from directories and from membership lists of various cooperating organizations (Sukhumsuvun and Resurreccion, 1988).

Several methods are employed to obtain data on consumer preferences, attitudes, and opinions. The most effective is probably the personal interview, because respondents are usually more motivated and involved. However, this method is also more expensive to conduct. Telephone interviews may likewise be conducted to reduce costs, particularly when the respondents are not easy to contact, but this method results in poorer cooperation and a lower completion rate than the personal interviews. Further, telephone interviews limit the type and range of questions that may be asked. It is critical, when using telephone interviews, that the questionnaire be pretested thoroughly.

Another approach is the use of the self-administered mailed questionnaire (Resurreccion and Heaton, 1987). In addition to the test products, the consumer is given instructions and a questionnaire to be completed after the product is used. The questionnaire is then mailed back to the researcher. This offers the advantage of reduced cost. Cooperation may be lowered by ambiguous instructions and questions. With appropriate instructions and a pretested questionnaire, response rates of at least 90% can be obtained (Resurreccion and Heaton, 1987).

The advantage of the home use test is that products are tested under actual use conditions and responses of the entire household, not only that of the consumer panelist, on use patterns, pricing, and attitudes may be obtained. The disadvantages are that the researcher can exert little or no control over the test, more time is required to conduct the test, and it is more expensive to run. Nonresponse rates are greater than those obtained with the other methods available, and only one or a maximum of two samples may be tested efficiently during the test period. Finally, this method requires additional methodological considerations such as planning for containers and labels to be used and preparation of product use instructions that are complete and unambiguous.

4. Scanner-generated

Scanner-generated data differ from data collected using any other method. Information about consumer attitudes and comparisons of food products are not collected. Rather, information about whether a consumer purchases or does not purchase a product is collected. Chapter 14 addresses in more detail the topic of scanner-generated data. This data is collected in supermarkets that scan UPC codes. The data reflect actual purchases. This source of data would provide a basis for a follow-up study on consumer attitudes to tie to an earlier study using one of the previously described data sources, that is, one will be able to compare consumer preferences with actual purchases.

B. Types

Once the method of data collection has been decided, the type of data to be collected must be determined. This section addresses two key types of data: product characteristics and consumer characteristics.

1. Product characteristics

The affective dimension of food is probably the most studied component of consumer evaluation of food. The types of questions that can be asked are classified according to structure and scaling types. Structure may be either open-ended or closed. Product evaluation by consumers involves the discipline of psychophysics. The researcher is translating affect resulting from a product or physical stimulus into a meaningful recorded response. Sensory evaluation techniques are used to correlate these physical scales of stimulus with the psychological value given by the respondent.

Scaling techniques can be broken down to several basic categories namely, nominal, ordinal, interval, and ratio scales. A nominal scale may be used to name various items that must be differentiated from one another. For example, one could ask respondents to list names of foods they like. Ordinal scales not only identify items by name but reveal whether each item has a quality greater than, equal to, or less than the other items. Simple paired preference tests or rank order tasks involve ordinal scaling techniques. In the interval scale, the distance between points on the scale is assumed to be equal and the scale has an arbitrary zero point. Interval scales are considered to be truly quantitative scales. The ratio scales exhibit the same properties as interval scales. In addition, there is a constant ratio between points and an absolute zero (Stone and Sidel, 1985).

The simplest application for preference is the paired preference test, which involves two samples presented simultaneously or sequentially. One or more pairs may be tested per panel session. The consumer is requested to express a preference based on a specified criterion. A forced choice may or may not be imposed. If it is, the consumer must indicate a preference for one sample over the other. Results are obtained in terms of the relative frequencies of choice of the samples (Institute of Food Technologists, 1981).

An extension of the paired preference test in which three or more samples are presented simultaneously is the ranking test. The consumers are asked to assign an order to the samples according to their preference. As in the paired preference method, rank order evaluates samples only in relation to one another (Institute of Food Technologists, 1981). Liking or disliking a sample cannot be determined adequately by this method.

Scale ratings reflect the consumers perceived intensity of a specified attribute under a given set of conditions. Various rating scales have been developed. The most common affective scale is the ordered metric scale or numbered category scale. This is an ordinal scale that assumes interval character because of the nature of assignment of numbers to the individual points.

The 9-point hedonic scale, pioneered by the Quartermaster Food Institute for the Armed Forces (Peryam and Pilgrim, 1957) is used to measure the level of liking the food products. The method is used widely and relies on consumer capacity to report, directly and reliably, feelings of like and dislike. Several variations of the scale reduce the number of points to seven or five and modify the terms used (Schutz, 1980).

2. Consumer characteristics

Once the product characteristics data are collected, consumer characteristics are needed to complete the data required for the modeling. According to consumer demand theory, the demand for a product is a function of consumer income, product price, substitute and complement prices, and environmental variables (Phlips, 1974). The environmental variables also sometimes are referred to as taste and preferences of the consumer. These variables, for example, could be age, education, employ-ment status, health, family size, eating habits, or marital status. These characteristics are used to control for differences in preferences, as well as to segment the market (i.e., establish market "niches").

C. Reliability Issues

Several correlation-based statistical techniques are used to assess the validity and reliability of a scale or index (Bohrnstedt, 1970), but their use does not guarantee the validity or reliability of a measurement instrument. Reliability is the extent to which the subjects will provide consistent or repeatable results in the absence of true change in the object being measured (Wimberley, 1980). A high degree of reliability is desirable, but does not guarantee validity. Validity is a more complex issue because it is not measured directly, but a valid measure also should be reliable.

Four basic methods for assessing reliability are (1) retest, (2) alternative form, (3) split-halves, and (4) internal consistency methods (Carmines and Zeller, 1979). The retest method is the easiest to conduct and involves giving the same test to the same group of people at a different time. The reliability may be calculated as the correlation between scores on the same test administered at two points in time. Drawbacks of this test include the expense of test–retest methods and test reactivity, that is, the test itself may induce a change in subject response during the retest. A naive interpretation of the test–retest correlations may underestimate the degree of reliability, and may overestimate reliability if subject memory of the first test influences retest responses. Giving the test in an alternative form is another method; again, the response correlation is the measure of reliability. However, this method is limited by the practical difficulty of constructing alternative yet parallel forms of the test. The split-halves method, unlike the other two methods, can be conducted on one occasion. The total set of items is divided into halves and the scores are correlated to obtain an estimate of reliability. The halves can be considered ap-proximations of alternative forms. Internal consistency coefficients estimate relia-bility without requiring the splitting or repeating of items. These require only a single test administration and provide unique estimates of reliability. The most popular of these coefficients is Chronbach's alpha (Chronbach, 1951; White et al. 1988a).

Validity is the extent to which any measuring instrument measures what it is intended to measure. The three basic types of validity are (1) criterion-related validity, (2) content validity, and (3) construct validity. Criterion-related validity (also called predictive validity) validates a test by showing that it accurately predicts some form of behavior external to the measuring instrument itself (Mullen *et al.*, 1984). Content validity depends on the extent to which empirical measurement reflects a specific domain of content. Construct validity is concerned with the extent to which a particular measure relates to other measures that are consistent with theoretically derived hypotheses concerning the concepts (or constructs) that are being measured.

III. Modeling

Developing new products can be costly and time consuming. Companies continually are seeking new approaches to achieving greater success in the marketplace. Theoretically, there is a set of attributes that, if present in a product, would lead to optimum acceptance (Schutz, 1983). The attainment of optimized attributes in a product does not guarantee success in the marketplace. Properties such as brand name and loyalty, advertising and promotion, price, quality control, and economic factors in the marketplace are major contributors to product success.

The measurement of consumer acceptance is not new. Historically, results were tabulated according to product acceptance scores, and mean values were obtained. When demographic information is obtained, it is cross-tabulated with acceptance scores (McWatters *et al.*, 1990).

In the past 20 years, there has been a remarkable expansion in the use of statistical techniques to examine food acceptance data. Several multivariate methods have been used in the assessment of food quality, including principal component analysis, factor analysis, cluster analysis, multidimensional scaling, and regression (Resurreccion, 1988). Qualitative choice models and conjoint analysis have been used in analyzing food product acceptance. However, applications to fresh fruits and vegetables have been minimal because of the lack of familiarity with these approaches. Thus, the rest of this section provides an overview of these approaches so readers may gain a better understanding of them and gain the ability to determine which approach may be best for them in their research on consumer acceptance of fresh fruits and vegetables. Then readers will be able to use the references to gain a deeper understanding of the approach appropriate to their research.

A. Principal Component and Factor Analysis

Principal component analysis (PCA) provides a method for extracting structure from the variance–covariance or correlation matrix (Federer *et al.*, 1987). PCA constructs linear combinations of the original data with maximal variance. PCA is used to reduce the matrix to a smaller number of components with as little loss of information as practicable (Resurreccion, 1988). PCA has been used in the study of cabbage (Martens, 1985) and cauliflower (Martens *et al.*, 1983).

Factor analysis (FA) is a technique most commonly used for data reduction and simplification in food acceptance studies. It is used to reduce a large number of variables to a smaller set of new variables called factors. The factors are often, but not always, independent of one another (Ennis *et al.*, 1982). Once a model has been computed for a particular set of experimental variables, each set of measures can be converted to factor scores that correspond to the extracted factors. Factor scores have been demonstrated to reflect the difference between samples more clearly than any individual variable can. Pilgrim and Kamen (1959) used factor analysis to group foods by using food preference data from consumers. Factor analysis was used by Sukhumsuvun and Resurreccion (1988) to identify the underlying perceptions of 200 Thai consumers of peanuts and peanut products and to identify new peanut products that exhibit good potential for being accepted by Thais. Resurreccion (1986), in earlier work, used this technique to study consumer perceptions of broccoli and 41 other fresh and processed vegetable products. Paguio and Resurreccion (1987) compared perceptions of dairy products and vegetables of low and middle income mothers of preschool children. The technique has been used in the study of apples (McLellan and Massey, 1984), dry beans (Hosfield *et al.*, 1984), green beans (Powers *et al.*, 1977), and tomatoes (Resurreccion and Shewfelt, 1985).

B. Cluster Analysis

Cluster analysis is a general term for any procedure that groups variables or cases according to some measure of similarity (Ennis *et al.*, 1982). The variables in a cluster are associated highly with one another, whereas those in different clusters are relatively distinct from one another (Cardello and Maller, 1987).

Resurreccion *et al.* (1987) and Godwin *et al.* (1978) described analyses that differ greatly in theory and practice. When analyzing data on tomatoes, Resurreccion (1988) used cluster analysis to establish groups of closely related measures. In another study on beans Resurreccion *et al.* (1987) evaluated 72 snap bean samples for 17 sensory and physicochemical measurements of quality to determine relationships between these measures. Cluster analysis was used also by Powers *et al.* (1978) in the study of blueberries.

C. Multidimensional Scaling

Multidimensional scaling (MDS) refers to a class of techniques that use proximities among objects. A proximity is a number that indicates how similar or different the objects are perceived to be. MDS is used to investigate product characteristics used by panelists in evaluating product similarities. The technique, using measures such as similarity, positions the products in a geometric configuration in such a way that significant features of the data on these products are revealed by the geometrical relationships among the points (Poste and Patterson, 1988).

The use of MDS has been proposed to identify product attributes by sensory means (Schiffman *et al.*, 1981). Moskowitz (1985) outlined the use of multidimensional scales for two problems in odor perception. The method was used to study odor quality of carrots and to determine individual's perceptions of odorants over a course of several sessions.

D. Regression

Multiple regression is a statistical technique that is essential to relating objective to instrumental data on food quality (Resurreccion, 1988). A series of predictor variables (quality measures) is employed to predict some dependent variable such as overall acceptance. In food quality evaluations, the most common use of multiple regression analysis is to predict the magnitude of some sensory attribute (such as overall acceptance) on the basis of a series of objective measures of the food (Cardello and Maller, 1987). Regression analysis was used in the study of five sensory parameters and 21 objective measures on carrots (Simon *et al.*, 1980), in which overall preference was enhanced by sugars and diminished by volatiles. Regression analysis was used also to study oranges (Basker, 1977). Response surface plots of regression equations were used by Santos *et al.* (1987) in a consumer study to optimize the levels of flavor and added color in an imitation cheese spread prepared from peanuts.

E. Qualitative Choice Models

Qualitative choice modeling is a refinement of multiple regression. As mentioned earlier in the data discussion section, scaling techniques are used in many consumer acceptance studies to quantify consumer response concerning attitude toward a particular food product. Since these responses are discrete (qualitative), the use of multiple regression (i.e., ordinary least squares) would result in biased and inefficient parameter estimates (Judge *et al.*, 1982). The use of binary qualitative dependent variable models can resolve problems associated with multiple regression estimates and has become more common in qualitative research investigations of individual preferences in the agricultural economics field (Hill and Kau, 1973; Rahm and Huffman, 1984; Fletcher and Terza, 1986).

If the study is structured so a consumer can express more than two outcomes and these outcomes can be ranked, a multiordered response model must be used for estimation. The reader is referred to Amemiya (1981) and Maddala (1983) for a more detailed and theoretical discussion of this set of models, as well as of the general qualitative choice models. The multiordered response model is appropriate when the dependent variable (i.e., consumer response) has more than two outcomes and the outcomes can be ranked or ordered. In 1986, this model was used by Carley and Fletcher (1986) to explain factors that affected decisions made by dairy farmers given a combination of alternative management practices. This model was used also by Fletcher *et al.* (1990,1991) to explain consumer attitudes toward and willingness to pay for a new fried food prepared from cowpea flour. A brief description of the multiordered response model in terms of consumer buying intention follows.

Theoretically, consumer buying intention follows a continuous function. This continuous function is assumed to be related linearly to a set of observed consumer and product characteristics. However, this continuous function is not observable. Rather, consumer buying intention is observed at discrete levels. Given this framework, the model is solved using the maximum likelihood procedure, as described by Maddala (1983).

Qualitative choice analysis allows the interpersonal comparison of utility, given the assumption that a random utility function exists. This approach provides the ability to compute selection probabilities and obtain consumer profiles.

F. Conjoint Analysis

Conjoint analysis methodology generally is acknowledged to be an important and useful consumer research technique for preference assessment (Carmone *et al.,* 1978). Although the foundation of conjoint methodology was laid in the 1920s, application of conjoint measurement was rare until the 1960s. In the 1960s, a search by mathematical psychologists for a new technique to quantify judgmental data resulted in a number of theoretical contributions and algorithmic developments. It is, however, agreed generally that the first detailed consumer-oriented paper by Green and Rao (1971) on conjoint measurement introduced the methodology to the marketing studies. Since its introduction in the marketing literature, conjoint analysis has been applied to a variety of multivariate problems.

The basic principle of conjoint analysis is that a product is composed of attributes and consumer preferences for products can be assessed effectively by estimating the importance of product attributes to consumers. This principle of conjoint analysis finds theoretical support in economics in the approach to consumer theory proposed by Lancaster (1971). Lancaster suggests that consumers derive utility not from goods themselves, but rather from the attributes or characteristics that the good possess (e.g., size, quality, price, nutritional value, or safety, in the case of fresh fruits and vegetables).

Conjoint analysis involves decomposing a composite good into its constituent attributes, surveying respondents about relative preferences for alternative attribute bundles, and quantifying marginal rates of substitution between attributes. For a detailed discussion of the various steps in conjoint analysis, the alternative methods of implementing each of the steps, and several conceptual and technical questions, readers are referred to Green and Srinivasan (1978) and Timmermans (1984).

The fundamental decision to be made initially by any user of conjoint measurement concerns the selection of a model of preference. Green and Srinivasan (1978) have identified four types of multiattribute preference models: the vector model, the ideal point model, the part-worth (additive and multiplicative) model, and the mixed model. Of these, the part-worth function model has received wide acceptance due to its ready interpretability and flexibility. Green and Srinivasan (1978) suggest that the flexibility of the preference model is greater as we proceed from the vector to the ideal point to the part-worth function models; however, the reliability of the estimated parameters is likely to improve in the reverse order.

After deciding on a model of preference and before actually collecting data from the respondents, the user of conjoint analysis must consider a number of technical issues associated with eliciting consumer relative preferences. One of the major factors affecting eliciting the views of the respondents is the selection of attributes that are relevant to consumers in forming their preferences. The identification of relevant attributes can be undertaken by several alternative means. Some of the alternatives suggested are (1) pretest, (2) consumer questioning, (3) Kelly's repertory grid test, and (4) focus group interviews. Green and Srinivasan (1978) point out that the more difficult task is to reduce the number of attributes to a manageable size so the estimation procedures are reliable but still account for consumer preferences sufficiently.

Using too many attributes to account for consumer preferences increases the number of judgments required from each respondent and may lead to superficial responses. Conversely, the inclusion of two few attributes may lead to inaccurate gauging of preferences. To keep the number of attributes manageable, a well-structured presentation of alternatives should be followed. Presentation strategies such as "fractional factorial" (Green, 1984) allow the researcher to use only a relatively small proportion of the total possible attributes.

The two most commonly used data collection alternatives for conjoint analysis are the two-factor-at-a-time procedure and the full-profile approach (Green and Srinivasan, 1978). In the two-factor-at-a-time procedure, attributes are considered in pairs; the respondent is required to rank order all possible combinations of levels generated from a pair of attributes. In contrast, the full-profile approach uses stimuli described in terms of all critical attributes, and the respondent is required to rank order all possible combinations of attribute levels (Segal, 1982). These profile rankings provide the input data from which respondent use for the various levels of each attribute are collected.

Since its introduction in the marketing literature in the early 1970s, conjoint analysis has been used extensively to understand consumer preferences. Cattin and Wittink (1982) have provided a comprehensive survey of the commercial use of conjoint analysis in marketing research. Several developments have occurred in conjoint analysis, and its application has been extended to other disciplines. However, application of conjoint analysis to assessing consumer attribute preferences of an agricultural product such as fresh fruit or vegetable has been infrequent.

In an agricultural application of conjoint analysis, Manalo (1990) assesses the importance of consumer preferences for apple attributes in apple purchasing decisions. An apple was defined, with the help of preliminary interviews with several consumers, to possess five attributes: size, color, price, crispiness, and flavor. Since each of these attributes had either two or three levels, a total of 108 possible attribute combinations was generated. By using an orthogonal array experimental design, Manalo (1990) identified only a subset (18 combinations) of the total number of attribute combinations. The data collection method used the full-profile approach and a random sample of 208 apple consumers was asked to rank the 18 combinations using 1 and 18 to indicate highest and lowest preference, respectively. Ranking data provided by the respondents were modeled by using an additive part-worth function model to estimate relative importance of product attribute combinations to the consumer. The results suggested that the most preferred combinations of attributes is represented by an apple that is large, red, priced at $0.79/lb, crisp, and sweet.

IV. Hypothetical Illustration of a Consumer Acceptance Test

No single procedure is superior to all others in establishing correlations between physicochemical measures and sensory assessment of quality (Powers *et al.*, 1977). The ultimate choice of method must be made by the individual investigator based on the research objectives. This section provides an illustration of how one may address a specific research objective of measuring consumer acceptance of a particular fresh produce item.

Let us assume that an agribusiness firm wants to investigate consumer acceptance of and the potential market for branded genetically engineered tomatoes with a home grown fresh flavor. The genetically engineered tomatoes can be different from bulk tomatoes in taste, and may cost more at the marketplace. It is, therefore, important for the agribusiness firm to investigate consumer preference, to identify the target population, and to measure willingness to pay for this new product to avoid ineffective product development and promotional programs.

In measuring consumer acceptance of and willingness to pay for branded genetically engineered tomatoes, the researcher is interested initially in collecting primary data. To collect such data, a group of consumers must be recruited that represents the target market for the product. The consumers should be users of fresh tomato products and should be qualified on the basis of preset demographic characteristics such as age, education, or income. For illustration purposes, let us assume that 400 consumers are to be recruited from a consumer database that is maintained specifically for consumer product acceptance studies.

Assuming that the testing is to be conducted in a sensory laboratory that is equipped with facilities for product preparation and presentation, a questionnaire must be designed carefully. The questionnaire should identify a set of attributes (e.g., color, flavor, appearance) common to both bulk tomatoes and genetically engineered tomatoes and be designed to enable the participants to rate and compare the samples. A 9-point hedonic scale could be used to rate the product attributes. Further, the questionnaire must include questions concerning the likelihood of consumer acceptance and their willingness to pay for genetically engineered tomatoes. It is important that the questionnaire be pretested to minimize ambiguity.

The recruited consumers then should be invited to the sensory laboratory and be received by trained staff at the reception area. Prior to the actual sensory tests, demographic information should be collected from the participants and necessary instructions should be administered. Participants then should be led to the test area that is equipped with partitioned booths to prevent communication among participants. Sample cut-up tomatoes placed in sample cups and labeled with three digit random numbers should be presented to the participants along with a copy of the questionnaire. The test participants should be asked to taste the samples before filling in the questionnaire.

After the collection and tabulation of the sensory data, preliminary statistical analysis could be conducted to identify the potential consumers of genetically engineered tomatoes and their willingness to pay for this product relative to bulk tomatoes. If product acceptance decision is cross-tabulated against product attribute ratings and demographic information, product and consumer characteristics that influence the genetically engineered tomato consumption decision could be classified.

A qualitative choice model then could be developed and estimated econometrically to investigate impacts of various product and consumer characteristics on consumer willingness to pay for genetically engineered tomatoes. For estimation purposes, the responses to the "willingness to pay" question of the potential consumers of genetically engineered tomatoes could be collapsed into a specific number of categories and a multiordered response model could be used. For example, the categories of willingness to pay could represent "not willing to pay a higher price than bulk tomatoes," "willing to pay a price premium up to 10%," and "willing to pay a price premium of more than 10%."

The estimation results from the multiordered response model will help identify statistically significant factors that influence consumer willingness to pay for genetically engineered tomatoes. Further, by computing marginal probabilities from the estimated model, one could measure the change in probability of willingness to pay with respect to a change in product and consumer characteristics. Marginal probabilities for consumer characteristics from the estimated model also could be used to segment the market and to identify the target population by willingness to pay categories.

Bibliography

Amemiya, T. (1981). Qualitative response models: A survey. *J. Econ. Lit.* **14,** 1483–1536.

American Society for Testing and Materials (1968). "Manual of Sensory Testing Methods STP 434." American Society for Testing and Materials, Philadelphia.

Amerine, M. A., Pangborn, R. M., and Roessler, E. B. (1965). "Principles of Sensory Evaluation of Food." Academic Press, New York.

Atlanta Journal, The (1991). What to buy? Shoppers digest new food choices. May 7.

Basker, D. (1977). Changes in the organoleptic quality of shamouti orange during their ripening season. *J. Food Sci.,* **1,** 147–156.

Bohrnstedt, G. W. (1970). Reliability and validity assessment in attitude measurement. *In* "Attitude Measurement" (G. F. Summers, ed.). pp. 80–99. Rand McNally & Company, Chicago.

Cardello, A. V., and Maller, O. (1987). Psychophysical bases for the assessment of food quality. *In* "Objective Methods in Food Quality Assessment"(J. G. Kapsalis, ed). p. 61. CRC Press, Inc., Boca Raton, Florida.

Carley, D. H., and Fletcher, S. M. (1986). An evaluation of management practices used by southern dairy farmers. *J. Dairy Sci.* **69,** 2458–2464.

Carmines, E. G., and Zeller, R. A. (1979). "Reliability and Validity Assessment." Sage University Paper No. 17. Sage Publ., Beverly Hills.

Carmone, F. J., Green, P. E., and Jain, A. K. (1978). The robustness of conjoint analysis: Some Monte Carlo results. *J. Marketing Res.* **15,** 300–303.

Castleberry, S. B., and Resurreccion, A. V. A. (1986). Effectively communicating the attribute of "quality": An exploratory examination. In "Marketing in an Environment of Change" (R. L. Key, ed.), pp. 228–232. Southern Marketing Association, Charleston, South Carolina.

Castleberry, S. B., and Resurreccion, A. V. A. (1989). Communicating quality to consumers. *J. Consumer Marketing* **6(3),** 21–28.

Cattin, P., and Wittink, D. R. (1982). Commercial use of conjoint analysis: A survey. *J. Marketing* **46,** 44–53.

Cronbach, L. J. (1951). Coefficient alpha and the internal structure of tests. *Psychometrika* **16,** 297–334.

Ennis, D. M., Boelens, H., and Bowman, P. (1982). Multivariate analysis in sensory evaluation. *Food Technol.* **36(11),** 83–90.

Federer, W. T., McCulloch, C. E., and Miles-McDermott, N. J. (1987). Illustrative examples of principal components analysis. *J. Sensory Studies* **2,** 37–54.

Fletcher, S. M., and Terza, J. V. (1986). Analyzing farmers' selection of available marketing alternatives using the multivariate probit model. *Canadian J. Agr. Econ.* **34,** 243–252.

Fletcher, S. M., McWatters, K. H., and Resurreccion, A. V. A. (1990). Analysis of consumer attitudes toward new fried food prepared from cowpea flour. *J. Food Dist. Res.* **21(2),** 75–82.

Fletcher, S. M., McWatters, K. H., and Resurreccion, A. V. A. (1991). Analysis of consumers' willingness to pay for a new food prepared from cowpea flour. *J. Consumer Stud. Home Econ.* (In press.)

Godwin, D. R., Bargmann, R. E., and Powers, J. J. (1978). Use of cluster analysis to evaluate sensory–objective relations of processed green beans. *J. Food Sci.* **43,** 1229–1234.

Green, P. E. (1984). Hybrid models for conjoint analysis: An expository review. *J. Marketing Res.* **21,** 155–169.

Green, P. E., and Rao, V. R. (1971). Conjoint measurement for quantifying judgmental data. *J. Marketing Res.* **8,** 355–363.

Green, P. E., and Srinivasan, V. (1978). Conjoint analysis in consumer research: Issues and outlook. *J. Consumer Res.* **5,** 103–123.

Hill, L., and Kau, P. (1973). Application of multivariate probit to a threshold model of grain drying purchasing decisions. *Am. J. Agr. Econ.,* **55,** 19–27.

Hosfield, G. L., Ghader, A., and Uebersax, M. A. (1984). A factor analysis of yield and sensory and physio-chemical data from tests used to measure culinary quality in dry edible beans. *Can. J. Plant Sci.* **64,** 285–293.

Institute of Food Technologists (1981). Sensory evaluation guide for testing food and beverage products. *Food Technol.* **35(11),** 50–59.

Judge, G. G., Hill, R. C., Griffiths, W. E., Lutkepohl, H., and Lee, T. C. (1982). "Introduction to the Theory and Practice of Econometrics." John Wiley & Sons, New York.

Lancaster, K. (1971). "Consumer Demand: A New Approach." Columbia University Press, New York.

McLellan, M. R., and Massey, L. M., Jr. (1984). Effect of postharvest storage and ripening of apples on the sensory quality of processed applesauce. *J. Food Sci.* **49,** 1323–1326.

McWatters, K. H., Resurreccion, A. V. A., and Fletcher, S. M. (1990). Response of American consumers to akara, a traditional West African food made from cowpea paste. *Int. J. Food Sci. Tech.* **25,** 551–557.

Maddala, G. S. (1983). "Limited-Dependent and Qualitative Variables in Econometrics." Cambridge University Press, New York.

Manalo, A. B. (1990). Assessing the importance of apple attributes: An agricultural application of conjoint analysis. *Northeastern J. Agr. Res. Econ.* **19,** 118–124.

Martens, M. (1985). Sensory and chemical quality criteria for white cabbage studied by multivariate data analysis. *Norwegian Food Res.* **18(2),** 100–104.

Martens, M., Martens, H., and Wold, S. (1983). Preference of cauliflower related to sensory descriptive variables by partial least squares (PLS) regression. *J. Sci. Food Agricult.* **34,** 715–724.

Moskowitz, H. R. (1976). Applications of multidimensional scaling to the psychological evaluation of odors. *In* Correlating Sensory Objective Measurements—New Methods for Answering Old Problems (J. J. Powers and H. R. Moskowitz, eds.). pp. 97–110. ASTM STP 594. American Society for Testing and Materials, Philadelphia.

Mullen, B. J., Krantzler, N. J., Grivetti, L. E., Schutz, H. G., and Meiselman, H. L. (1984). Validity of a food frequency questionnaire for the determination of individual food intake. *Am. J. Clin. Nutr.* **39,** 136–143.

Paguio, L. P., and Resurreccion, A. V. A. (1987). Preschool mother's perception of dairy products and vegetable uses. *Nutr. Rep. Int.* **35(1),** 79–86.

Peryam, D. R., and Pilgrim, F. J. (1957). Hedonic scale method of measuring food preferences. *Food Technol.* **11(9),** 9–14 (supplement).

Phlips, L. (1974). "Applied Consumption Analysis." North-Holland/American Elsevier, New York.

Pilgrim, F. J., and Kamen, J. M. (1959). Patterns of food preferences through factor analysis. *J. Marketing* **24,** 68–72.

Poste, L. M., and Patterson, C. F. (1988). Multidimensional scaling—sensory analysis of yoghurt. *Can. Instit. Food Sci. and Technol.* **21(3):**271–278.

Powers, J. J., Godwin, D. R., and Bargmann, R. E. (1977). Relations between sensory and objective measurements for quality evaluation of green beans. *In* Flavor Quality: Objective Measurement (R. A. Scanlan, ed.). pp. 51–70. ACS Symposium Series, Washington, D. C.

Powers, J. J., Smit, C. J. B., and Godwin, D. R. (1978). Relations among sensory and objective attributes of canned rabbiteye (*Vaccinium ashei* v. *reade*) blueberries. *Lebensm.-Wiss. Technol.* **11,** 275–278.

Rahm, M. R., and Huffman, W. E. (1984). The adoption of reduced tillage: The role of human capital and other variables. *Am. J. Agr. Econ.* **66(4),** 405–413.

Resurreccion, A. V. A. (1986). Consumer use patterns for fresh and processed vegetable products. *J. Consumer Stud. Home Econ.* **10,** 317–332.

Resurreccion, A. V. A. (1988). Applications of multivariate methods in food quality evaluation. *Food Technol.* **42(11),** 128–136.

Resurreccion, A. V. A., and Heaton, E. K. (1987). Sensory and objective measures of quality of early harvested and traditionally harvested pecans. *J. Food Sci.* **52,** 1038–1040,1058.

Resurreccion, A. V. A., and Shewfelt, R. L. (1985). Relationship between sensory attributes and objective measurements of postharvest quality of tomato. *J. Food Sci.* **50,** 1242–1256.

Resurreccion, A. V. A., Shewfelt, R. L., Prussia, S. E., and Hurst, W. C. (1987). Relationships between sensory and objective measures of postharvest quality of snap bean as determined by cluster analysis. *J. Food Sci.* **52:**113–123.

Resurreccion, A. V. A., Chompreeda, P., and Beuchat, L. R. (1991a). "Sensory Quality Evaluations of Peanut-Supplemented Chinese-Type Noodles by Consumers." Paper presented at the Annual Meeting of the Institute of Food Technologists, June 1–5, Dallas, Texas.

Resurreccion, A. V. A., Dull, G., Smittle, D. A., and Thai, C. (1991b). "Comparison between Trained Judges' and Consumer Panelists' Evaluations of Honeydew Melons of Varying Stages of Maturity." Paper presented at the 88th Annual Meeting of the Southern Association for Agricultural Scientists Food Science and Human Nutrition Section, February 2–6, Fort Worth, Texas.

Santos, B. L., Koehler, P. E., and Resurreccion, A. V. A. (1987). Sensory analysis of a peanut-based imitation cheese spread. *J. Food Qual.* **10,** 43–56.

Santos, B. L., Resurreccion, A. V. A., and Garcia, V. V. (1989). Quality characteristics and consumer acceptance of a peanut-based imitation cheese spread. *J. Food Sci.* **54,** 468–472.

Schiffman, S. S., Reynolds, M. L., and Young, F. W. (1981). Introduction to Multidimensional Scaling—Theory, Methods, and Applications. Academic Press, New York.

Schutz, H. G. (1980). "Food-Related Attitudes and Their Measurement." Paper presented at the Symposium on Attitude Theory and Measurement in Food and Nutrition Research, June 15–17, Pennsylvania State University, University Park, Pennsylvania.

Schutz, H. G. (1983). Multiple regression approach to optimization. *Food Technol.* **37(11),** 46–48,62.

Schutz, H. G., Moore, S. M., and Rucker, M. H. (1977). Predicting food purchase and use by multivariate attitudinal analysis. *Food Technol.* **31(8),** 85–92.

Segal, M. N. (1982). Reliability of conjoint analysis: Contrasting data collection procedures. *J. Marketing Res.* **19,** 139–144.

Simon, P. W., Peterson, C. E., and Lindsay, R. C. (1980). Correlations between sensory and objective parameters of carrot flavor. *J. Agr. Food Chem.* **28,** 559–562.

Stone, H., and Sidel, J. L. (1985). "Sensory Evaluation Practices." Academic Press, San Diego.

Sukhumsuvun, S., and Resurreccion, A. V. A. (1988). Food habits and eating patterns of Thai nationals in the United States. *Nutr. Rep. Int.* **37(5),** 913–922.

Supermarket News (1989). "Production Introduction Dip 5.6%: Researcher." January 16.

Timmermans, H. J. P. (1984). Decompositional multi-attribute preference models in spatial choice analyses: A review of some recent developments. *Prog. Human Geog.* **8,** 189–221.

White, F. D., Lillard, D. A., and Resurreccion, A. V. A. (1987). "Effect of Warmed-Over Flavor on Consumer Acceptance and Purchase of Precooked Ready-to-Eat Beef Products." Paper presented at the 47th Annual Institute of Food Technologists Meeting, June 16–19, Las Vegas, Nevada.

White, F. D., Resurreccion, A. V. A., and Lillard, D. A. (1988a). Consumer characteristics, food related attitudes, and consumption of meat and meat products. *Nutr. Res.* **8,**1333–1344.

White, F. D., Resurreccion, A. V. A., and Lillard, D. A. (1988b). Effect of warmed-over flavor on consumer acceptance and purchase of precooked top round steaks. *J. Food Sci.* **53(5)**, 1251–1257.

Wimberly, R. C. 1980. Issues in the measurement of food and nutrition attitudes. *In* Attitude theory and measurement in food and nutrition research. Proceedings of a symposium, June 15–17, (1980). Pennsylvania State University, University Park, Pennsylvania.

CHAPTER 8

MODELING QUALITY CHARACTERISTICS

Chi N. Thai

I. Introduction

Throughout this book, we are using the Kramer and Twigg (1970) definition of quality as "the composite of those characteristics that differentiate individual units of a product and have significance in determining the degree of acceptability by the buyer." This statement demonstrates the importance of the buyer or consumer, especially when one considers that consumption of fruits and vegetables in the United States has increased in the past two decades (Shewfelt, 1990). The demand for high quality products is occurring worldwide also (Sarig, 1989). Thus, to remain successful in a more competitive market, quality maintenance and control are essential to the fruit and vegetable sector of the economy. To deliver optimal quality products to the consumer, we must understand, predict, and possibly control quality changes in a reliable manner. In other words, we need accurate tools for predicting quality changes.

II. General Approach to Modeling

In the broadest sense, Rothenberg (1989) defined modeling as "the cost-effective use of something in place of something else for some cognitive purpose." He characterized any model with three essential attributes.

1. *Reference:* The model refers to something in the real world (its referent).
2. *Purpose:* The model has an intended cognitive purpose with respect to its referent.
3. *Cost-effectiveness:* The model is more cost-effective to use for this purpose than the referent itself.

Models can be classified in many ways. Neelamkavil (1987) presented a detailed and layered classification of models. All models are categorized as "physical," "symbolic," or "mental." "Symbolic" models were classified further into "mathematical" and "nonmathematical" models. "Mathematical" models are of particular interest since they can be used to optimize and control processes involving product quality changes. Clements (1989) discussed various methodologies of mathematical modeling and noted that the Checkland (1981) "soft systems methodology" can add richness and capability to current modeling methodologies.

Contrary to most authors, Rothenberg (1989) viewed simulation as a kind of modeling rather than a kind of model, since the purpose of simulation is comprehension, planning, prediction, and control of the referent by observing the behavior of its model. Shannon (1975) and Bratley *et al.* (1983) can be consulted for general issues on crucial components of a simulation project: data gathering, model building and validation, statistical design and estimation, computer programming, and implementation. Rand (1983) reviewed the methods most pertinent to the modeling of food processing and storage. Kapsalis (1987) presented objective methods and an integrative approach to be used in food quality measurement and simulation.

III. Models of Fruit and Vegetable Quality

Literature on quality of fruits and vegetables is too vast to gather into this one chapter, but a volume by Jen (1989a) can serve as a starting reference for workers in this field. Jen (1989b) listed the three major quality factors of fruits and vegetables as color, texture, and flavor. Applying the previously described approach to modeling, our particular referent is the consumer perception of these quality factors. Our cognitive purpose is to create a predictive model of sensory perception of fruit and vegetable quality as the produce progresses through the postharvest system. Chapter 7 provides details on measuring and modeling consumer acceptance. This chapter is concerned with complementary issues and has the goal of presenting the feasibility of a quality assurance and control system for fresh fruits and vegetables.

Williams (1981) listed three types of interrelationships that need to be investigated in modeling human perception of quality.

1. interrelationships of preference (sensory) data
2. relationships between preference data and objective descriptive (physical) data
3. interrelationships of objective descriptive data

Regression is the mathematical method used most often for relationships of the first two types, but other statistical methods are used also (Ross, 1987; Chapter 7). Neural networks are being shown to be promising modeling tools for relationship types 1 and 2. (Thai *et al.,* 1990c; Thai and Shewfelt, 1991c). Neural computing techniques were developed from research in neurophysiology and cognitive processes (Hecht-Nielsen, 1990; Simpson, 1990). A neural network is an interconnection of a finite number of computing elements called nodes or neurons which can "learn" to generalize from a given set of matching data inputs and outputs using different learning algorithms (Maren *et al.,* 1990). The "knowledge" of the network is represented by the relative strengths of its interconnections (analogous to regression parameter estimates). For type 3 relationships, a single model has shown its applicability for the last 30–40 years and is based on chemical kinetics concepts and the Arrhenious model for temperature effect on the rate of quality change (Saguy and Karel, 1987). This model is usually defined by the following equations:

$$\frac{dQ}{dt} = k \, Q^n \tag{1}$$

$$k = k_0 \, e^{-(E_a/RT)} \tag{2}$$

where Q is the quality index, t is time, n is the order of the kinetic reaction, k is the rate parameter, k_0 is the constant independent of temperature, E_a is the activation energy, R is the ideal gas constant, and T is absolute temperature. It must be emphasized that Eqs. 1 and 2 are only convenient mathematical structures found to be applicable in the modeling of specific quality attributes, whereas the actual biochemical reactions involved in the ripening of fruits and vegetables are quite complex and most often not quantified accurately yet (Tucker and Grierson, 1987).

Throughout the years, methodological advances have been made using this basic model in all phases of the simulation project, from experimental design to data gathering and analysis, to model fitting and implementation in the industrial environment. Selected relevant studies are presented in the following sections to give the reader an overview of methodologies used for food quality attribute quantification and modeling.

A. Experimental Design and Data Collection

1. Input data variability

Although it is well accepted that fruits and vegetables follow definite developmental stages during maturation, it is known also that these products cannot be harvested at the precise maturity level as quantified by analytical instruments, because human operators, even if well trained, cannot insure such discrimination because of human sensory limitations and fatigue factors. The experimenter or modeler thus must select only groups of products with a narrowly defined maturity level (or assume it to be so) to be used in subsequent laboratory experiments. This "group" approach is quite popular, but the variance caused by differences in measured initial maturity levels would mask out the true kinetics of the maturation process. An alternative method to data collection and analysis is to follow the same "individual" sample in the groups being tested throughout the experiment. Consequently, repeated quality measurements are performed on the individual specimen, and therefore are required to be nondestructive. Color can be monitored this way routinely. Nondestructive firmness testers have become available to the research establishment only recently (Meredith *et al.*, 1990; Delwiche and Sarig, 1991). Flavor is even more difficult to measure nondestructively. (See Chapter 11 for a more complete view of nondestructive evaluation methods.) Thai and Shewfelt (1991d) and Thai *et al.* (1990b) applied this "individual" approach to modeling color development in peaches and tomatoes, resulting in large improvements in data fitness over the "group" approach ($R^2 = 0.82 - 0.98$ for "individual" approach compared with $R^2 = 0.17 - 0.69$ for "group" approach). Thai *et al.* (1990b) also presented a procedure for converting a "group" model to an equivalent "individual" model.

2. Test temperature profile

When modeling kinetic processes, the standard method is to use different constant storage temperatures, but nonisothermal methods have been used also to overcome the thermal lag problem with such temperature profiles as inverse logarithmic (Rogers, 1963), inverse linear (Eriksen and Stelmach, 1965), linear increasing (Rhim *et al.*, 1989), and sinusoidal (Nunes and Swartzel, 1990b). If one can access such programmable storage facilities, these methods do simplify the collection of data. We are not aware of any application of these methods to fresh fruit and vegetable quality modeling.

B.   Data Analysis and Model Fitting

1.   Flavor models

Flavor is associated most closely with sensory evaluation (see Chapters 5, 7). Stone *et al.* (1991) recommended that sensory analysis should "begin with the consumer rather than with a preconceived notion as to what is best for the consumer." Flavor studies of fruits and vegetables are quite numerous, but these data usually are analyzed with correlation statistics only, and occasionally are presented as regression models (see Chapter 7 for other statistical techniques). These studies also tend not to have sufficient data ranges or details from which to build models. A few selected studies follow. Watada *et al.* (1985) correlated soluble solids and acidity to light transmission properties of whole apples. Dhanaraj *et al.* (1986) studied the effect of orchard altitude on apple starch content and juiciness. Temperature effects on acidity and soluble solids of apples after low oxygen storage were studied by Chen *et al.* (1989). Asparagus ascorbic acid and soluble solids were quantified by Hudson and Lachance (1986) at different steps of the marketing system, and their changes under calcium treatment were studied by Drake and Lambert (1985). Broccoli acidity contents were measured by Hudson *et al.* (1986) and by Perrin and Gaye (1986) for different postharvest conditions. Dull *et al.* (1989) used near-infrared transmission properties to quantify soluble solids of whole cantaloupes, whereas Forbus and Senter (1989) used delayed light emission to quantify cantaloupe chlorophyll and soluble solids. Matsumoto *et al.* (1983) and MacRae *et al.* (1989) presented comprehensive nutrient data on kiwifruit, but made no attempt to derive quantitative models. As for mangoes, Vasquez-Salinas and Lakshminarayana (1985) studied the effects of storage temperatures on changes in ascorbic acid, carotenoids, and soluble solids, and Medlicott *et al.* (1987) investigated the effects of ethylene and acetylene on ripening. Stone *et al.* (1986) compared changes in ascorbic acid and thiamine in okra under blanching and dehydration conditions. Onion pungency and sugar content changes during summer storage were studied by Hurst *et al.* (1985). Brix and total acidity for oranges were measured for different seasons and under different postharvest conditions by Basker (1977). Consumer-relevant sensory and objective variables have been identified by cluster analysis for snap beans (Resurreccion *et al.,* 1987) and tomatoes (Resurreccion and Shewfelt, 1985). Chemical and sensory properties have been quantified for peaches (Shewfelt *et al.,* 1987), tomatoes (Watada and Aulenbach, 1979), and small fruits such as strawberries, blueberries, and grapes (Morris *et al.,* 1988).

2.   Texture models

The main textural attribute in fruits and vegetables is firmness, which is commonly measured by puncture and deformation tests (see Chapter 5). Thus, the maximum puncture force or the maximum deformation distance under a constant load are the standard variables to model; most authors agreed to use a model of first-order kinetics with the Arrhenius equation (Rao and Lund, 1986; Bourne, 1989; Labuza and Breene, 1989). Chapter 5 also reported some successes in linking these physical firmness measurements to sensory perceptions of firmness using regression models. Other techniques based on acoustics (Armstrong and Brown, 1991; Farabee and

Stone, 1991; Rosenfeld *et al.*, 1991) or on impact responses (Delwiche *et al.*, 1991; Pitts *et al.*, 1991) have been used to quantify fruit physical firmness nondestructively (see Chapter 11 for other technologies). However, it is not yet shown that these new techniques are related to human perception of firmness. Formidable tasks still remain in developing new nondestructive firmness tests and instrumentation that would correspond better to the human hand and mouth.

3. Color models

The external color of fruits and vegetables is only the outward manifestation of complex internal biochemical reactions (Friend and Rhodes, 1981; Gross, 1987; Goodwin, 1988). Although these reactions have been identified, concrete kinetic information is not yet available. Thus, colorimeter readings usually are used to characterize and model color development. Color is a three-dimensional entity (Hunt, 1987); the two most popular color measurement systems used for fruits and vegetables are the Hunter "Lab" and the CIE L*a*b* systems. Chapter 5 discusses the merits of each system; for convenience, Lab will be used to refer to either system in this chapter. The standard variables used in modeling color have been the ratio a/b (Thorne and Alvarez, 1982; Yang and Chinnan, 1988) and L, a, and b as independent variables (Shewfelt *et al.*, 1988; Delwiche, 1989). However to model human color perception, the preferred system is based on hue angle, $H[\tan^{-1}(b/a)]$, chroma, $C[(a^2 + b^2)^{1/2}]$, and value, L (Little, 1975; Setser, 1984; Chapter 5). Thai *et al.* (1990b) showed that, for tomato, H was the primary variable that changed with time and temperature, whereas C and L were secondary components dependent only on H. For peaches, Thai and Shewfelt (1990) showed that only the ground color hue angle component was of importance in the modeling of both physical color development and human color judgment. Note that not all color models fit Eq. 1, since tomato color development in hue angle was shown to behave sigmoidally with time (Thorne and Alvarez, 1982; Thai *et al.*, 1990b).

4. Determination of kinetic parameters

For constant temperature experiments, the classic method still in use (Taoukis and Labuza, 1989a) for determining the kinetic parameters of Eqs. 1 and 2 involved two steps. The first step is to use an integral form of Eq. 1 and regress the quality index Q (or a mathematical transformation of Q to linearize the integral form of Eq. 1 with respect to time) with respect to time t for each test temperature to obtain the corresponding rate parameter k. The second step regresses those k values with respect to the inverse of temperature $1/T$ to obtain E_a and k_0 using Eq. 2. The second method requires the combination of Eqs. 1 and 2 into a new equation expressing the quality index Q as a function of initial Q_0, time t, temperature T, E_a, and k_0. Nonlinear regression then is performed on this expanded equation to obtain E_a and k_0 directly (Arabshahi and Lund, 1985; Cohen and Saguy, 1985; Haralampu *et al.*, 1985). This second method has the advantage of increased degree of freedom, yielding a lower confidence interval on E_a. Additionally, Davies and Hudson (1981) suggested a way to reduce the high correlation between E_a and k_0 by moving the origin of the $1/T$ axis from 0 to the mean of observed $1/T$. For the special case of first-order kinetics, the thermal death time (TDT) method is used also, mainly in food sterilization and processing. The basic difference is that using TDT, the rate parameter k is proportional to temperature whereas the Arrhenius k is proportional to the inverse of temperature. Ramaswamy *et al.* (1989) discussed problems inherent in the interconversion of TDT and Arrhenius data and provided reasonable means to transform from one system to the other.

For varying temperature experiments or model usages, the basic assumptions are that quality kinetic changes are additive and commutative (i.e., independent of the sequence of temperatures applied), and that temperature change has no effect by itself. From these concepts, two general methods are devised. One is based on the time derivatives of the measured quality data, and is known as the time–temperature tolerance (TTT) hypothesis (Van Arsdel, 1957; Thorne and Alvarez, 1982; Rhim *et al.*, 1989). This differential method has the disadvantage of being very sensitive to experimental errors due to the use of time derivatives. The second method is integral and uses the measured data directly. Three implementations have been offered for this second approach.

1. Senum and Yang (1977) used rational approximations of the integral of the combined Eqs. 1 and 2.
2. Nunes and Swartzel (1990a) used a concept of equivalent time t_E and temperature T_E (the Equivalent Point Method).
3. Thai *et al.* (1989) proposed the Time Shift Method to account for the fact that the individual fruits or vegetables in a test group are most likely to be at slightly different stages of maturity at the start of an experiment.

5.   Recent modeling paradigms

Zhang *et al.* (1989) proposed a quality evaluation model based on fuzzy set theory, wherein sharp explicit boundaries between grades were replaced by continuous fuzzy membership functions. Zhang and Litchfield (1991) used this approach to compare the overall quality of different brands of pickles, and to determine the most important factors for a particular market when introducing a new cookie product.

Thai and Shewfelt (1991c) used the recent technique of neural networks, which can be considered a nonlinear mapping technique, to determine the relevant contributors to sensory evaluation of tomato and peach color.

C.   Model Implementation

1.   Use of time–temperature integrators

Full-history time–temperature integrators (TTI) (Wells and Sigh, 1988a; Taoukis and Labuza, 1989b) have shown their applicability and reliability in monitoring food quality indirectly by quantifying the accumulated effects of time–temperature histories encountered during storage and transport of a product. Full-history TTIs are active at all temperatures, whereas partial-history TTIs are only active in defined ranges of temperatures. With this approach, first the correspondence between the product quality kinetics and the sensor kinetics must be established and quantified;

it then is assumed to be permanent. Thus, when applying this technology to fresh
fruits and vegetables, we must determine the range of variations in product quality
kinetics across cultural practices and seasons (Thai and Shewfelt, 1991a, d), and
the corresponding variations in TTI kinetics.

2. Real-time modeling approach

From studies of thermal color kinetics of tomatoes across several seasons, Thai and
Shewfelt (1991a) noted that, although the same model formulations applied across
seasons, the numerical values of the kinetic parameters were significantly different,
yielding a maximum prediction error of 1.5 days. Thus, a "real-time" modeling
approach is proposed whereby the product kinetic parameters are determined during
the product transit through the postharvest system. This approach would require
complete temperature histories to be recorded as well as a limited number of
monitoring points (before reaching the retail level) at which actual quality measurements would be made and used in regressing the appropriate model equations
to obtain optimal values of the kinetic parameters. The completed model would
then be used for prediction from that point onward to the retail level.

3. Model integration

Thai *et al.* (1990a) described a prototype decision support system for peaches and
tomatoes, wherein models for consumer preferences of color and firmness were
integrated with physical quality and temperature models to allow a user with information on initial qualities of a product to find the optimum sale date(s) for a
given storage temperature, or to find the optimum storage temperature(s) for a target
sale date. Wells and Singh (1988b) simulated three inventory issue policies for
perishable food and showed that the "shortest remaining shelf life" policy provided
products with equal or higher average quality than the other two policies: "first-in,
first-out" and "last-in, first-out". Schoorl and Holt (1986) described a quality management system based on deterioration energy of horticultural produce that included
physiological, pathological, and mechanical damages.

D. An Illustration: Modeling Tomato Color and Firmness

Previous sections in this chapter essentially presented a review of current methodologies. Using tomatoes as an example, this section provides more details on
one modeling approach being developed.

1. Multifacet modeling approach

As a tomato fruit ripens, it changes in purchase (color and firmness) as well as
consumption (mouthfeel and flavor) quality characteristics. Thus, one can state also
that maturity, color, firmness, and flavor are different facets of the same maturation
process. This concept is an important criterion to remember when modeling the
maturation process of the tomato.

   Current technologies do not allow monitoring of tomato consumption characteristics nondestructively, a necessary condition for applying the "individual" ap-

proach (see Section III.A.1). Thus, only tomato purchase characteristics can be modeled. In storage tests of tomatoes at different constant temperatures, corresponding color changes were monitored in individual fruits with a tristimulus colorimeter using the Hunter (Lab) scale. Firmness on the same fruit was measured also using the nondestructive technique of Meredith *et al.* (1990), which actually measured the coefficient of restitution, R, of a fruit when it was dropped from a height of about half an inch onto a metallic plate equipped with force transducers (Thai *et al.*, 1989; Thai and Shewfelt, 1991b).

Color, having three components (L,a,b), was a more complex attribute than firmness and, thus, was modeled first. Just as color and firmness are different facets of the same maturation process, L, a, and b are different facets of the perceived color of tomatoes. Thus, it was necessary to assess the changes in all three components as storage time progressed. This approach resulted in the three-dimensional analysis tool proposed by Thai *et al.* (1990b). Figure 1 represents three-dimensional plots of observed L, a, and b for only three typical tomatoes labeled "0", "4", and "5" (from a set of 140 "mature-green to breaker" fruits). In Fig. 1, the upper set of curves represented true three-dimensional plots of respective L, a, and b data sets, starting from the upper left with observations on day 0, connected to respective observations on day 1, and so forth to the observations for day 9 in the lower right. These three curves had different starting points, but all converged on the same "red ripe" point. The lower set of curves in Fig. 1 represented projections of these three-dimensional curves onto the a,b plane. It was clear that they formed nearly one continuous curve. These two trends were found to apply for all other tomatoes tested. Interestingly, the resulting curve on the a,b plane turned out to be fixed, meaning that parameters a and b were not independent attributes as previously described (Thorne and Alvarez, 1982; Shewfelt *et al.*, 1988), because a fixed curve meant that we could find a unique function f, so $b = f(a)$. For this case, since these physical color models were to be used later as inputs to color sensory models describing consumer preferences for tomato color, we chose to use the HCL color

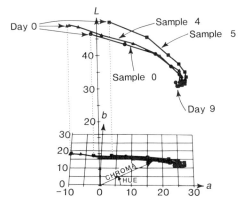

Figure 1   Three-dimensional plots of observed data (L,a,b) for tomato labeled 0, 4, and 5. [Reprinted, with permission, from Thai, C. N., Shewfelt, R. L., and Garner, J. C. (1990b). Tomato color changes under constant and variable storage temperatures: Empirical models. *Trans. ASAE* **33(2)**, 607–614.]

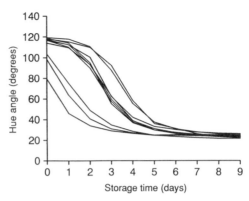

Figure 2 Plot of hue angle H vs. storage time at 20°C for 10 tomato fruits. [Reprinted, with permission, from Thai, C. N., Shewfelt, R. L., and Garner, J. C. (1990b). Tomato color changes under constant and variable storage temperatures: Empirical models. *Trans. ASAE* **33(2)**, 607–614.]

space instead (Little, 1975; Setser, 1984). Since H and C were parameters derived from a and b, this dependency property also applied to H and C. Chapter 5 further shows that hue usually is the main variable, whereas chroma and value are only modifying attributes of hue. Observed data also supported this hypothesis, as shown in Figs. 2, 5, and 6, respective plots of hue and storage time, chroma and hue, and value and hue for 10 typical fruits. In these figures, each curve represents the changes experienced by the same fruit.

2. Hue H submodel

Figure 2 shows that starting hues H_0 varied from 120° to 80° and that individual hue changes behave sigmoidally. Most notably, a single sigmoidal curve seems to be able to fit each tomato curve simply by shifting it along the time axis. Thorne and Alvarez (1982) also reported that tomato color behaved sigmoidally, but they modeled the ratio a/b. Modifying their equation yielded

$$\ln \left(\frac{H_{mg} - H}{H - H_{rr}} \right) = \beta + \alpha t \tag{3}$$

where H, H_{mg}, and H_{rr} are, respectively, the hue angle for the sample, mature green, and red ripe fruits; t is time, and β and α are parameters to be determined for each storage temperature. Equation 3 describes a sigmoidal curve that behaves in the manner shown in Fig. 2, but is still set for the "group" approach in which differences in starting hue H_0 are not distinguished. Noting that when $t = 0$, $H = H_0$, Eq. 3 provides another form for the parameter β:

$$\beta = \ln \left(\frac{H_{mg} - H_0}{H_0 - H_{rr}} \right)$$

Thus, Eq. 3 becomes

$$\ln \left(\frac{H_{mg} - H}{H - H_{rr}} \right) = \ln \left(\frac{H_{mg} - H_0}{H_0 - H_{rr}} \right) + \alpha t \tag{4}$$

Table I
Estimates and Standard Errors for Parameters of Equation 5[a]

Temperature treatment (°C)	Parameters[b]			Goodness of fit (R^2)
	H_{mg}	H_{rr}	α	
15	120.74	27.21	0.9641	0.9790
	(0.093)	(0.530)	(0.0163)	
20	119.61	25.21	1.3579	0.9823
	(0.274)	(0.393)	(0.0303)	
25	119.15	28.24	1.2432	0.9859
	(0.134)	(0.345)	(0.0197)	

[a]Reprinted with permission from Thai, C. N., Shewfelt, R. L., and Garner, J. C. (1990b). Tomato color changes under constant and variable storage temperatures: Empirical models. *Trans. ASAE* **33**(2), 607–614.
[b]Values in parentheses are standard errors.

With further algebraic manipulations of Eq. 4, H can be expressed explicitly as

$$H = \frac{H_{mg} + H_{rr}\left(\dfrac{H_{mg} - H_0}{H_0 - H_{rr}}\right) e^{\alpha t}}{1 + \left(\dfrac{H_{mg} - H_0}{H_0 - H_{rr}}\right) e^{\alpha t}} \tag{5}$$

Next, the procedure NLIN (SAS INSTITUTE, INC., 1985) with the Marquardt search method was used to find the optimal parameters H_{mg}, H_{rr}, and α for Eq. 5. Table I lists results found for parameters H_{mg}, H_{rr}, α, and the coefficient of determination R^2 for storage temperatures of 15, 20, and 25°C. Table II lists similar results when Eq. 3 was used, that is, when the "group" approach was chosen. The "group" approach yielded poorer R^2 and larger standard errors for the parameter

Table II
Estimates and Standard Errors for Parameters of Equation 3[a]

Temperature treatment (°C)	Parameters[b]				Goodness of fit (R^2)
	H_{mg}	H_{rr}	a	β	
15	135.91	25.09	0.5592	−0.5859	0.6179
	(55.126)	(5.779)	(0.2574)	(1.2461)	
20	117.88	24.17	0.9141	−1.0687	0.6941
	(22.938)	(2.322)	(0.2570)	(0.8386)	
25	132.21	26.86	0.7276	−0.5683	0.6631
	(43.567)	(2.979)	(0.2522)	(1.0465)	

[a]Reprinted with permission from Thai, C. N., Shewfelt, R. L., and Garner, J. C. (1990b). Tomato color changes under constant and variable storage temperatures: Empirical Models. *Trans. ASAE* **33**(2), 607–614.
[b]Values in parentheses are standard errors.

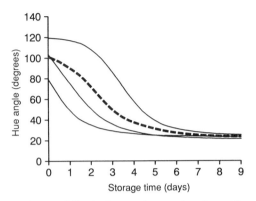

Figure 3 "Group" approach to modeling in the experimenter time frame. The solid curves are plots of hue angle data for 3 tomatoes. The dashed curve is the best fit curve.

estimates than did the "individual" approach. These results can be explained by examining Figs. 3 and 4. The solid curves in Fig. 3 are plots of hue angle data for three tomatoes, representing a typical range of hue data. The dashed curve in Fig. 3 is the best fit curve, from least squares, for the "group" approach to this data set using Eq. 3. One can see easily that the data variations above and under this best fit curve result in a poor R^2. When the initial hue angle H_0 was introduced by the "individual" approach into Eq. 5, the net effect was that each fruit data set was "ranked" first according to its initial H_0 value, that is, each data curve was shifted appropriately in parallel with the time axis, as shown in Fig. 4. Then the best least squares curve (dashed) was fit to this data set using Eq. 5, resulting in a better R^2. Notice also that the vertical axis, marking the point at which $t = 0$, is no longer drawn in Fig. 4, because it becomes variable as the vertical projection to the time axis from wherever H_0 happened to be. In other words, the "individual" approach allows adjustments for each fruit time frame through parameter H_0, whereas the "group" approach views all fruit under a common time frame that happens to be the experimenter time frame.

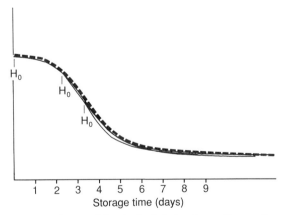

Figure 4 "Individual" approach to modeling in the fruit time frame. See text for explanation.

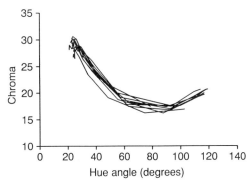

Figure 5 Plot of chroma C vs. hue angle H at 20°C for 10 tomato fruits. [Reprinted, with permission, from Thai, C. N., Shewfelt, R. L., and Garner, J. C. (1990b). Tomato color changes under constant and variable storage temperatures: Empirical models. *Trans. ASAE* **33(2)**, 607–614.]

Results for parameter α in Table I show clear evidence of the influence of temperature T on α. Thus, α in Eq. 5 is replaced by a linear function of T, $\alpha = cT + d$, yielding Eq. 6.

$$H = \frac{H_{mg} + H_{rr} \left(\dfrac{H_{mg} - H_0}{H_0 - H_{rr}} \right) e^{(cT+d)t}}{1 + \left(\dfrac{H_{mg} - H_0}{H_0 - H_{rr}} \right) e^{(cT+d)t}} \tag{6}$$

where parameters H_{mg}, H_{rr}, c, and d must be redetermined using the original data set, as recommended in Section III,B,4, with a resulting R^2 of 0.9721 (Thai *et al.*, 1990b).

3. Chroma C submodel

In Fig. 5, the individual curves tend to weave around a central curve and are not shifted versions of one unique curve, as for H. Similar plots were made for all other temperature treatments and the same trend was confirmed. These observed changes in chroma with respect to hue then simply were modeled with polynomial equations, with $R^2 = 0.9379$ (Thai and Shewfelt, 1991a).

4. Value L submodel

In Fig. 6, value L decreases to a minimum L_{min} of 25–26 at the "red-ripe" point, irrespective of the initial L_0 (corresponding to high hue angles). This observation means that tomatoes with higher L_0 decrease their L value at a faster rate than those with lower L_0. Therefore, scaling techniques were used to model the changes of L with respect to H (see Thai *et al.*, 1989, for more details on this technique). The final value L submodel is described by Eq. 7, with $R^2 = 0.9473$.

$$L = L_0 \left(\frac{1.0 - \delta e^{-\gamma H}}{1.0 - \delta e^{-\gamma H_0}} \right) \tag{7}$$

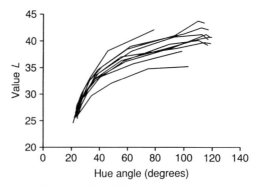

Figure 6 Plot of value L vs. hue angle H at 20°C for 10 tomato fruits. (Reprinted with permission of the ASAE.)

where (L_0, H_0) and (L, H) were, respectively, the measurements of value L and hue H at $t = 0$ and at any time t later, and γ and δ are parameters listed in Thai and Shewfelt (1991a).

5. Firmness submodel

Firmness always has been considered an independent attribute and, thus, usually was linked directly to time and temperature (Thorne and Alvarez, 1982; Thai *et al.*, 1989). However, the approach used in modeling tomato color components H, C, and L also should apply to firmness, since firmness is simply another facet of the maturation process. In other words, is there a relationship between firmness and hue angle H?

Firmness was measured in two ways: first with a nondestructive impact test to

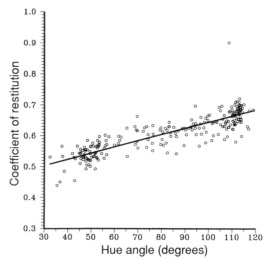

Figure 7 Plot of coefficient of restitution R vs. hue angle H. [Reprinted, with permission, from Thai, C. N., and R. L. Shewfelt. (1991b). Tomato color and firmness models for varying temperature storage. *ASAE Technical Paper* No. 916566.]

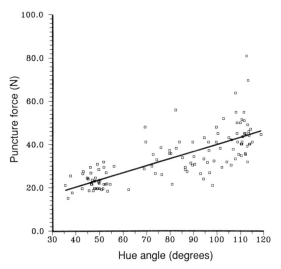

Figure 8 Plot of peak puncture force F vs. hue angle H . [Reprinted, with permission, from Thai, C. N., and R. L. Shewfelt. (1991b). Tomato color and firmness models for varying temperature storage. *ASAE Technical Paper* No. 916566.]

get the coefficient of restitution R at a randomly chosen test area on the fruit, then with an Instron force apparatus to get the peak puncture force F at the same test area (Thai and Shewfelt, 1991b). Figure 7 is a plot of R and hue angle H; Fig. 8 is a plot of F (in Newtons) and H. Notice the obvious linear trends (also reported by Campbell *et al.*, 1990) in both figures, but variations in R were consistent at all levels of hue, whereas F exhibited larger variations at large hue values (i.e., for green tomatoes). This relationship indicates that the nondestructive impact test may be more reliable because of its consistency.

Thai and Shewfelt (1991b) also performed another experiment in which color and R were measured daily on the same fruit stored under different constant temperatures T (15–25°C). They reported the following relationship:

$$R = R_0 + (aT + b)(H - H_0) \qquad (8)$$

where R_0 corresponded to initial hue and H_0, $R^2 = 0.8239$, and a and b are constants.

In the previous sections, we showed that, given a properly chosen primary quality attribute (such as hue angle for tomato), we could model the other attributes (i.e., chroma, value, and firmness) in relation to that primary variable. Currently we are working on extending these modeling principles to other fruits and vegetables.

IV. Conclusions

This chapter showed that there have been many attempts at designing quality control systems, each from a different viewpoint. Thus, although the economic incentive exists and current scientific and technological achievements allow us to implement fresh fruit and vegetable quality control systems, the large number of commodities to consider and the existing nonuniform approach to experimentation and analysis are hampering the actual implementation of such systems.

In the future, the use of modeling also can be incorporated appropriately into quality measuring equipment such as on-line sorters, which may then sort fruits and vegetables for anticipated quality attributes rather than sorting using only present quality attributes. Further, using machine learning concepts, perhaps we can design equipment that updates its own knowledge base about fruit and vegetable postharvest behavior to adapt to changes due to varietal or production differences.

Bibliography

Arabshahi, A., and Lund, D. B. (1985). Considerations in calculating kinetic parameters from experimental data. *J. Food Process Eng.* **7**, 239–251.

Armstrong, P. R., and Brown, G. K. (1991). "Apple Firmness Sorting Using a Nondestructive Acoustic Technique." ASAE Technical Paper No. 91-6044. American Society of Agricultural Engineers, St. Joseph, Michigan.

Basker, D. (1977). Changes in the organoleptic quality of shamouti oranges during their ripening season. *J. Food Qual.* **1**, 147–156.

Bourne, M. C. (1989). Applications of chemical kinetic theory to the rate of thermal softening of vegetable tissue. *In* "Quality Factors of Fruits and Vegetables: Chemistry and Technology" (J. J. Jen, ed.), pp. 98–110. American Chemical Society, Washington, D.C.

Bratley, P., Fox, B. L., and Schrage L. E. (1983). "A Guide to Simulation." Springer-Verlag, New York.

Campbell, A. D., Huysamer, M., Stotz, H. U., Greve, L. C., and Labavitch, J. M. (1990). Comparison of ripening processes in intact tomato fruit and excised pericarp discs. *Plant Physiol.* **94**, 1582–1589.

Checkland, P. B. (1981). "Systems Thinking, Systems Practice." John Wiley & Sons, Chichester.

Chen, P. M., Varga, D. M., Mielke, E. A., and Drake, S. R. (1989). Poststorage behavior of apple fruit after low oxygen storage as influenced by temperatures during storage and in transit. *J. Food Sci.* **54(4)**, 993–996.

Clements, R. R. (1989). "Mathematical Modelling: A Case Study Approach." Cambridge University Press, Cambridge.

Cohen, E., and Saguy, I. (1985). Statistical evaluation of Arrhenius model and its applicability in prediction of food quality losses. *J. Food Process. Preserv.* **9**, 273–290.

Davies, O. L., and Hudson, H. E. (1981). Stability of drugs: Accelerated storage tests. *In* "Statistics in the Pharmaceutical Industry" (C. R. Buncher and J. Y. Tsay, eds.), pp. 355–395. Marcel Dekker, New York.

Delwiche, M. J. (1989). Maturity standards for processing clingstone peaches. *J. Food Process Eng.* **10**, 269–284.

Delwiche, M. J., and Sarig, Y. (1991). A probe impact sensor for fruit firmness measurement. *Trans. ASAE* **34(1)**, 187–192.

Delwiche, M. J., Singh, N., Arevalo, H., and Mehlschau, J. (1991). "A Second Generation Fruit Firmness Sorter." ASAE Technical Paper No. 91-6042. American Society of Agricultural Engineers, St. Joseph, Michigan.

Dhanaraj, S., Krishnaprakash, M. S., Arvindaprasad, B., Ananthakrishna, S. M., Krishnaprasad, C. A., and Narasimham, P. (1986). Effect of orchard elevation on maturity and quality of apples. *J. Food Qual.* **9(3)**, 129–142.

Drake, S. R., and Lambert, E. F. (1985). Influence of calcium treatment, spear size, and storage time on fresh and frozen asparagus quality. *J. Food Qual.* **8(2/3)**, 101–111.

Dull, G. G., Birth, G. S., Smittle, D. A., and Leffler, R. G. (1989). Near infrared analysis of soluble solids in intact cantaloupe. *J. Food Sci.* **54(2)**, 393–395.

Eriksen, S. P., and Stelmach, H. (1965). Single-step stability studies. *J. Pharm. Sci.* **54**, 1029–1034.

Farabee, M. L., and Stone, M. L. (1991). "Determination of Watermelon Maturity with Sonic Impulse Testing." ASAE Technical Paper No. 91-3013. American Society of Agricultural Engineers, St. Joseph, Michigan.

Forbus, W. R., Jr., and Senter, S. D. (1989). Delayed light emission as an indicator of cantaloupe maturity. *J. Food Sci.* **54(4)**, 1094–1095.

Friend, J., and Rhodes, M. J. C. (eds.) (1981). "Recent Advances in the Biochemistry of Fruits and Vegetables." Academic Press, London.

Gross, J. (1987). "Pigments in Fruits." Academic Press, New York.

Goodwin, T. W. (ed.) (1988). "Plant Pigments." Academic Press, London.

Haralampu, S. G., Saguy, I., and Karel, M. (1985). Estimation of Arrhenius model parameters using three least squares methods. *J. Food Process. Preserv.* **9**, 129–143.

Hecht-Nielsen, R. (1990). "Neurocomputing." Addison-Wesley, Reading, Massachusetts.

Hudson, D. E., and Lachance, P. A. (1986). Ascorbic acid and riboflavin content of asparagus during marketing. *J. Food Qual.* **9**, 217–224.

Hudson, D. E., Cappellini, M., and Lachance, P. A. (1986). Ascorbic acid content of broccoli during marketing. *J. Food Qual.* **9**, 31–37.

Hunt, R. W. G. (1987). "Measuring Color." Halsted Press, New York.

Hurst, W. C., Shewfelt, R. L., and Schuler, G. A. (1985). Shelf-life and quality changes in summer storage onions (*Allium cepa*). *J. Food Sci.* **50(3)**, 761–763, 772.

Jen, J. J. (1989a). "Quality Factors of Fruits and Vegetables: Chemistry and Technology." American Chemical Society, Washington, D.C.

Jen, J. J. (1989b). Chemical basis of quality factors in fruits and vegetables: An overview. *In* "Quality Factors of Fruits and Vegetables: Chemistry and Technology" (J. J. Jen, ed.), pp. 1–9. American Chemical Society, Washington, D.C.

Kapsalis, J. G. (1987). "Objective Methods in Food Quality Assessment." CRC Press, Boca Raton, Florida.

Kramer, A., and Twigg, B. A. (1970). "Fundamentals of Quality Control for the Food Industry." AVI, Westport, Connecticut.

Labuza, T. P., and Breene, W. M. (1989). Applications of "active packaging" for improvement of shelf-life and nutritional quality of fresh and extended shelf-life foods. *J. Food Process. Preserv.* **13**, 1–69.

Little, A. C. (1975). Off on a tangent. *J. Food Sci.* **40**, 410–411.

MacRae, E. A., Lallu, N., Searle, A. N., and Bowen, J. H. (1989). Changes in the softening and composition of kiwifruit (*Actinidia deliciosa*) affected by maturity at harvest and postharvest treatments. *J. Sci. Food Agric.* **49**, 413–430.

Maren, A. J., Harston, C. T., and Pap, R. M. (1990). "Handbook of Neural Computing Applications." Academic Press, New York.

Matsumoto, S., Obara, T., and Luh, B. S. (1983). Changes in chemical constituents of kiwifruit during post-harvest ripening. *J. Food Sci.* **48(2)**, 607–611.

Medlicott, A. P., Sigrist, J. M. M., Reynolds, S. B., and Thompson, A. K. (1987). Effects of ethylene and acetylene on mango fruit ripening. *Ann. Appl. Biol.* **111**, 439–444.

Meredith, F. I., Leffler, R. G., and Lyon, C. E. (1990). Detection of firmness in peaches by impact force response. *Trans. ASAE* **33(1)**, 186–188.

Morris, J. R., Makus, D. J., and Main, G. L. (1988). Small fruit quality as affected by point of origin— Distant versus local. *J. Food Qual.* **11(3)**, 193–204.

Neelamkavil, F. (1987). "Computer Simulation and Modeling." John Wiley & Sons, Chichester.

Nunes, R. V., and Swartzel, K. R. (1990a). Modeling thermal processes using the equivalent point method. *J. Food Eng* **11**, 103–117.

Nunes, R. V., and Swartzel, K. R. (1990b). Modeling chemical and biochemical changes under sinusoidal temperature fluctuations. *J. Food Eng* **11**, 119–132.

Perrin, P. W., and Gaye, M. M. (1986). Effects of simulated retail display and overnight storage treatments on quality maintenance in fresh broccoli. *J. Food Sci.* **51(1)**, 146–149.

Pitts, M. J., Cavalieri, R. P., and Drake, S. (1991). "Evaluation of the PFT Apple Firmness Sensor." ASAE Technical Paper No. 91-3017. American Society of Agricultural Engineers, St. Joseph, Michigan.

Rao, M. A., and Lund, D. B. (1986). Kinetics of thermal softening of foods—A review. *J. Food Process. Preserv.* **10**, 311–329.

Ramaswamy, H. S., Van de Voort, F. R., and Ghazala, S. (1989). An analysis of TDT and Arrhenius methods for handling process and kinetic data. *J. Food Sci.* **54(5)**, 1322–1326.

Rand, W. M. (1983). Development of analysis of empirical mathematical kinetic models pertinent to food processing and storage. *In* "Computer-Aided Techniques in Food Technology" (I. Saguy, ed.), pp. 49–70. Marcel Dekker, New York.

Resurreccion, A. V. A., and Shewfelt, R. L. (1985). Relationships between sensory attributes and objective measurements of postharvest quality of tomatoes. *J. Food Sci.* **50(5)**, 1242–1256.

Resurreccion, A. V. A., Shewfelt, R. L., Prussia, S. E., and Hurst, W. C. (1987). Relationships between sensory and objective measures of postharvest quality of snap beans as determined by cluster analysis. *J. Food Sci.* **52(1)**, 113–116, 123.

Rhim, J. W., Nunes, R. V., Jones, V. A., and Swartzel, K. R. (1989). Determination of kinetic parameters using linearly increasing temperature. *J. Food Sci.* **54(2)**, 446–450.

Rogers, A. R. (1963). An accelerated storage test with programmed temperature rise. *J. Pharm. Pharmcol.* **15**, 101T–105T.

Rosenfeld, D., Shmulevich, I., and Rosenhouse, G. (1991). "Three-Dimensional Simulation of Acoustic Response of Fruits for Firmness Sorting." ASAE Technical Paper No. 91-6046. American Society of Agricultural Engineers, St. Joseph, Michigan.

Ross, E. W., Jr. (1987). A quantitative approach to standardization in food quality assessment. *In* "Objective Methods in Food Quality Assessment" (J. G. Kapsalis, ed.), pp. 38–58. CRC Press, Boca Raton, Florida.

Rothenberg, J. (1989). The nature of modeling. *In* "Artificial Intelligence, Simulation, and Modeling" (L. E. Widman, K. A. Loparo, and N. R. Nielsen, eds.), pp. 75–92. John Wiley & Sons, New York.

Saguy, I., and Karel, M. (1987). Index of deterioration and simulation of quality losses. *In* "Objective Methods in Food Quality Assessment" (J. G. Kapsalis, ed.), pp. 234–260. CRC Press, Boca Raton, Florida.

Sarig, Y. (1989). Quality oriented post-harvesting technologies for horticultural crops. *In* "Potentialities of Agricultural Engineering in Rural Development" (M. H. J. Wang, ed.), pp. 382–389. International Academic Publishers, New York.

SAS Institute Inc. (1985). SAS® user's guide: statistics, version 5 ed. SAS Institute Inc., Cary, North Carolina.

Schoorl, D., and Holt, J. E. (1986). Post-harvest energy transformations in horticultural produce. *Agric. Syst.* **19**, 127–140.

Senum, G. I., and Yang, R. T. (1977). Rational approximations of the integral of the Arrhenius function. *J. Thermal Anal.* **11**, 445–447.

Setser, C. S. (1984). Color: Reflections and transmissions. *J. Food Qual.* **6(3)**, 183–197.

Shannon, R. E. (1975). "Systems Simulation: The Art and Science." Prentice Hall, Englewood Cliffs, New Jersey.

Shewfelt, R. L. (1990). Quality of fruits and vegetables. *Food Technol.* **44(6)**, 99–106.

Shewfelt, R. L., Meyers, S. C., Prussia, S. E., and Jordan, J. L. (1987). Quality of fresh-market peaches within the postharvest handling system. *J. Food Sci.* **52(2)**, 361–364.

Shewfelt, R. L., Thai, C. N., and Davis, J. W. (1988). Prediction of changes in color of tomatoes during ripening at different constant temperatures. *J. Food Sci.* **53(5)**, 1433–1437.

Simpson, K. (1990). "Artificial Neural Systems: Foundations, Paradigms, Applications, and Implementations." Pergamon Press, New York.

Stone, H., McDermott, B. J., and Sidel, J. L. (1991). The importance of sensory analysis for the evaluation of quality. *Food Technol.* **45(6)**, 88–95..

Stone, M. B., Toure, D., Greig, J. K., and Naewbanij, J. O. (1986). Effects of pretreatment and dehydration temperature on color, nutrient retention, and sensory characteristics of okra. *J. Food Sci.* **51(5)**, 1201–1203.

Taoukis, P. S., and Labuza, T. P. (1989a). Applicability of time–temperature indicators as shelf life monitors of foods products. *J. Food Sci.* **54(4)**, 783–788.

Taoukis, P. S., and Labuza, T. P. (1989b). Reliability of time–temperature indicators as food quality monitors under nonisothermal conditions. *J. Food Sci.* **54(4)**, 789–792.

Thai, C. N., and Shewfelt, R. L. (1990). Peach quality changes at different constant storage temperatures: Empirical models. *Trans. ASAE* **33(1)**, 227–233.

Thai, C. N., and Shewfelt, R. L. (1991a). Seasonal variability of tomato color thermal kinetics. *Trans. ASAE* **34(4)**, 1830–1835.

Thai, C. N., and Shewfelt, R. L. (1991b). "Tomato Color and Firmness Models for Varying Temperature Storage" ASAE Technical Paper No. 91-6566. American Society of Agricultural Engineers, St. Joseph, Michigan.

Thai, C. N., and Shewfelt, R. L. (1991c). Modeling sensory color quality of tomato and peach: Neural networks and statistical regression. *Trans. ASAE* **34(3)**, 950–955.

Thai, C. N., and Shewfelt, R. L. (1991d). Redglobe peach color kinetics under step-varying storage temperatures. *Trans. ASAE* **34(1)**, 212–216.

Thai, C. N., Shewfelt, R. L., and Garner, J. C. (1989). "Tomato Color and Firmness Changes under Different Storage Temperatures." ASAE Technical Paper No. 89-6596. American Society of Agricultural Engineers, St. Joseph, Michigan.

Thai, C. N., Pease, J. N., and Shewfelt, R. L. (1990a). A decision support system for delivering optimal quality peach and tomato. *In* "Proceedings of the First CLIPS Users' Group Conference" (J. C. Giarratano and C. Culbert, eds.), pp. 386–396. National Aeronautics and Space Administration, Houston.

Thai, C. N., Shewfelt, R. L., and Garner, J. C. (1990b). Tomato color changes under constant and variable storage temperatures: Empirical models. *Trans. ASAE* **33(2)**, 607–614.

Thai, C. N., Resurreccion, A. V. A., Dull, G. G., and Smittle, D. A. (1990c). "Modeling Consumer Preferences with Neural Networks." ASAE Technical Paper No. 90-7550. American Society of Agricultural Engineers, St. Joseph, Michigan.

Thorne, S., and Alvarez, J. S. (1982). The effect of irregular storage temperatures on firmness and surface color in tomatoes. *J. Sci. Food Agric.* **33**, 671–676.

Tucker, G. A., and Grierson, D. (1987). Fruit ripening. *In* "The Biochemistry of Plants: A Comprehensive Treatise" (D. D. Davies, ed.), Vol. 12, pp. 265–318. Academic Press, San Diego.

Van Arsdel, W. B. (1957). The time–temperature tolerance of frozen foods. I. Introduction—The problem and the attack. *Food Technol.* **11**, 28–33.

Vasquez-Salinas, C., and Lakshminarayana, S. (1985). Compositional changes in mango fruit during ripening at different storage temperatures. *J. Food Sci.* **50(6)**, 1646–1648.

Watada, A. E., and Aulenbach, B. B. (1979). Chemical and sensory qualities of fresh market tomatoes. *J. Food Sci.* **44(3)**, 1013–1016.

Watada, A. E., Massie, D. R., and Abbott, J. A. (1985). Relationship between sensory evaluations and nondestructive optical measurements of apple quality. *J. Food Qual.* **7(3)**, 219–226.

Wells, J. H., and Singh, R. P. (1988a). A kinetic approach to food quality prediction using full-history time–temperature indicators. *J. Food Sci.* **53(6)**, 1866–1893.

Wells, J. H., and Singh, R. P. (1988b). A quality-based inventory issue policy for perishable foods. *J. Food Process. Preserv.* **12**, 271–292.

Williams, A. A. (1981). Relating sensory aspects to quality. *In* "Quality in Stored and Processed Vegetables and Fruit" (P. W. Goodenough and R. K. Atkin, eds.), pp. 17–33. Academic Press, London.

Yang, C. C., and Chinnan, M. S. (1988). Computer modeling of gas composition and color development of tomatoes stored in polymeric film. *J. Food Sci.* **53(3)**, 869–872.

Zhang, Q., and Litchfield, J. B. (1991). Applying fuzzy mathematics to product development and comparison. *Food Technol.* **45(7)**, 108–115.

Zhang, Q., Breen, A., and Litchfield, J. B., (1989). "A Fuzzy Comprehensive Evaluation Model for Food Products Grading." ASAE Technical Paper No. 89-6613. American Society of Agricultural Engineers, St. Joseph, Michigan.

CHAPTER 9

VISUAL INSPECTION AND SORTING: FINDING POOR QUALITY BEFORE THE CONSUMER DOES

Frank Bollen, Stanley E. Prussia, and Amos Lidror

Postharvest Handling: A Systems Approach

In any agricultural production system, a proportion of the product does not meet the quality demanded by the primary buyer. Nonconforming product must be left in the field, disposed of, or sold to a different outlet. In some cases, pickers can perform quality determinations in the field, and only harvest salable product. However, generally the product is subjected to a subsequent sorting operation, the primary quality determining step.

Establishing optimum design and operating conditions for a sorting process has been recognized as critical in determining the potential performance of the products shipped and has received considerable research attention in the past (Section II). Unfortunately, this optimization tends to be treated independently of other parts of the postharvest system.

The second aspect of the analysis of the sorting operation is the ability of humans to inspect and remove defective product. Sorter productivity and accuracy are a function of the design and operational parameters of the process, the quality of the incoming product, and the grade standards required (Section III).

This chapter discusses the design and operation of sorting equipment as well as techniques that have been developed to analyze the sorting operation. Information is presented to indicate how these techniques may be used to contribute to improvements in entire postharvest systems.

I. Background

The postharvest system spans all the operations that occur from harvest of a product to final purchase by a consumer. As previously discussed in Chapter 4, at one end of this system are the producers, who have grown and nurtured a product that they hope to sell at the highest possible return to their operation. At the other end are consumers (see Chapter 7), who will only purchase a product that they consider a good value.

A. Marketing Factors

Consumers will vary their quality perception for many reasons, for example, time of year, supply of product, supply of other products, and end use. Although initial product quality is determined by the producer, the dynamic parameters of price and quality requirements are established largely by the consumer. Any intermediary buyers (e.g., wholesalers) between the producer and the consumer also will have quality requirements.

The modern market has many established criteria; for producers to be competitive they must meet the specified requirements. Buyers will pay premium prices for fruit of uniform size and color. Fruit should not be misshapen or bruised and should be free of blemishes, diseases, and insects. For exporters, many international and national quarantine regulations must be met for insects, animals, and diseases. Product that will be stored for a length of time prior to marketing must meet criteria for maturity, firmness, and damage levels.

To assist both the seller and the buyer, many public agencies and marketing organizations have developed standards for the grades of most horticultural crops. The documents commonly are called "grade standards" and include one or more sets of specifications and tolerances. Compliance with a specific set of requirements in the grade standard enables the lot to be sold as a shipment labeled with the specified grade, for example, U.S. Extra Fancy, Fancy, No. 1, or Utility. When required, a third party grade inspector evaluates samples from each lot shipped by a seller before certifying that the lot complies with the grade standard specified.

As discussed in Chapter 2, the sorting operation must be viewed in the context of the overall postharvest system. It is important to understand how cultural practices and uncontrolled inputs such as weather cause variation of quality of products that enter the packinghouse. Likewise, at the time of shipping, it is important to be able to predict the quality of shipments as they progress through the rest of the postharvest system, as described in Chapter 8.

A major function of a packinghouse is to transform the highly variable product received from the harvesting operation into uniform lots of product for shipments that comply with the requirements of the buyer.

B. Sorting Terminology

Clarifying distinctions made between terms such as sorting and separation, sorter and grader, and grade inspection and quality control inspection is useful.

Separation is the removal of nonusable material from usable product. An early operation in packinghouses is the separation of debris and nonedible items from the flow of marketable product. Separations are made with mechanical devices such as sizers, blowers, and washers, and by employees who remove sticks, leaves, debris, and other nonedible material.

Sorting is the segregation of edible or marketable product into distinct quality categories. Sorting of the marketable items is accomplished by both mechanical (sizers or color sorters) and human (visual or tactical) means. Sorters also may perform some separation tasks such as removing debris and nonmarketable materials missed by earlier separation operations. The equipment used for sorting should be referred to as a sorting line, not a grading line.

People who perform sorting operations should be referred to as sorters, not graders. Graders are the third party inspectors who evaluate whether or not the packed lot complies with requirements of a grade standard for a predetermined grade classification.

Sample inspection for quality control (Chapter 13) is more precise than, and differs from, the inspection of product necessary for sorting. The importance of the sorting operation cannot be overstated, since variations in this operation will affect returns for most other parts of the postharvest handling system.

C. Sorting Equipment

Many attempts have been made to automate the sorting operation using machine vision (Drury and Sinclair, 1983). The primary impediments to progress are the complexities of defining algorithms to describe the numerous defects that must be identified by an image processor. One review (Chen and Sun, 1991) indicates that progress has been made for selected products and features. The economics of machine vision continue to hinder wider applications, especially for packinghouses with short operating seasons. Some systems have been developed that combine human skills of vision and defect detection with mechanical removal (MacRae, 1985).

Most sorting operations still are performed by human visual inspection of the product and manual removal of items with defects. Humans have special abilities for identifying defects and for determining if they exceed prescribed threshold criteria. This combination of rapid detection and flexibility in selecting thresholds for rejection suggest that postharvest systems will always include some form of human visual sorting operation. However, special attention is needed to avoid inconsistent performance.

A typical sorting operation consists of a continuous flow of product passing in front of one or more stationary sorters (Fig. 1). Normally, the task of the sorter is to remove items that do not meet the specifications for the lot being shipped. Nonconforming items are placed into a discard flow and items meeting other specifications are placed on separate conveyers that may flow to packing areas for lesser quality markets.

The design of sorting equipment has considerable effect on efficiency of the sorter in detecting and removing defective items that cause off grades. Consideration must be given to how the product is presented to the sorters. For efficient operation, the product must be easy to see, reach, and remove, and reject chutes or belts must be readily accessible.

For horticultural crops, a sorting conveyor is the customary means for presenting products to inspectors; four common techniques are illustrated in Fig. 2. When possible, items are rotated, which makes it easier for sorters to see defects over the entire product surface. A common sorting table for items with a spherical or cylindrical shape consists of a roller conveyor supporting the items between consecutive rollers. In many designs, the rollers are dragged over a flat surface. This friction surface rotates the rollers and, consequently, the product. The drive surface may be stationary, or driven to enable variable control of product rotation speed.

Figure 1   Typical sorting operation.

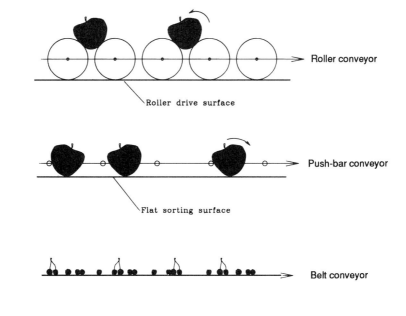

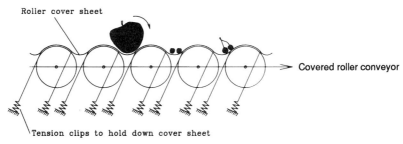

Figure 2 Typical methods for transporting product past sorters.

Alternatively, the "push-bar conveyor" pushes fruit over a flat surface. The push-bars are a set of stainless steel or plastic tubes separated by at least the diameter of the product. Drive chains attached to both ends of the bars pull them along the table surface, causing them to rotate the product as it is pushed past the sorters. The sorting surface may be fixed, but a flat belt is used often to enable the rotation speed of the product to be varied.

The shape or size of some products precludes the option of easily rotating them; in these cases a flat belt conveyor is used. This technique is common for small fruit such as cherries and berries. The major disadvantage of this type of presentation is that the product can be viewed only from above; any defects on the underside are not visible unless the fruit is moved manually.

Fruits can be rotated by cascading them from one belt to a second, where the product is reinspected, or by placing curtains over the table which drag over the product and rotate it. The height of the curtains can be adjusted to enable most of the product to be rotated by 180°.

A combination design that uses a roller conveyor with a flat surface is shown in Fig. 2 also. This approach uses a fixed roller cover sheet that is held tightly over a roller conveyor with a series of tension clips. The movement of the rollers under the sheet causes a ripple effect, that propagates along the table. The product is transported in the valleys, and rotates as it moves forward.

D. Visual Perception

The ability of humans to perceive a visual image depends on both physical and cognitive factors (Prussia, 1991). Changes in the color and intensity of light change the image received by the eye. The method of presenting the product to the sorters also has an important effect on perception. If product speed (either translation or rotation) is too fast, it is not possible to fixate properly on a defect; hence, it is not possible to reach a decision about whether or not the item should be rejected.

Any vision difficulties adversely influence detection of defects. Visual acuity decreases with age, but can be increased by increasing the brightness of the test object. A 60-year-old worker requires about twice the brightness level that a 20-year-old worker requires for equal visual acuity (Luckiesh, 1944). Vision examinations for sorters are useful for determining problems with visual acuity, peripheral vision, and color blindness. Also, the inability to concentrate for long periods of time results in a relaxing of vigilance, which is an important factor of visual perception.

II. Design and Operation of Sorting Equipment

The general sorting operation has developed along with the various postharvest systems used for different crops. Most design and operating conditions have been determined by trial and error for parameters such as table width, table speed, number of sorters, and speed of product rotation. Each product also places slightly different requirements on the system.

The interrelationship between these physical design parameters, the productivity and accuracy of the sorter, and the quality of the product are only beginning to be understood (Prussia and Meyers, 1989), yet the result of the sorting operation has significant effect on much of the rest of the postharvest system.

Relatively few studies have been undertaken in this area. In addition, sorting research has tended to be concentrated on large spherical and ellipsoidal products such as potatoes, apples, and citrus. Tables I and II summarize design and operational parameters reported by various researchers. In every case the product studied was large and able to be rotated on a roller conveyor or push-bar sorting table.

Research has concentrated on selected parts of the system with specific objectives, such as

1. optimizing design and operating parameters to achieve high accuracy of reject removal (Malcolm and DeGarmo, 1953; Hunter and Yaeger, 1970)
2. optimizing the design and operating parameters to achieve sorter productivity (Bollen, 1986a)

Table I

Optimal Design and Operational Performances Parameters Established by Various Researchers

Product	Defect type	Defects (%)	Design parameters				Operational parameters			Research objective	Experimental sample time (min)	Reference
			Table width (m)	Translation speed (m/min)	Rotation speed (rev/min)	Product rotation (rev/m)	Sorters per table	Fruit per row on table	Fruit/s per person			
Lemons	Actual	25–50	0.5	6.6	22–81	3–13	1	4	4.5	To achieve high accuracy of reject removal	1–2	Malcolm and DeGarmo (1953)
Oranges	Actual	20–60	0.5	5.5	54	9.8	1	4	3.7	To achieve high accuracy of reject removal	1–2	Malcolm and DeGarmo (1953)
Potatoes	Actual	15–40	0.5	9.0	36	3.3	1	3	4.5	To achieve high accuracy of reject removal	1–2	Malcolm and DeGarmo (1953)
Potatoes	Simulated	10–20	0.4[a]	6.7–11.0		5.8	1	4–5	7.0–11.6[b]	To achieve high accuracy of reject removal		Hunter and Yaeger (1970)
Kiwifruit	Actual	10–40	0.65	7.0	20–40	4.9	2–4	4	1.6	To achieve high sorter productivity at a standard removal accuracy	60–240	Bollen (1986a)
Spheres	Simulated	10–30	0.5	6.4–8.5		5.3	1	3–5	4.2–5.6	To achieve high accuracy of reject removal	1–2	Malcolm and DeGarmo (1953)
Ellipsoids	Simulated	10–30	0.5	6.4–8.5		2.5	1	3–5	4.2–5.6	To achieve high accuracy of reject removal	1–2	Malcolm and DeGarmo (1953)

[a]This value is per sorter.
[b]The lower value is at 20% defect level, the higher at 10% defect level.

Table II
Design and Operational Parameters Used by Sorting Researchers

Product	Defect type	Defects (%)	Design parameters				Operational parameters			Research objective	Experimental sample time (min)	Reference
			Table width (m)	Translation speed (m/min)	Rotation speed (rev/min)	Product rotation (rev/m)	Sorters per table	Fruit per row on table	Fruit/s per person			
Apples	Simulated culls III II	5 5[a] 10	0.24	3.4		4.9	1	2	1.0	Evaluation of semiautomatic sorting, using keypad to identify various grades as they pass station	20	Stevens and Gale (1970)
Spheres	Simulated	30	0.41	6.6	131	19.9	1	4	5.3	Optimum parameters for samples moving directly toward sorter	5–6	Meyers (1988)
Spheres	Simulated	30	0.41	3.8	19	5.0	1	5	3.0	Determination of optimum preview length for samples flowing directly to sorter	1–2	Prussia (1985)
Oranges	Actual	10–50		9.0	44	4.9	4	3–10	2.0	Modeling effect of defect rate on inspection rate at optimum conditions	20	Pasternak, Lidror, and Engel (1989)

[a]Three simulated defect grades.

3. evaluating a specific piece of equipment (Stevens and Gale, 1970; Prussia, 1985; Meyers, 1988)
4. determining the relationship between sorter productivity and the quality of the product (Lidror *et al.*, 1978; Pasternak *et al.*, 1989)

Table I summarizes the conclusions of the research with objectives 1 and 2. To conduct the experiments on objectives 3 and 4, the researchers had to make assumptions about, or otherwise determine, the optimum design and operation parameters for their investigation. Table II summarizes the designs used by the various researchers; these designs can be assumed to be near optimal, or at least typical for the particular product of interest.

Observations on the design and operation of sorting tables can be made as follows.

A. Size of Table

A sorting table should be designed at a height that is comfortable for the sorter to reach product on both sides of the table, and to deposit rejects on the appropriate belt or in the appropriate chute.

The design philosophy is to minimize hand movements to enable rapid location and removal of defective items. Also, hand movements should occur within a comfortable envelope of space. Dreyfus (1967) suggests that the sorters be positioned so that an angle of 45° is measured at the shoulder.

Sorting tables must accommodate people of various heights. One solution is to install false floors to lift short people up to a suitable working height. Some managers provide sorters with stools to reduce fatigue, which means the table height must allow for the stools. A height of 1.2 m is a practical height that will allow variations in application.

DeGarmo and Woods (1953) recommend that sorting tables should not exceed a width of 0.45 m. In further research with lemons, oranges, and potatoes, Malcolm and De Garmo (1953) suggest a width of 0.5 m. Hunter and Yaeger (1970) suggest a width of 0.4 m/sorter for potatoes.

As seen in Tables I and II, most research has been based on one sorter per table. However, for ellipsoidal produce it is important that both ends be viewed. In such cases, the sorters must be placed on both sides of the table, but must be able to reach completely across. Having sorters on both sides of tables is probably more efficient for items of any shape, since in most cases it is important to use at least two sorters per table.

When using more than one sorter, the table must be wider than 0.5 m, but should be no wider than arm's reach. Studies have shown that tables wider than 0.75 m have significant detrimental effects on sorting efficiency (Bollen, 1986b). This width does not include the reject belts mounted on the side of the tables.

Sorters have difficulty sorting fruit that lies directly below them, since it is easier to look further away with less head movement. For example, on a table 0.8 m wide, a person scanning from the opposite side to the middle of the table describes a 16° arc with their eyes and head. Scanning from the middle to the near side describes a 31° arc. Thus, product near the sorters tends to be ignored. Sorting is improved if sorters can be constrained by the table design to stand slightly away from the product. This can be achieved by installing reject belts down both sides of each table or by constructing tables with 100-mm wide sides.

B. Translation Speed

Translation speed is the velocity at which products pass the sorter. If the feed rate for incoming items is constant, then changes in translation speed will vary the amount of product on the table at a given time. In others words, translation speed controls the number of fruit per row. If the table rotates the fruit using a static friction drive, then changing translation speed also varies the rotational speed of the product.

Having a variable speed controller for the sorting table is becoming common practice. Changing the translation speed must be done with caution since it is unsettling for sorters if speed is adjusted frequently. However, sorters have the ability to adapt to a wide range of steady speeds. The limiting factors appear to be overflowing the table with product when operating at a low speed, and rotating the fruit too fast at a high translation speed. Most researchers suggest speeds of 6.5–9.0 m/min (Table I).

C. Product Loading

The quantity of product often is described in terms of product density on the table (kg/m^2 or fruit/m^2) or in terms of number of fruit per row. Loading can be regulated by adjusting the translation speed or the product feed rate. Loading should be regulated to insure the capability of the sorters to maintain a desired accuracy, and to insure that sufficient product can be handled when incoming quality levels require a high reject rate. Product loading is generally between 3 and 5 fruit per row (Table I), irrespective of table width for high quality products.

D. Rotational Speed

To achieve effective sorting, the product must be rotated in front of the sorter. It is desirable to rotate the fruit completely at least twice within the immediate field of view. The maximum rotational speed at which sorters can operate effectively is determined partly by the types of defects being removed (see Tables I, II) but, in general, rotational speeds above 50 rev/min are detrimental.

To achieve rotary motion on a roller conveyor, the rollers are dragged over stationary rails by chains connected to the ends of the rollers (Fig. 2). The roller rotation causes the product to rotate in the opposite direction. If desired, a belt can be installed beneath the rollers, or the rollers may be driven by a chain, to vary the rotation speed independently of translation speed.

With the belt stationary, the moving product will have maximum reverse rotational speed. Rotational speed will be zero when the belt is run at the same speed as the roller conveyor. When the belt speed is increased further, the fruit will start to rotate forward. A similar effect can be achieved with a push-bar conveyor (Fig. 2). If the flat surface consits of a variable speed flat belt then different rotational speeds and directions are possible. If forward rotation is used, the maximum rotational speed should be reduce below 50 rev/min, as translational and rotational speeds are additive at the product surface.

E. Sorter Position

The most efficient sorting operations require two sorters per table for a single line carrying products with low levels of defects. A good design allows accommodation of additional sorters in the event of high defect rates in the product. Sorters should be on opposite sides of the conveyor. However, sorting productivity is reduced if they stand directly opposite one another, since they tend to compete for the same product and do not use the full width of the table properly. Research with kiwifruit on a table 0.8 m wide showed that the proportion of the defective fruit removed was 96% when the sorters were staggered and fell to 68% when the sorters were standing directly opposite each other (Bollen, 1986b).

Research with simulated fruit (Meyers *et al.*, 1990a) reported a 23% improvement in defect detection for sorters positioned at the end of an inspection conveyor compared with sorters positioned at the sides. Approximately two-thirds of the improvement was shown to result from the ability to see more of the surface area when at the end than when at the side. The researchers suggested that the remaining one-third of the improvement could involve unknown ergonomic factors.

F. Lighting

Correct lighting is critical for an efficient sorting operation. It improves defect detectability and reduces eye strain. Low intensity light makes perception of contrasts difficult. A study on lighting for fruit sorting by Nicholson (1985) recommended a uniform light level of at least 1000 lux at the table.

Fluorescent tubes are used most commonly. If they are mounted 1.5 m above the table, there is minimum glare and the whole area is well lit. For wide tables, it is good practice to mount the tubes perpendicular to the table, since this insures uniform light levels across the whole table. For narrow tables, tubes mounted parallel to the table provide suitably uniform light. When it is necessary to mount lights at or below eye level to avoid shadowing, the lights should be fitted with deflectors and diffusers to direct a diffuse light onto the table where it is required, and not into the eyes of sorters.

Nicholson also suggested that the surroundings should be well lit. When sorters look up from the table, their eyes adjust to the light intensity of the background. Background light of a similar intensity helps reduce eye strain. Neutral-colored walls help reflect diffuse light back to the table. Sorting products on white belts can produce glare or high reflectivity of the incident light. Dark dull belts can ease eye strain and improve visibility of the product.

If determining product color is important, then it is necessary to use lights that produce a spectrum similar to that of daylight. In the extreme case, green light falling on a red surface will make the surface appear dark to the eye, since most of the light at these wavelengths will be absorbed by the surface. Making accurate decisions based on dark images is difficult. Unfortunately, "cool white" fluorescent tubes have a high intensity but a blue bias, which makes products appear excessively green. Therefore special tubes are required.

G. Location of Reject Chutes and Conveyors

Removing rejects or segregating products on a table is physically tiring work, and it is important to reduce hand movements to a minimum. For some crops, it has become habitual to throw rejects across the table. Some sorters prefer this design, but it is more energy efficient to have narrow chutes directly in front of the sorters or immediately beside them. Conveyors installed above the table usually require extra hand movements and the conveyor can shadow the table, which also hinders sorting.

H. Defect Type

The types of defects have significant effects on the optimum operating parameters. Some of the simulated products shown in Tables I and II could be sorted at very high throughput rates of 5.3 fruits/sec (Meyers *et al.*, 1990b), 4.2–5.6 fruits/sec (Malcolm and De Garmo, 1953), and 7.0–11.6 fruits/sec (Hunter and Yaeger, 1970). These simulated defects tended to be limited to one or two types, and usually were all of similar size. A real sorting operation encounters a large range of defect types; sorters must make decisions on the severity of each. This additional decision process results in a significant slowing in the potential product throughput.

For the examples shown in Tables I and II, typical throughput rates are reported as 2.0 fruits/sec (Pasternak *et al.*, 1989), 1.0 fruits/sec (Stevens and Gale, 1970), and 1.6 fruits/sec (Bollen, 1986b). Malcom and De Garmo (1953) reported throughputs of 3.7–4.5 fruits/sec for actual defects on lemons, oranges, and potatoes. However, their trial run time was only 1–2 min, and sorters may be able to sustain these levels of activity only for short periods.

III. Analysis of Sorting Operations

To analyze a sorting operation, it is necessary to establish parameters, such as efficiency or accuracy, that may be useful for comparing performance under different conditions. Because of the complexity of sorting, no standard analysis has been established.

When analyzing a sorting operation, the information directly available includes the throughput, the rate of removal of product, the proportion of defects removed, and the proportion of good fruit removed with the rejects. Sometimes a breakdown by defect type is also possible. This information has been used by various investigators to predict the performance of operations, to provide sorting system design information, and to provide operational information and management tools.

A. Sorting Performance

The performance of sorting systems can be described in many different ways depending on the type of information of interest. However, terminology is often confusing.

Simple measures of sorter accuracy have been quoted by many investigators. Often these are called sorting efficiencies. However, they generally relate only to the ability of the sorter to detect and remove a particular product and are not related to the product throughput. Stevens and Gale (1970) discussed inspection efficiency in terms of the proportion of sorting errors in an observed product flow. Malcolm and DeGarmo (1953) defined their inspection efficiency as the proportion of defective product that was removed from a determined quantity of incoming defects.

Sorting performance also may be described primarily as a throughput variable, for example, (fruit/sec)/sorter, as in Tables I and II, or (kg/sorter)/hr. Often throughput is correlated with the level of incoming defects, and decreases as some function of increasing defect level. The sorter throughput parameter does not describe the sorter accuracy, and assumes that a prescribed packout quality and an allowable level of good fruit in the reject flow is being maintained.

Peleg (1985) presents several quality criteria indices to describe the performance of the sorting operation. His following description of sorting includes both a sorting accuracy and a product throughput variable. The efficiency is defined as

$$E_i = \Sigma\, (P_{gi}G_i/P_iQ) \tag{1}$$

for i separate quality grades, where E is efficiency, Q is throughput rate of incoming product, G_i is outflow rate sorted into the ith grade, P_i is the proportion of i in the total incoming product flow Q, and P_{gi} is the proportion of i in the outflowing grade G_i. The most generalized definition includes weighting for the relative monetary value of the different grades of product. The weighted sorting efficiency for an entire operation is defined as

$$E_{\text{w}} = \Sigma\, (P_{gi}G_i/P_iQ)\, W_i \tag{2}$$

where the weighting function is

$$W_i = K_iP_i/\Sigma(K_iP_i) \tag{3}$$

and K_i is the cost fraction of grade i (must be $\leqslant 1.0$). In a simple sorting operation, in which the product is either packed or discarded ($i = 1.2$; $K_1 = 1$, $K_2 = 0$), Eq. 2 is reduced to

$$E_{\text{w}} = P_{g1}G_1/P_1Q \tag{4}$$

The weighted efficiency E_{w} is equal to the probability of a correctly sorted product being placed in the correct grade of outflowing product.

Signal detection theory is another method that may be used to describe performance and is detailed in Section III.D. This analysis method takes into account both the proportion of defects and the proportion of good quality product removed for specified operating conditions.

B. Modeling Sorting Operations

In the systems approach to postharvest handling, it is necessary to predict how a particular operation might function under various conditions. It may be necessary, for example, to predict productivity and staffing levels for various throughput or quality conditions. Many attempts have been made to analyze and describe the sorting operation mathematically; some of these models can be useful in a systems analysis.

Manual sorting operations can be subdivided into the following tasks:

1. visual search and detection of product with defects
2. decision about whether the defects are severe enough to be rejected
3. physically moving a hand to grasp and remove the particular item

These tasks are similar to many industrial operations, for which much research has been directed at each part as well as at the overall operation (Ferree and Rand, 1927,1928,1931).

Two approaches to modeling the system are described in the following sections. The first approach is to develop an empirical model that fits some experimental data and normally is applied to the whole sorting operation. Another approach enables separate analyses of the physical and psychological factors that influence sorting accuracy.

C. Empirical Models

The objective of developing an empirical model is to fit a mathematical relationship to some observed or experimental data (Portiek and Saedt, 1974). This model then can be used to predict outcomes from other situations.

By examining the momentary condition of a sorting process, it can be observed that a certain amount of work must be input to achieve a reduction in the quantity of defective product. From studies on the sorting of citrus fruit (Lidror *et al.*, 1978), the equation developed was

$$-kdP_p = P_q dt \tag{5}$$

where t is the sorting work input in terms of inspection time (min/1000 fruit), k is a constant factor that is a function of product type, P_p is the proportion of defective fruit in the product outflow, and P_q is the proportion of defective fruit in the incoming product flow. The solution to the equation, which was correlated highly in over 305 experiments, was

$$t = k(P_q - P_p)/P_q \tag{6}$$

with k = 18 for citrus. A similar expression is defined by Groocock (1986) as inspection effectiveness for industrial quality management.

In another study on the sorting performance of oranges, Pasternak *et al.* (1989) determined that the process was described effectively by

$$P_p = P_q A e^{-Bt} \tag{7}$$

where A and B are constants that are a function of defect type.

Significant differences were noted between sorting slight defects and sorting severe ones. The model was developed using two sets of constants:

$$A = 0.958 \quad \text{and} \quad B = 0.0266 \quad \text{for slight defects}$$
$$A = 0.762 \quad \text{and} \quad B = 0.0784 \quad \text{for severe defects}$$

These values were determined using a sorting system considered to be operating under optimum sorting conditions.

Equation 7 can be used in a systems analysis to predict the quantity of defective fruit that will pass on to the packing operation, for a given input defect level and sorting rate. However, this equation does not take into account the quantity of good quality fruit rejected.

Similar relationships could be established for other sorting operations. Different values of the constants A and B can be generated easily using curve fitting techniques with observed sorting data.

D. Signal Detection Theory

Signal detection theory (SDT) was developed to quantify the effectiveness of systems used for detecting communications signals from background noise (reviewed by Egan, 1975). The ability to detect a signal can be described by two nondimensional parameters, d' and β. The first parameter gives a measure of the detectability of the signal and the second represents the criterion used to identify the signal. SDT was applied later in psychology experiments to determine human ability to distinguish a visual signal from a background visual noise (Tanner and Swets, 1954).

A growing body of literature is available on SDT applications to visual and other sensory perceptions, to vigilance, and to various industrial inspection tasks (Jaraiedi *et al.*, 1986). The sorting operation is analogous to situations in communications engineering, psychology, and psychophysics. SDT offers interesting possibilities as an analysis tool for fruit and vegetable sorting operations. Prussia (1991) presents SDT principles, examples, exercises, and suggested laboratory experiments intended to help students apply SDT to practical sorting questions.

Psychology analysts have retained the original communications terminology to describe SDT; this convention is used in this discussion (Green and Swets, 1966). When attempting to detect a signal, there are two possibilities: a signal-plus-noise stimulus (SN) and a noise-with-no-signal stimulus (N). The two possible responses to a stimulus, "yes" and "no", indicate the observer's belief that the signal is present or absent. That either response may be in error is always a possibility.

For the general sorting operation, a flow of product passes in front of the sorter, who removes the defective product; the good product is conveyed to the packing area. The mixture of both good and defective product can be considered SN; the good product can be considered N.

The ability of the sorter to make "yes" and "no" decisions correctly is influenced by the physical parameters of the operation, for example, speed, rotation, fruit density, number of defective fruit, types of defects, and lighting. Decisions are influenced also by psychophysical factors such as sorter sensitivity to perceptual stimulus, sorter alertness, and sorter motivation to give one response or the other.

Table III
Responses to SN and N Stimuli

Stimulus	"Yes" response	"No" response	Sum of probabilities
SN	$p(H)$	$p(M)$	1.0
N	$p(FA)$	$p(CR)$	1.0

A major contribution of SDT is the ability to separate the physical from the psychophysical influences.

The conditional probability of responding "yes" when a signal is present is termed the hit rate, $p(Hit)$ or $p(H)$; the conditional probability of responding "yes" when a signal is not present is termed the false alarm rate, $p(False\ alarm)$ or $p(FA)$. A third possibility is the conditional probability of responding "no" when a signal is present, which is called a $p(Miss)$ or $p(M)$. The last conditional probability is that of responding "no" when a signal is not present, which is called $p(correct\ rejection)$ or $p(CR)$. All four responses to the SN and N stimuli are shown graphically in Fig. 3 and Table III. Since the probabilities are conditional, all possibilities can be described using only the two probabilities $p(H)$ and $p(FA)$, since $p(M) = 1.0 - p(H)$ and $p(CR) = 1.0 - p(FA)$. Conventional SDT techniques assume that the SN and N probabilities are distributed normally and are of equal variance (Fig. 3).

For a particular system in which the physical and operational parameters and the product characteristics do not change, SN and N distributions do not change. The difference between the normal deviates of the means is described by the parameters d', the detectability (Freeman, 1973)

$$d' = z(H) - z(FA) \tag{8}$$

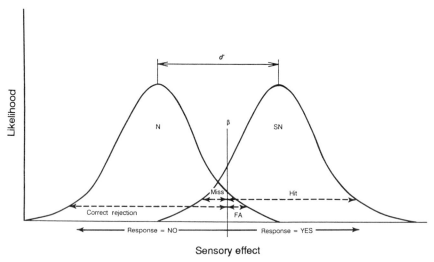

Figure 3 Probability distributions for "Noise-with-no-signal" (N) and "Signal-plus-noise" (SN). The four possible outcomes are shown, as are representations for detectability, d', and criterion, β.

z is the standard deviation value or z-score for a normal variant. The relative value of d' is also important. The easier the detection of defects, the further apart the two distributions will be, and the higher d'. A sorting table with a higher d' indicates that the sorters at that table have the potential to detect and remove more of the defects than those at a table with a lower d'.

The detection performance at a sorting table may be evaluated visually by using a graph called the receiver operating characteristic (ROC) graph (Fig. 4), on which $p(H)$ is plotted against $p(FA)$.

Experimentally, $p(FA)$ may be varied in some manner, for example, by varying the instructions to the sorters or by adopting some incentive payment scheme. By varying $p(FA)$, it is possible to generate an ROC curve. Since the physical parameters of the system (speed or rotation) have been unchanged, this curve is characteristic of that operation, which is shown as a unique d' line for each set of conditions.

Each sorting system has a curve for each value of $p(H)$ and the corresponding $p(FA)$; these thus are characterized by a detectability curve. The power of SDT is that, after one pair of $p(H)$ and $p(FA)$ values has been determined, it is possible to generate the complete curve.

The second useful descriptor is the criterion likelihood ratio, β, which represents the probability that a decision was based on SN stimuli relative to the probability that the decision was based on N stimuli. For a particular cutoff β, shown in Fig. 3, the ratio of the two normal density functions simplifies to (Egan, 1975)

$$\beta = \exp\{[z(FA)]^2/2 - [z(H)]^2/2\} \qquad (9)$$

The physical representation of the criterion β is a description of the cutoff position that the sorter sets in his or her own mind (Fig. 3). Any stimulus above this criterion

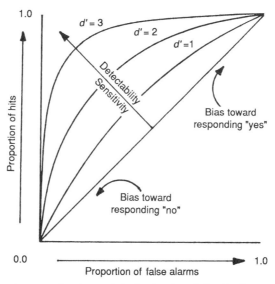

Figure 4 Typical receiver operating characteristic plot showing relationships between p(Hit) and p(False Alarm) as a family of detectability (d') curves.

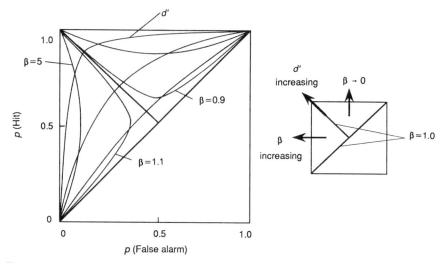

Figure 5   Receiver operating characteristic plot showing the contour lines for criterion, β, superimposed on a family of detectability curves d'. d' is a plot of the relationship between the probability of hits and the probability of false alarms.

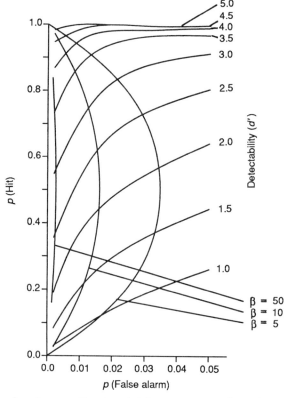

Figure 6   Section of receiver operating characteristic space common for sorting operations showing the contour lines for criterion, β. superimposed on a family of detectability curves, d'.

is called a signal, regardless of whether it is SN or only N. The criterion β may be varied by the sorter, either on instruction or as a result of some other influence (e.g, incentive payments). A deliberate change in criterion β results in differing hit and false alarm rates for an otherwise constant sorting system.

A low value for β indicates that a considerable amount of N is being accepted as SN, so p(FA) is high. This tendency is termed a liberal criterion. A high β implies a low p(FA) and is called a conservative criterion. Figure 5 shows a ROC plot with constant values of β plotted as a family of curves added to the family of d' curves.

Each pair of p(H) and p(FA) has a corresponding d' and β. For a particular system, d' is constant and the values of p(H) and p(FA) will follow this line, depending on the value of β. The system parameters thus are separated conveniently. The physical description is encapsulated by the value of d' and the psychological factors are described by β.

In a commercial sorting operation, false alarm rates are normally very low, so it is not possible to generate a full ROC curve experimentally. However, it is necessary to ensure that the theory holds true for the portion of ROC space of interest, as shown if Fig. 6.

Analyses using SDT highlight the importance of determining both the hit rates and the false alarm rates in the assessment of any sorting operation, which rarely has been considered in past research. The resulting detectability and criterion values have several useful applications.

1.   Physical design and operational parameters

The detectability, d', can be used to compare the design and operating characteristics of different sorting systems. For example, fewer false alarms (discarded good items) result when physical changes increase d' and the hit rate, p(H), remains fixed. Operating or design criteria for a particular system also can be optimized by generating d' values using real product for various alterations and modifications.

The advantage of SDT over traditional techniques of evaluating sorting operations is that d' can be determined by different sorters with products of differing quality in various systems, and still allow comparisons. That this analysis is subject to some restrictions should be noted. For example, product quality does have some effect.

2.   Sorter criteria

By calculating β for a sorter, it is possible to determine whether that sorter has a conservative or liberal approach, which could serve as a useful management tool. A conservative criterion also results when the sorters only remove the worst product.

In a commerical operation, the objective usually is to maintain a consistent quality of the packed product. If the incoming product quality is poor, it is necessary for the hit rate to be high, whereas if the quality is good, a low hit rate is adequate. The sorters thus are required to vary their own criterion, depending on the incoming quality. Analysis of individual criterion values will determine how effectively each sorter is able to adjust.

3. Systems analysis

SDT mathematically describes the relationship between hit and false alarm rates. Once a value of d' has been determined experimentally, it may be used in a model of the packinghouse to predict hit and false alarm rates. The analyst also has the ability to use β to vary the behavior of the sorters and predict the impacts of such changes on the whole system.

IV. Economics of Sorting Operations

In many cases, the ability to predict performance of a sorting operation with SDT will be useful for management operations on a daily basis. In addition, the usefulness of the model can be extended if it is possible to predict the economic value of potential changes. Economic models provide a useful tool during planning and design phases, as well as for managing an operation. Economic models are also the most suitable way to integrate the sorting operation into the postharvest system.

One method of optimizing a sorting operation is using the sorting efficiency defined by Peleg (1985) (Eq. 4) which requires the proportion of defective product to be known or estimated in both the incoming and the sorted product flows, and dollar values to be assigned to the various grades. The efficiency calculated is weighted according to product values, so it is possible to compare various scenarios. The total value of sorted product, V, then can be calculated as

$$V = E_w T(v) \tag{10}$$

where T is the throughput for sale (kg/hr, ton/season, and so forth) and v is the unit value of the product ($/kg, $/ton, and so forth). The returns, represented by total value, then can be used in addition to the costs to evaluate various alternatives or used as part of a wider system model.

For any marketed product, there will be a payout schedule that is a function of the amount that is defective. Therefore, the higher the hit rate that can be achieved, the better the returns, as shown in Fig. 7. Every "false alarm" represents a lost item thus, the value of returns decreases with increasing $p(\text{FA})$. Actually, this scale should be continuous, but is shown as a discrete scale for clarity in the following discussions.

For a predetermined payment schedule, a payout matrix can be established, as represented in Fig. 7. This payout must be established for a known incoming fruit quality to specify a hit rate to produce the desired output quality. For example, for a required output quality of 95% good product, the sorter must maintain a hit rate of 0.53 if 10% defects are incoming, but must have a hit rate of 0.92 if 40% defects are entering.

If the relationship between $p(\text{FA})$ and $p(\text{H})$ for a particular system is known, or can be determined, this information can be used to determine the relationships between the sorting operation and the returns for product sold, thus combining economic and operational parameters into the same model. The $p(\text{H})$–$p(\text{FA})$

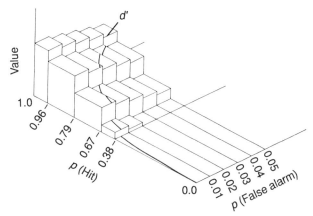

Figure 7 Payoff matrix with receiver operating characteristic plot and d' curve superimposed.

relationship may be determined from historical data of sorter performance or it may be predicted using SDT.

For the example shown in Fig. 7, the management might want to know the consequences of advising their sorters to "reduce the number of good fruit in the reject bin" (reduce false alarms) or "reduce the number of defects in the outgoing product" (increase hit rate). If the sorters concentrate on reducing good fruit entering the reject bin, the consequence will be a reduction in the hit rate of defective product; similarly, the consequence of reducing rejects passing into the final pack will be an increase in good fruit entering the reject bin. This situation, as previously discussed, is represented by moving along the d' line shown in Fig. 7.

If a buyer of quality product pays according to the following schedule

Grade I	<2% defects
Grade II	<10% defects
Grade III	<15% defects
Grade IV	<25% defects
Will not buy	>25% defects,

and the packinghouse operator has incoming product with 35% defects then sorters must achieve a hit rate of 0.96 to ensure Grade I product, 0.79 for Grade II, 0.67 for Grade III, and 0.38 for Grade IV. These hit rates are shown in Fig. 7.

If the operation is maintaining an average hit rate of 0.8, then, referring to Fig. 7, a "reduction in the number of good fruit in the reject bin" by lowering p(FA) to 0.015 would result in a reduction of p(H) to below 0.79. The payout schedule cutoff is represented by p(H) = 0.79; therefore, any reduction in the false alarm rate will result in a considerably lower return to the packinghouse.

The second scenario is to "reduce the number of defects in the outgoing product." If the hit rate was increased to 0.9, the objective would be achieved. However, the increase in p(H) will also result in an increase in false alarms to above 0.02; thus, the return to the packinghouse will be reduced by the loss of salable fruit.

For the situation illustrated, any changes in the instruction to the sorters will result in a reduction in overall packinghouse returns.

An operator contemplating an upgrade for the system could also apply this technique. The expected performance of some new equipment might be an increase in hit rate from 0.82 to 0.95 at a false alarm rate of 0.018. A new d' curve may be established for the upgraded system, representing $p(H) = 0.92$ and $p(FA) = 0.02$. Then, if the sorters can be instructed suitably, it will be possible for the operator to achieve a hit rate of over 0.95 at a false alarm rate of 0.02. This increases the value of the product; to determine whether this increase will be a profitable return on the capital invested in the upgrade is a simple matter.

V. Summary

This chapter has discussed the pivotal importance of the sorting operation to the postharvest system. It has outlined the approaches researchers have adopted in the past to analyze the operation, and detailed some of their important recommendations. It has introduced new techniques that are being developed to enable the operation to be described and understood in the context of the overall postharvest system.

Bibliography

Bollen, A. F. (1986a). "Sorting Fruit: Systems Design and Operation." Postharvest Manual. New Zealand Agricultural Engineering Institute/Ministry of Agriculture and Fisheries. Hamilton, New Zealand. Unpublished.

Bollen, A. F. (1986b). How efficient is your packhouse—and its sorting table. *N. Z. Kiwifruit* **March,** 20–21.

DeGarmo, E. P., and Woods, B. M. (1953). "Introduction to Engineering Economy." MacMillan, New York.

Chen, P., and Sun, Z. (1991). A review of non-destructive methods for quality evaluation and sorting of agricultural products. *J. Agric. Eng. Res.* **49,** 85–98.

Dreyfus, H. (1967). "Measurement of Man: Human Factors in Design." Whitney Publications, New York.

Drury, C. G., and Sinclair, M. A. (1983). Human and machine performance in an inspection task. *Human Factors* **25(4),** 391–399.

Egan, J. P. (1975) "Signal Detection Theory and ROC Analysis." Academic Press, Orlando, Florida.

Ferree, C. E., and Rand, G. (1927) *Trans. Illum. Eng. Soc.* **22(1),** 79.

Ferree, C. E., and Rand, G. (1928) *Trans. Illum. Eng. Soc.* **23(5),** 507.

Ferree, C. E., and Rand, G. (1931) *Trans. Illum. Eng. Soc.* **26(8),** 820.

Freeman, P. R. (1973). "Tables of d' and BETA." Cambridge University Press, Cambridge.

Green, D. M., and Swets, J. A. (1966). "Signal Detection Theory and Psychophysics." John Wiley & Sons, New York.

Groocock, J. M. (1986). "The Chain of Quality, Market Dominance Through Product Superiority (Inspection Effectiveness)." John Wiley & Sons, New York.

Hunter, J. H. and Yaeger, E. C. (1970). "Use of a Float Roll Table in Potato Grading Operations." Maine Agricultural Experimental Station Bulletin No. 690. Orono, Maine.

Jaraiedi, M., Toth, G. J. Aghazadeh, F., and Herrin, G. D. (1986). Modeling of decision making behavior of inspectors. *Trends Ergonomics* **III,** 351–357.

Lidror, A., Roll, Y., and Siv, S. (1978). Citrus fruit sorting parameters influencing manual work rate. *Proc. Int'l. Soc. Citriculture,* **20,** 115–116.

Luckiesh, M. (1944). "Light, Vision, and Seeing." Van Nostrand, New York.

MacRae, D. C. A review of developments in potato handling and grading. *J. Agr. Eng. Res.* **31**, 115–138.

Malcolm, D. G. and DeGarmo, E. D. (1953). Visual inspection of products for surface characteristics in grading operations. USDA Marketing Research Report No. 45. U.S. Govt. Printing Office, Washington, D.C.

Meyers, J. B. (1988). Improving dynamic visual inspection performance. Masters Thesis, University of Georgia, Athens, Georgia.

Meyers, J. B., Jr., Prussia, S. E., Thai, C. N., Sadosky, T. L. and Campbell, D. J. (1990a). Visual inspection of agricultural products moving along sorting conveyors. *Trans. ASAE* **33(2)**, 367–372.

Meyers, J. B., Jr., Prussia, S. E., and Karwoski, C. J. (1990b). Signal detection theory for optimizing dynamic visual inspection performance. *Appl. Eng. Agric.* **6(4)**, 412–417.

Nicholson, J. V. (1985). Color and light for the inspection table. SIRTEC Publication No. 1. New Zealand Department of Scientific and Industrial Research. Wellington, New Zealand.

Pasternak, H., Lidror, A., and Engel, H. (1989). The effect of the incidence of defect on orange inspection time. *Can. Agric. Eng.* **31(2)**, 131–134.

Peleg, K. (1985). "Produce Handling, Packaging, and Distribution." AVI, Westport, Connecticut.

Portiek, J.H., and Saedt, A. P.H. (1974). An analytic and a descriptive model for the potato sorting process. *J. Agr. Eng. Res.* **19**, 189–198.

Prussia, S. E. (1985). "Visually Sorting at Ergonomically Designed Workstations." ASAE Paper No. 85-1618. American Society of Agricultural Engineers, St. Joseph, Michigan.

Prussia, S. E. (1991). Dynamic visual inspection. *In* "Module 9 of Human Factors" (D. L. Roberts and W. I. Becker, eds.) American Society of Agricultural Engineers, St. Joseph, Michigan.

Prussia, S. E. and Meyers, J. B., Jr. (1989). Ergonomics for improving visual inspection at fruit packinghouses. *Acta Hort.* **258**, 357–364.

Stevens, G. N. and Gale, G. E. (1970). Investigations into the feasibility of semi-automatic quality inspection of fruit and vegetables. *J. Agr. Eng. Res.* **15(1)**, 52–64.

Tanner, W. P., and Swets, J. A. (1954). A decision-making theory of visual detection. *Psychol. Rev.* **61**, 283–288.

CHAPTER 10

LATENT DAMAGE: A SYSTEMS PERSPECTIVE

Yen-Con Hung

From the moment a fruit or vegetable is detached from the parent plant, it becomes susceptible to internal and external factors that can lead to premature loss of quality and waste. Traditional approaches to reducing losses of horticultural crops focused on cultural practices during crop production and extension of shelf life during postharvest handling. Managers at many stages of the postharvest system tend to be more concerned about maintenance and control of quality attributes of perishable goods than about extending the shelf life (Thai *et al.*, 1986).

One way to maintain quality attributes of perishable goods is to minimize mechanical damage. Mechanical damage increases the rate of deterioration and may occur during harvesting, loading and unloading, transport, sorting and grading operations, and bulk storage (Shewfelt, 1986). Efforts to reduce mechanical damage require knowledge of the susceptibility of the product and of the times at which impact actually occurred. However, damage detection has been complicated by the lack of visible evidence when physical damage has been incurred (Peleg, 1986; Thai *et al.*, 1986; Prussia *et al.*, 1987).

Variyam *et al.* (1988) suggest that preharvest conditions may have a greater impact on postharvest quality than does the postharvest handling system. According to a current model of plant growth and development, crop yield and quality can be viewed as the integration of flows of energy, water, and nutrients through the plant (see Chapter 4). Many other environmental and managerial factors also may determine the quality of marketable products. However, the effect of many of these factors on the final quality of horticultural products does not become evident until harvest or distribution.

The quality of any fruit or vegetable at any point in the production or postharvest system is a function of its history at that time. The phrase "latent damage" was coined by Peleg (1985) and later defined by Shewfelt (1986) as "damage incurred at one step but not apparent until a later step" to describe the result of nonvisible quality degradation.

I. Types of Latent Damage

Bruising is the most obvious example of latent damage, since the impact can occur during rough handling but the discoloration and textural breakdown do not become visible immediately. Bruise development is a relatively slow process. Initial discoloration of potatoes appears in approximately 4 hr; completion of the reaction requires about 24 hr (Belknap *et al.*, 1990). Klein (1987) reported that bruises on apples were developed completely 12 hr after impact. Prussia *et al.* (1987) noted that impacts at 0.043 Joule (J) level produced very little visible damage to apples immediately after impact (Fig. 1). However, the percentage of fruits with minor visible damage increased from 8.3% immediately after impact to 75% after 12 wk storage at 4°C. Apples that had minor visible bruises shortly after impact had an increase in the severity of bruises after storage (Fig. 1). Bruising on potatoes is also not visible until the tuber is peeled (Mondy *et al.*, 1987). Burton and Schulte-Pason (1985) reported that visible damage in the form of bruising for blueberries,

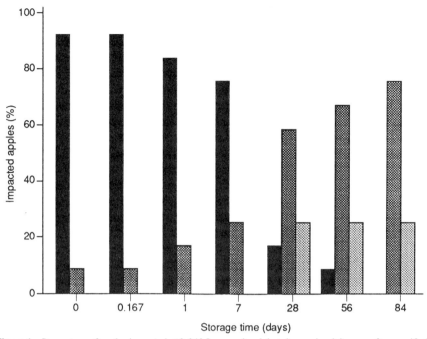

Figure 1 Percentage of apples impacted at 0.043 J energy level that shows visual damage after specified storage times. Filled bars, no visible damage; hatched bars, minor visible damage; dotted bars, visible damage. (Reprinted with permission from Prussia *et al.*, 1987.)

sweet and tart cherries, tomatoes, and apples is not always immediately apparent, and much of it can be internal.

Another example of latent damage is quiescent infection. Preharvest infection of fruits and vegetables may occur through direct penetration of the skin, infection through natural openings of the products, and infection through damage. The infection then is arrested and remains quiescent until after harvest, when the resistance of the host decreases and conditions become favorable for growth (Wills *et al.*, 1989). The latent phenomenon is caused by a transient resistance of the host to the extensive development of the pathogenic microorganism (Verhoeff, 1974). Latent infections have been documented for anthracnose of mango and papaya, crown rot of banana, and stem-end rot of citrus (Wills *et al.*, 1989).

Physiological disorders resulting from inadequate nutrition or preharvest stress conditions represent another form of latent damage. These disorders may be caused by an adverse environment or a nutritional deficiency during growth and development. Examples of adverse environments are frost, high temperature, limited water, or high salt concentrations. Plants also require a balanced mineral intake for proper development, so a deficiency in any essential mineral will lead to maldevelopment of the plant as a whole. Fruits and vegetables often show various browning symptoms that have been attributed to mineral deficiencies (Wills *et al.*, 1989).

Postharvest stress disorders constitute another form of latent damage. It is normal to store or transport high-moisture produce under refrigeration to repress metabolism and maintain fresh flavor and nutritive value (Uritani, 1978). However, tropical and subtropical plants will suffer from chilling injury during cold storage. A safe length of time for storage of fruits and vegetables at a particular chilling temperature is difficult to predict, because chilling injury symptoms often are developed and become detectable or visible only after these products have been removed from cold storage to a warmer temperature (Wang, 1989; Chapter 12). Chilling injury results in a loss of quality, manifested as pitting, discoloration, internal breakdown, uneven or incomplete ripening, off flavor, and weakening of the tissues, which renders the commodity very susceptible to decay by postharvest pathogens (Paull, 1990; Saltveit and Morris, 1990).

Superficial scratches have been identified as a major problem in mechanically harvested citrus fruits (Burkner and Chesson, 1972). Many of the superficial injuries are barely visible immediately after harvest but scars later develop, apparently as a result of increased moisture loss from and around the injured tissue (Golomb *et al.*, 1984).

II. Importance of Latent Damage

Postharvest deterioration is a serious problem for the producer and the distributor of fresh produce and may influence adversely the availability and cost of these commodities to the consumer (Eckert, 1978). From an economic standpoint, costs are associated with handling and transporting defective items that eventually will be discarded. Thus, early detection of latent damage can reduce costs, help identify causes of the damage, and help prevent its manifestation (see Chapter 2). Aside from the obvious loss of edible product, postharvest deterioration has other important but less evident consequences: (1) transmission of decay from an item in a package or container to otherwise healthy items, (2) reduced postharvest life of the product because of accelerated ripening or senescence triggered by ethylene released from affected items in a container or storage room (Wild *et al.*, 1976), and (3) possible contamination of the edible product with a mycotoxin elaborated by the disease-inducing microorganism (Buchanan *et al.*, 1974).

Apple bruising is a frequent cause of loss of value of marketed and processed apples. In addition to the loss of value of the raw materials, it is frequently necessary to sort bruised apples from the acceptable product. Dewey and Schueneman (1971) reported that 12.2% of the Michigan apples that were handpicked and stored for fresh-market exhibit have bruises classed as grade defects. Bartram *et al.* (1983) studied two packinghouse operations and found that 89% of the 'Golden Delicious' apples were bruised after completing the packing operations, compared with 74% before starting the operation. Cappellini *et al.* (1987) reported that 67% of the 4453 apple shipments that arrived at the New York terminal markets between 1972 and 1984 showed bruise damage. Stanton (1961) noted an average of 24 bruises over 0.5 in. in diameter per 100 'McIntosh' apples from nine New York farms. Apples with bruises larger than that must be identified and removed in the grading operation.

Many supermarket chains exhibit a growing trend to include "vine-ripe" tomatoes in the produce section as a premium-priced selection. However, greater care must be taken to minimize mechanical damage characterized by bruising, cuts, punctures, and abrasions and latent damage known as internal bruising (Sargent *et al.*, 1989). Detection of internal bruising is normally not possible by inspection personnel in the packinghouse, since it is latent damage that becomes apparent during ripening. However, the consumer will notice the disorder at slicing, which may affect future sales.

Wright and Billeter (1975) found that bruising was the dominant loss factor for fancy and extra fancy 'Delicious' apples at the wholesale and retail levels, constituting 75–90% of the total recorded loss. Blue mold rot was by far the most dominant disorder of these apples. Although blue mold decay manifests itself at wholesale or retail levels, it may not originate there. The *Penicillium expansum* fungus is considered a weak pathogen, requiring mechanically injured or weakened tissue in which its airborne spores can germinate and develop. A microscopic break in the skin or an injured lenticel provides an avenue for the fungus to enter the fruit (English *et al.*, 1946).

The unpredictability of time–temperature combinations required to produce chilling injury symptoms makes this damage difficult to control. A limited period of cold storage may not prove harmful, but its safe duration cannot be determined readily since injury often is detectable only after the food product has been removed from cold storage for distribution. The symptoms of chilling injury will affect product quality, often to the point where usefulness is sacrificed. Chilling injury can occur in the field or during harvesting, storage, and/or distribution of susceptible fruits and vegetables, resulting in loss of sensory quality and consumer acceptability. The practical consequences include economic losses, reduced product storage life, and an increased dependence on imported produce (Jackman *et al.*, 1988).

Although widely recognized as a problem, only Peleg (1985), Prussia (1985), and a few other authors have described the concept of latent damage. Damaged fruit that is undetected at a packinghouse requires the same cost inputs as undamaged fruit for cooling, packaging, transporting, distributing, retailing, and other marketing. Therefore, early removal of items that may become unacceptable later in the postharvest system results in economic benefits (Prussia *et al.*, 1987). Another reason to learn more about latent damage is to improve loss assessment accuracy for complete postharvest systems.

An understanding of latent damage is critical in identifying the handling operations that cause mechanical damage within a postharvest system (Campbell *et al.*, 1986). When using normal procedures, the cause of damage typically is assigned to the operation that shows increased damage between its input and its output. Latent damage can cause inflated damage assessments for operations with significant holding times while underestimating damage at those sites with short cycles. Thai (1987) has indicated that accurate identification of operations causing latent damage requires excessively large sample sizes. Thus, methods for early detection would assist more accurate assignment of responsibility to the postharvest step(s) at which damage is incurred.

A primary goal of postharvest handling is quality maintenance of the product from harvest to consumption. A better understanding of latent damage should lead to better inventory control and management by manipulation of conditions and the development of quality management schemes, as well as to loss prevention by identifying the sources of damage and taking corrective action. In addition, transportation must be considered carefully, including early culling of product that will develop damage during subsequent handling and storage. Each of these factors has an economic impact that cannot be ignored; successful research must integrate the technical and the economic aspects carefully.

III. Implications

Physical damage to plant tissue is followed often by physiological responses (MacLeod *et al.*, 1976). These responses include localized increased respiration at the site of injury, stress ethylene production, accumulation of secondary metabolites, and cellular disruption leading to decompartmentalization of enzymes and substrates (Rolle and Chism, 1987).

Increased ethylene production after impact damage has been reported for freshly harvested apples (Lougheed and Franklin, 1974) and mature-green tomatoes (MacLeod *et al.*, 1976). Yang and Pratt (1978) reported that the biosynthetic pathway for ethylene in wounded tissue is similar to the pathway for ethylene synthesis in ripening fruit. Ethylene promotes senescence and accelerated deterioration, which directly or indirectly affects the color, flavor, and texture in fruits and vegetables (Kader, 1985). The magnitude of ethylene production increases with frequency of impact. Impact damage also results in increased respiration rates one day after injury. Sustained increases in respiration of intact fruit after damage by impact has been observed in oranges (Eaks, 1961) and freshly harvested apples (Lougheed and Franklin, 1974). The increased ethylene production and other possible physiological changes are conducive to faster ripening of damaged fruits. MacLeod *et al.* (1976) concluded that physiological and physical changes induced by the damaged tissue alter the composition and reduce the quality of internal appearance of bruised mature-green tomatoes, although acceptable external quality may be retained during ripening. Other physiological changes in damaged tissue have been reported, both whole plant (Hale and Orcutt, 1987) and for harvested product (Wang, 1989).

A. Bruising

Bruising has been defined as damage to plant tissue by external forces causing physical change in texture and/or eventual chemical alteration of color, flavor, and texture (Mohsenin, 1986). In addition to apples, pears, peaches, tomatoes, cherries, and blueberries, bruise damage during harvest and postharvest handling has been reported for avocadoes, papayas, and pineapples (Timm and Brown, 1990). Holt and Schoorl (1977) observed that cells near the point of compression were distorted, resulting in cell wall distention. When the elastic limit of the cell wall has been exceeded, the cell walls rupture, releasing the contents of the cell into the air-filled intercellular spaces. Tissue softening is accomplished by polygalacturonases, pectin methylesterases, and cellulases that degrade cell wall material (Ben-Arie and Naomi,

1979). The browning reaction is mediated by polyphenol oxidases, which oxidize phenolic compounds to quinones that are unstable and polymerize as melanotic compounds with high molecular weights (Vamos-Vigyazo, 1981).

Bruises require time to develop fully. Initial discoloration appears in approximately 4 hr, with completion of the reaction in about 24 hr (Belknap *et al.*, 1990). The time course required for bruise development may depend on increased activity of the enzymes involved in the biosynthesis or metabolism of substrates for the oxidation reaction, or of polyphenol oxidase itself (Belknap *et al.*, 1990).

The development of bruises in fruits is dependent on the tissue structure of each fruit (Garcia *et al.*, 1988). Fruit tissue with a low air-space volume, for example, peach tissue, is susceptible to deep bruises that are often not visible on the skin surface (Kunze *et al.*, 1975). However, fruits such as apples with a high volume of air-filled interstitial spaces will have bruises from the contact surface toward the center of the fruit until all the impact energy is either dissipated by cell breakage or stored by elastic membrane distention (Holt *et al.*, 1981).

Potato tissue responds to physical injury by initiating a number of biochemical and physiological changes. Belknap *et al.* (1990) reported a large transient increase in phenylalanine ammonia lyase (PAL) activity in bruised potatoes; the maximal PAL activity was observed 48 hr after bruise induction. Physical impact also induced a marked decrease, followed by recovery, in the messenger RNA level for patatin, a primary storage protein in the potato tuber. The change in PAL activity was associated with a transient increase in messenger RNA encoding PAL. The researchers suggested that the stress resulting in bruising elicits a wound response similar to those observed in other injuries to plant tissues. Bruised potato tissue also contained more phenols, total glycoalkaloid, and lower ascorbic acid than unbruised tissue (Mondy *et al.*, 1987). Bruising in tomatoes has been described as a breakdown of the locular gel from the normal clear pink color to a cloudy yellowish "stringy, gelatinous tissue" (Hatton and Reeder, 1963).

When impacting peaches on a flat rigid surface, the damaged tissue usually appears at some depth within the fruit where the shear stress is maximum (Horsfield *et al.*, 1972; Kunze *et al.*, 1975). Vergano *et al.* (1991) observed that impacted tissues exhibit tear lines and voids in peaches. Surface manifestation of the bruise damage may not be apparent because bruised flesh was separated from the skin by an unbruised area. Miles and Rehkugler (1973) concluded that shear stress was the most significant failure parameter in compression tests. According to the theory of elasticity, the maximum shear stress resulting from an applied load occurs at a distance of about one-half the radius of the contact area beneath the surface (Fridley *et al.*, 1968).

When impacted, fruits such as citrus, tomatoes, cherries, and apples show an immediate increase in the evolution of CO_2 (Robitaille and Janick, 1973; MacLeod *et al.*, 1976; Parker *et al.*, 1984). Burton and Schulte-Pason (1985) demonstrated that increasing CO_2 evolution was proportional to the severity of damage for blueberries and sweet and tart cherries. The increase in CO_2 evolution is not from enhanced normal respiratory activity but from the decarboxylation of malic acid spilled from damaged cells at the site of the bruise (Klein, 1983). However, freshly harvested apples did not emit increased amounts of CO_2 with progressive bruising.

B. Quiescent Infections

Preharvest infection of fruit and vegetables may occur through direct penetration of the skin, natural openings of the products, or openings resulting from mechanical damage. Some infections can be arrested and remain quiescent until after harvest when the resistance of the host decreases and conditions become favorable for growth (Wills *et al.*, 1989). Many fungi that cause considerable loss of produce are unable to penetrate intact skin, but readily invade through any break in the skin. The damage is often microscopic, but is sufficient for pathogens present in the field or packinghouse to gain access to the produce.

Blue mold, caused by *Penicillium expansum* L. ex Thom, causes the principal postharvest decay of apples, accounting for 80 to 95% of the apple decays in United States markets (Cappellini *et al.*, 1987). The relationship between bruising and decay incidence of apples has been studied; however, little work has been done to show the relationship between the amount and severity of bruising that might occur during packinghouse operations and subsequent blue mold decay development (Burton *et al.*, 1987). Several studies have indicated that bruising leads to blue mold infection and that lenticel infection of apples is associated with bruising (Baker and Heald, 1936; English *et al.*, 1946). Burton *et al.* (1987) found that very little decay developed at inoculated bruised sites immediately after harvest unless damage was severe; however, after 4 months in storage, 'Golden Delicious' and 'McIntosh' apples showed decay at inoculated bruise sites caused by drops of only 5 cm. Impacts of this magnitude are common in commercial packinglines.

Conidia of *Botrytis cinerea* are abundant in the atmosphere of strawberry and raspberry plantations during flowering in the spring. These spores may germinate in a water drop on the petals of the flower, then move from the diseased flower part into the receptacle (strawberry) and form a latent infection. However, the site of disease will not develop until the berries are harvested several months later (Jarvis, 1962). Similarly, spores of *Collectotrichum gloeosporioides* germinate in water on the surface of avocado, banana, citrus, mango, and papaya fruits during the development of the fruit on the plant (Baker, 1938; Stanghellini and Aragaki, 1966; Brown, 1975). The germ tube of the fungus then forms a structure known as an appressorium, which will develop into an infection hypha when the fruits begin to ripen after harvest.

C. Preharvest Stress Disorders

Physiological disorders refer to the breakdown of tissue that is not caused by either invasion by pathogens or mechanical damage (Wills *et al.*, 1989). These disorders may be caused by adverse environments or a nutritional deficiency during growth and development.

The degree of browning in avocado fruit has been correlated with polyphenol oxidase activity (Kahn, 1975). Bower and Van Lelyveld (1985) demonstrated that soluble polyphenol oxidase activity increased in avocado fruit from water-stressed (drought) trees when they were ripened after being placed for 48 hr in restricted ventilation. These investigators suggested that stress history predisposes cell membranes to break down under restricted conditions, resulting in a higher browning potential.

Fruits and vegetables often show various browning symptoms that have been attributed to deficiencies in mineral constituents (Wills *et al.*, 1989). Plants require a balanced mineral intake for proper development, so a deficiency in any essential mineral will lead to maldevelopment of the plant as a whole (Chapter 4). Expression of these disorders can be prevented by correct nutritional balance during growth or, to some extent, by postharvest application (Wills *et al.*, 1989). Calcium has been associated with more deficiency disorders than other minerals. Some examples of these disorders include water core and internal breakdown in apple, blackheart in celery, soft nose in mango, cork spot in pear, and blossom-end rot and blackseed in tomato. Calcium will suppress respiration and several other metabolic sequences in plant tissues. Calcium also is associated with pectic substances in the middle lamella and may prevent disorders merely by strengthening structural components of the cell without alleviating the original cause of cell collapse. The strengthening of cell components may prevent or delay the loss of cell compartmentalization and the enzyme reactions that cause those disorder symptoms (Wills *et al.*, 1989).

D. Postharvest Stress Disorders

Chilling injury describes the physiological damage that occurs in many plants and plant products as a result of their exposure to low but nonfreezing temperatures (Jackman *et al.*, 1988). The chilling injury symptoms commonly seen have been summarized by Wang (1989). The symptom most commonly seen is surface lesions, such as pitting on cucumbers and squash, scald on papaya and citrus, and large sunken areas on peppers. Another symptom that is common also is internal discoloration on avocado, tomato seeds, and pineapple fleshy tissues.

Initial responses to lowered temperature usually are considered to be physical in nature and include such phenomena as membrane alterations and protein or enzyme dysfunction. Physiological changes such as cessation of protoplasmic streaming, alterations in respiration rates, and changes in ethylene biosynthesis then follow. If these changes continue over a period of time, structural integrity and overall quality of plant products can be lost (Morris, 1982). This disorder has been attributed to the lipophilic nature of the proteins (Yamaki and Uritani, 1973) and the phase transition of the lipids (Lyons and Raison, 1970; Raison and Lyons, 1971; Chapter 12). Increases in l-aminocyclopropane-l-carboxylic acid (ACC) levels or ethylene production have been shown to be a response of chilling-sensitive tissues to low-temperature stress (Wang and Adams, 1980,1982). The accumulation of ACC and ethylene that has been documented is correlated closely with the chilling susceptibility of different fruits and vegetables (Wang, 1989). Because the amount of ACC and ethylene in tissues from chilling-resistant species or in nonchilled tissues from chilling-sensitive species is insignificant, the endogenous ACC levels and ethylene production stimulated by chilling in the sensitive tissues can be used as an indicator of chilling injury.

Chilling injury is not the only postharvest disorder associated with stress. When some fruits such as tomato are stored above 30°C, they suffer from heat injury, inducing the disturbance of the normal ripeness. Mesocarp discoloration is a serious problem in low-temperature container-shipped avocados. Van Lelyveld and Bower (1984) reported that this symptom was a response to the restricted ventilation of the shipping containers, which caused a decreased O_2 content or an elevated CO_2 content. A certain proportion of mature-green 'd'Anjou' pears stored in controlled atmosphere storage is vulnerable to a disorder called "skin speckling" or "black speck" (Chen and Varga, 1989). This disorder is associated closely with low O_2 concentration over pathogens or chemicals in controlled atmosphere storage. Lee *et al.* (1990) found that the affected tissue contained higher percentages of dry matter and soluble proteins than the normal tissue. Their study indicated that a certain proportion of 'd'Anjou' pears might have been exposed to unfavorable preharvest environmental stresses; therefore, the fruit could no longer tolerate the subsequent low oxygen and chilling stresses during prolonged controlled atmosphere storage. A more detailed discussion of postharvest response to stress is provided in Chapter 12.

E. Superficial Scratches

Superficial scratches are a major problem in mechanically harvested citrus fruits (Burkner and Chesson, 1972). Many of the superficial injuries are barely visible immediately after harvest, but scars later develop, apparently as a result of increased moisture loss from and around the injured tissue (Golomb *et al.*, 1984). Therefore, inhibition of transpiration and improvement of healing conditions of the scarred tissue might preserve the marketability of slightly injured fruit. Golomb *et al.* (1984) found that sealing individual grapefruit in 0.015-mm thick high-density polyethylene (HDPE) sheets did not keep the injury scars from becoming visible but did improve active healing and inhibit further softening and spreading of scars, thus partly preserving fruit appearance. They also found that the highest PAL enzyme activity, a key in the process of lignification and wound healing, occurred in scratched HDPE-wrapped fruit stored at 30°C. Decay of scratched grapefruit also was reduced markedly by wrapping in HDPE and was attributed both to insulation against secondary contact infections and to provision of a humid atmosphere that, at warm temperature, enhances lignification and the healing process of superficial flavedo scars.

IV. Future Directions

The difficulty of studying latent damage is inherent in its name. Because the damage is not apparent when incurred, it is difficult to assess the amount of damage at an early stage and to identify the step(s) causing the damage accurately.

Developing a nondestructive method to detect damage at its early stage is important (see Chapters 11, 12). Damage detection based on physiological changes in fruits or vegetables is one of the possible approaches. A number of physiological measurements, for example, rate of electrolyte leakage and respiration, lipid fluidity, degree of unsaturation of membrane lipids, and rate of photosynthesis, have been suggested as indices of the severity of chilling injury. However, most of these methods of detection are either insensitive to early effects of chilling, time consuming, destructive, or inconvenient (Wang, 1989).

Several methods have been suggested that can assess chilling injury without the previously mentioned difficulties. One method is the analysis of chlorophyll fluorescence, since direct or indirect changes in photosynthetic metabolism are likely to affect this fluorescence (Smillie and Hetherington, 1983; Wilson and Greaves, 1990). Another method that has been used to assess chilling injury is delayed light emission from chlorophyll. Abbott and Massie (1985) reported that the maximum levels of delayed light emission of the chilled cucumber and bell pepper fruits were lower than those of nonchilled fruits, and suggested that the reduction in delayed light emission by chilling might be due to impairment of chloroplast activity rather than simple loss of chlorophyll. These methods provide a nondestructive and rapid indication of chilling injury for photosynthetic tissue, but not for nongreen items.

Progress also has been made in detecting nonvisible bruise damage. On impact, fruits show an immediate increase in the evolution of CO_2 (MacLeod et al., 1976). Pollack and Hills (1956) reported that the increase in CO_2 evolution was not caused by enhanced normal respiratory activity, but was due to the decarboxylation of malic acid spilled from damaged cells at the site of the bruise. An infrared CO_2 analyzer system was reported as a nondestructive and rapid means of monitoring the CO_2 evolution of blueberries, sweet and tart cherries, and apples to evaluate impact damage (Burton and Schulte-Pason, 1985). Upchurch et al. (1988) reported that reflectance in the 700–830-nm wavelength region showed promise for discrimination of bruised and nonbruised apples. Two- and three-wavelength models they proposed were successful in distinguishing bruised and nonbruised areas on unpeeled 'Red Delicious' apples with only 2.5–5% misclassification. Maw et al. (1989) found that X-ray computed tomography can detect ring separation on bruised sweet onions. NMR imaging also has been suggested to provide high resolution images of intact fruits and vegetables for detecting internal damages or defects such as bruises, dry regions, worm damage, and presence of voids (Chen and Kauten, 1988). However, each method still has limitations. Further improvement on the current detection methods still is needed.

If quality of fresh fruits and vegetables is to be improved using a systems approach, a better understanding of latent damage is needed. A systems approach seeks to understand the handling system, identify causes of problems, and find solutions. Latent damage obscures the cause and, thus, the solution of problems within the handling system. The most important area of research in latent damage is nondestructive detection.

Bibliography

Abbott, J. A., and Massie, D. R. (1985). Delayed light emission for early detection of chilling in cucumber and bell pepper fruit. *J. Amer. Soc. Hort. Sci.* **110**, 42–47.

Baker, K. F., and Heald, F. D. (1936). The effect of certain cultural and handling practices on the resistance of apples to *Penicillium expansum*. *Phytopathology* **26**, 932–948.

Baker, R. E. D. (1938). Studies in the pathogenicity of tropical fungi. 2. The occurrence of latent infections in developing fruits. *Ann. Botany (N.S.)* **2**, 919–931.

Bartram, R. J., Fountain, J., Olsen, K., and O'Rourke, D. (1983). Washington State apple condition at retail 1982–83. (Eating Quality). *Proc. Wash. State Hort. Assoc.* **79,** 36–46.

Belknap, W. R., Rickey, T. M., and Rockhold, D. R. (1990). Blackspot bruise dependent changes in enzyme activity and gene expression in Lemhi Russet potato. *Amer. Potato J.* **67,** 253–265.

Ben-Arie, R., Kislev, N., and Frenkel, C. (1979). Ultrastructural changes in the cell walls of ripening apple and pear fruit. *Plant Physiol.* **64,** 197–202.

Bower, J. P., and Van Lelyveld, L. J. (1985). The effect of stress history and container ventilation on avocado fruit polyphenol oxidase activity. *J. Hort. Sci.* **60,** 545–547.

Brown, G. E. (1975). Factors affecting postharvest development of *Colletotrichum gloeosporioides* in citrus fruits. *Phytopathology* **65,** 404–409.

Buchanan, J. R., Sommer, N. F., Fortlage, R. J., Maxie, E. C., Mitchell, F. G., and Hsieh, D. P. H. (1974). Patulin from *Penicillium expansum* in stone fruits and pears. *J. Amer. Soc. Hort. Sci.* **99,** 262–265.

Burkner, P. F., and Chesson, J. H. (1972). A shake and catch harvest system for desert grapefruit. *In* "International Conference on Tropical and Subtropical Agriculture, Honolulu, Hawaii, April 1972," pp. 92–96. American Society of Agricultural Engineers, St. Joseph, Michigan.

Burton, C. L., and Schulte-Pason, N. L. (1985). "Assaying Fruit Impact Damage Using an Infrared CO_2 Gas Analyzer." ASAE Tech. Paper No. 85–1564. American Society of Agricultural Engineers, St. Joseph, Michigan.

Burton, C. L., Pason, N. L. S., Brown, G. K., and Timm, E. J. (1987). "The Effect of Impact Bruising on Apples and Subsequent Decay Development." ASAE Tech. Paper No. 87–6516. American Society of Agricultural Engineers, St. Joseph, Michigan.

Campbell, D. T., Prussia, S. E., and Shewfelt, R. L. (1986). Evaluating postharvest injury of fresh market tomatoes. *J. Food Dist. Res.* **17(2),** 16–25.

Cappellini, R. A., Ceponis, M. J., and Lightner, G. W. (1987). Disorders in apple and pear shipments to the New York market, 1972–1984. *Plant Dis.* **71,** 852–856.

Chen, P., and Kauten, R. (1988). "Potential use of NMR for Internal Quality Evaluation of Fruits and Vegetables." ASAE Tech. Paper No. 88–6572. American Society of Agricultural Engineers, St. Joseph, Michigan.

Chen, P. M., and Varga, D. M. (1989),. "lack speak," a superficial disorder of 'd'Anjou' pears after prolonged CA storage. *In* "Controlled Atmospheres for Storage and Transport of Perishable Agricultural Commodities" (J. K. Fellman, ed.), pp. 145–156.

Dewey, D. H., and Schueneman, T. J. (1971). "Quality and Packout of Storage Apples: Their Effects on Costs and Return." Michigan Agricultural Experimental Station Research Report No. 147. Michigan State University, East Lansing.

Eaks, I. (1961). Techniques to evaluate injury to citrus fruit from handling practice. *Proc. Amer. Soc. Hort. Sci.* **78,** 190–196.

Eckert, J. W. (1978). Pathological diseases of fresh fruits and vegetables. *In* "Postharvest Biology and Biotechnology" (H. O. Hultin and M. Milner, eds.). Food & Nutrition Press, Westport, Connecticut.

English, H., Ryall, A. L., and Smith, E. (1946). "Blue Mold Decay of Delicious Apples in Relation to Handling Practices." U.S. Department of Agriculture Circular No. 151. U.S. Govt. Printing Office, Washington, D.C.

Fridley, R. B., Bradley, R. A., Rumsey, J. W., and Adrian, P. A. (1968). Some aspects of elastic behavior of selected fruits. *Trans. ASAE* **11,** 46–49.

Garcia, C., Ruiz, M., and Chen, P. (1988). "Impact Parameters Related to Bruising in Selected Fruits." ASAE Tech. Paper No. 88–6027. American Society of Agricultural Engineers, St. Joseph, Michigan.

Golomb, A., Ben-Yehoshua, S., and Sarig, Y. (1984). High-density polyethylene wrap improves wound healing and lengthens shelf-life of mechanically harvested grapefruit. *J. Amer. Soc. Hort. Sci.* **109,** 155–159.

Hale, M. G., Orcutt, D. M. (1987). "The Physiology of Plants Under Stress." John Wiley & Sons, New York.

Hatton, T. T., and Reeder, W. F. (1963). Effect of field and packinghouse handling on bruising of Florida tomatoes. *Proc. Florida State Hort. Soc.* **76,** 301–304.

Horsfield, B. C., Fridley, R. B., and Claypool, L. L. (1972). Application of theory of elasticity to the design of fruit harvesting and handling equipment for minimizing bruising. *Trans. AEAE* **15,** 746–753.

Holt, J. E., and Schoorl, D. (1977). Bruising and energy dissipation in apples. *J. Texture Stud.* **7**, 421–432.

Holt, J. E., Schoorl, D., and Lucas, C. (1981). Prediction of bruising in impacted multilayered apple packs. *Trans. AEAE* **24**, 242–247.

Jackman, R. L., Yada, R. Y., Marangoni, A., Parkin, K. L., and Stanley, D. W. (1988). Chilling injury. A review of quality aspects. *J. Food Qual.* **11**, 253–278.

Jarvis, W. R. (1962). The infection of strawberry and raspberry fruits by *Botrytis cinerea*. Fr. *Ann. Appl. Biol.* **50**, 569–575.

Kader, A. A. (1985). Ethylene-induced senescence and physiological disorders in harvested horticultural crops. *HortScience* **20**, 54–57.

Kahn, V. (1975). Polyphenol oxidase activity and browning of three avocado varieties. *J. Sci. Food Agric.* **26**, 1319–1324.

Klein, J. D. (1983). Physiological causes for changes in carbon dioxide and ethylene production by bruised apple fruit tissues. Ph. D. Thesis. Michigan State University, East Lansing.

Klein, J. D. (1987). Relationship of harvest date, storage conditions, and fruit characteristics to bruise susceptibility of apple. *J. Amer. Soc. Hort. Sci.* **112(1)**, 113–118.

Kunze O. R., Aldred, W. H., and Reeder, B. D. (1975). Bruising characteristics of peaches related to mechanical harvesting. *Trans. ASAE* **18**, 939–945.

Lee, S. P., Chen, P. M., Chen, T. H. H., Varge, D. M., and Mielke, E. A. (1990). Differences of biological components between the skin tissues of normal and black-speckled 'd'Anjou' pears after prolonged low-oxygen storage. *J. Amer. Soc. Hort. Sci.* **115**, 784–788.

Lougheed, E. C., and Franklin, E. W. (1974). Ethylene production increased by bruising of apples. *HortScience* **9**, 192–193.

Lyons, I. M., and Raison, J. K. (1970). Oxidative activity of mitochondria isolated from plant tissues sensitive and resistant to chilling injury. *Plant Physiol.* **45**, 385–389.

MacLeod, R. F., Kader, A. A., and Morris, L. L. (1976). Stimulation of ethylene and CO_2 production of mature-green tomatoes by impact bruising. *HortScience* **11**, 604–606.

Maw, B. W., Hung, Y.-C., Tollner, E. W., and Smittle, D. A. (1989). "Some Physical Properties of Sweet Onions." ASAE Tech. Paper No. 89–6007. American Society of Agricultural Engineers, St. Joseph, Michigan.

Miles, J. A., and Rehkugler, G. E. (1973). A failure criterion for apple flesh. *Trans. ASAE* **16**, 1148–1153.

Mohsenin, N. N. (1986). "Physical Properties of Plant and Animal Materials." Gordon and Breach, New York.

Mondy, N. I., Leja, M., and Gosselin, B. (1987). Changes in total phenolic, total glycoalkaloid, and ascorbic acid content of potatoes as a result of bruising. *J. Food Sci.* **52**, 631–633.

Morris, L. L. (1982). Chilling injury of horticultural crops: An overview. *HortScience* **17**, 161–162.

Parker, M. L., Wardowski, W. F., and Dewey, D. H. (1984). A damage test for oranges in a commercial packinghouse line. *Proc. Florida St. Hort. Soc.* **97**, 136–137.

Paull, R. E. (1990). Chilling injury of crops of tropical and subtropical origin. *In* "Chilling Injury of Horticultural Crops" (C. Y. Wang, ed.). CRC Press, Boca Raton, Florida.

Peleg, K. (1985). "Produce Handling, Packaging, and Distribution," AVI/Van Nostrand Reinhold, New York.

Pollack, R. L., and Hills, C. H. (1956). Respiratory activity of normal and bruised red tart cherries (*Purnus cerasus*). *Fed. Proc.* **15**, 328.

Prussia, S. E. (1985). Ergonomics of manual harvesting. *Appl. Ergonomics* **16**, 209–215.

Prussia, S. E., Hung, Y.-C., Shewfelt, R. L., and Jordan, J. L. (1987). "Latent Damage in Apples and Peaches." ASAE Tech. Paper No. 87–6520. American Society of Agricultural Engineers, St. Joseph, Michigan.

Raison, J. K., and Lyons, J. M. (1971). Hibernation: Alteration of mitochondrial membranes as a requisite for metabolism at low temperature. *Proc. Natl. Acad. Sci. U.S.A.* **68**, 2092–2094.

Robitaille, H. A., and Janick, J. (1973). Ethylene production and injury in apples. *J. Amer. Soc. Hort. Sci.* **98**, 411–412.

Rolle, R. S., and Chism, G. W., III (1987). Physiological consequences of minimally processed fruits and vegetables. *J. Food Qual.* **10**, 157–177.

Saltveit, M. E., Jr., and Morris, L. L. (1990). Overview on chilling injury of horticultural crops. *In* "Chilling Injury of Horticultural Crops" (C. Y. Wang, ed.). CRC Press, Boca Raton, Florida.

Sargent, S. A., Brecht, J. K., Zoellner, J. J., Chau, K. V., and Risse, L. A. (1989). "Reducing Mechanical Damage to Tomatoes during Handling and shipment." ASAE Tech. Paper No. 89–6616. American Society of Agricultural Engineers, St. Joseph, Michigan.

Shewfelt, R. L. (1986). Postharvest treatment for extending the shelf life of fruits and vegetables. *Food Technol.* **40(5),** 70–89.

Smillie, R. M., and Hetherington, S. E. (1983). Stress tolerance and stress-induced injury in crop plants measured by chlorophyll florescence in vivo. *Plant Physiol.* **72,** 1043–1050.

Stanghellini, M. E., and Aragaki, M. (1966). Relation of periderm formation and callous desposition to anthracnose resistance in papaya fruit. *Phytopathology* **56,** 444–449.

Stanton, F. (1961). "The Effect on Bruising of Different Methods of Harvesting McIntosh Apples." Agricultural Economics Research Mimeo No. 68. Cornell University Library, Ithaca, New York.

Thai, C. N. (1987). "Partial Latent Damage Detection Problems." ASAE Tech. Paper No. 87–6519. American Society of Agricultural Engineers, St. Joseph, Michigan.

Thai, C. N., Prussia , S. E., Shewfelt, R. L., and Davis, J. W. (1986). "Latent Damage Simulation and Detection for Horticultural Products." ASAE Tech. Paper No. 86–6552. American Society of Agricultural Engineers, St. Joseph, Michigan.

Timm, E. J., and Brown, G. K. (1990). "Impact Recorded on Avocado, Papaya, and Pineapple Packing Lines." ASAE Tech. Paper No. 90–6007. American Society of Agricultural Engineers, St. Joseph, Michigan.

Uritani, I. (1978). Temperature stress in edible plant tissues after harvest. *In* "Postharvest Biology and Biotechnology" (H. O. Hultin and M. Milner, eds.), pp. 136–160. Food & Nutrition Press, Westport, Connecticut.

Upchurch, B. L., Affeldt, H. A., Norris, K. A., and Throop, J. A. (1988). "Spectrophotometric Study of Bruises on Whole Red Delicious Apples." ASAE Tech. Paper No. 88–6566. American Society of Agricultural Engineers, St. Joseph, Michigan.

Vamos-Vigyazo, L. (1981). Polyphenol oxidase and peroxidase in fruits and vegetables. *CRC Crit. Rev. Food Sci. Nutr.* **15,** 49–127.

Van Lelyveld, L. J., and Bower, J. P. (1984). Enzyme reactions leading to avocado mesocarp discoloration. *J. Hort. Sci.* **59,** 257–263.

Variyam, J. N., Jordan, J. L., Beverly, R., Shewfelt, R. L., and Prussia, S. E. (1988). An application of models for survival data to postharvest systems evaluation. *Proc. Florida St. Hort. Soc.* **101,** 200.

Vergano, P. J., Testin, R. F., and Newall, W. C., Jr. (1991). Distinguishing among bruises in peaches caused by impact, vibration, and compression. *J. Food Qual.* **14,** 285–298.

Verhoeff, K. (1974). Latent infections by fungi. *Ann. Rev. Phytopath.* **12,** 99–110.

Wang, C. Y. (1989). Chilling injury of fruits and vegetables. *Food Rev. Int.* **5,** 209–236.

Wang, C. Y., and Adams, D. O. (1980). Ethylene production by chilled cucumbers (*Cucumis sativis* L.). *Plant Physiol.* **66,** 841–843.

Wang, C. Y., and Adams, D. O. (1982). Chilling-induced ethylene production in cucumbers (*Cucumis sativis* L.). *Plant Physiol.* **69,** 424–427.

Wild, B. L., McGlasson, W. B., and Lee, T. H. (1976). Effect of reduced ethylene levels in storage atmosphere on lemon keeping quality. *HortScience* **11,** 114–115.

Wills, R. B. H., McGlasson, W. B., Graham, D., Lee, T. H., and Hall, E. G. (1989). "Postharvest: A Introduction to the Physiology and Handling of Fruit and Vegetables," 3d Rev. Ed. Blackwell Scientific, Boston.

Wilson, J. M., and Greaves, J. A. (1990). Assessment of chilling sensitivity by chlorophyll fluorescence analysis. *In* "Chilling Injury of Horticultural Crops" (C. Y. Wang, ed.). CRC Press, Boca Raton, Florida.

Wright, W. R., and Billeter, B. A. (1975). "Market Losses of Selected Fruits and Vegetables at Wholesale, Retail, and Consumer Levels in the Chicago Area." U.S. Department of Agriculture Marketing Research Report No. 1017. U.S. Govt. Printing Office, Washington, D.C.

Yamaki, S., and Uritani, I. (1973). Mechanism of chilling injury in sweet potato. 10. Changes in lipid-protein interaction in mitochondria from cold-stored tissue. *Plant Physiol.* **51,** 883–888.

Yang, S. F., and Pratt, H. K. (1978). The physiology of ethylene in wounded plant tissue. *In* "Biochemistry of Wounded Plant Tissues" (G. Kahl, ed.), pp. 595–622. de Gruyter, Berlin.

CHAPTER 11

NONDESTRUCTIVE EVALUATION: DETECTION OF EXTERNAL AND INTERNAL ATTRIBUTES FREQUENTLY ASSOCIATED WITH QUALITY OR DAMAGE

E. W. Tollner, J. K. Brecht, and B. L. Upchurch

Advances in microelectronics and other integrated circuit technologies have increased the possibility of early detection of quality attributes using nondestructive testing approaches. Similarly, early detection of quality change precursors that will cause a systems problem is becoming increasingly possible with miniaturization and microelectronics. A definition of nondestructive evaluation has been stated as "a gaining of meaningful information which can be used in making judgments, both positive and negative, about the degree of excellence of a food without altering the physical and chemical properties of that food" (Dull *et al.*, 1980). That definition can apply both to maturity and to quality evaluation in the postharvest system. Strictly speaking, conventional approaches to measuring size and shape generally are considered nondestructive. However, for purposes of this review, nondestructive evaluation will refer primarily to energy-based approaches wherein energy is imparted to a product, modulated by the product, and then detected.

Maturity and quality are controversial issues among horticulturists and engineers. Dull (G. G. Dull, personal communication, 1991) defined fruit and vegetable maturity as a stage of physiological development that is characterized by objective physical and chemical parameters. Presently, quality evaluation carries an inherent human judgment of these quality parameters. To develop and apply nondestructive techniques for quality evaluation, the physiological parameters that normally are used to define a specific quality attribute for the product must be known. The challenge is to be able to measure those parameters nondestructively, once they are identified.

Classical energy-based nondestructive approaches are concerned with how the product affects energy propagation. In this approach, the physical or chemical characteristics of interest are measured for each individual of a population and are used to sort the product.

In the precursor detection approach, the sensor measures some feature of the surrounding immediate system environment that may indicate or affect product quality. The feature may be an exogenous factor (e.g., temperature, humidity, pressure) or a compound produced by the product, for example, ethylene. Pseudoproduct or biosensor methods embed a variety of miniaturized sensors into batches of product. These devices may transmit the sensed data remotely to a base station or may store the information on board for later retrieval.

In the context of the whole postharvest system, quality degradation can be described in terms of a schematic plot of some generic quality indicator against time in the postharvest system (Chapter 3). Suppose that, at some time in the product flow, damage inception occurs (e.g., impact or temperature shocks), defining the earliest time boundary of interest. Damage inception initiates an alternative trajectory of product quality, which results in an increasing disparity with the trajectory of no damage. Ultimately, difference in quality between the two paths can result in rejection of the damaged product by consumers, provided the difference has become obvious. The time of rejection by consumers defines the latest time boundary of interest. These features are shown schematically in Fig. 1. Postharvest systems strive to detect damage before product with recognizable damage reaches consumers. Because of the phenomenon of "latent damage," how or why particular damage may have arisen is often uncertain. Early detection of damage with nondestructive testing or early detection of environmental shock conditions using sensor platforms may establish better links between effect and cause of damage.

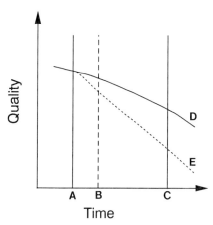

Figure 1 Hypothesized relationships associated with latent damage and damage-free quality trajectories. (A) Time of inception, not known precisely. (B) Time of nondestructive detection, ideally at time of damage inception. (C) Time of detection and rejection by consumer. (D) Damage-free trajectory. (E) Trajectory resulting from latent damage.

Nondestructive testing may allow effective damage detection to occur closer to the time of damage inception, as shown in Fig. 1. Placing sensor platforms in the product flow may be another strategy toward this end.

Successful application of nondestructive or precursor detection test methods to relate a sensed output to a sensory attribute requires that the sensed output be unique and monotonic, and that sensors be adequately sensitive to the attribute and stable. Powers *et al.* (1953) stated these requirements in part. Uniqueness is defined as a high degree of cause–effect correspondence between instrument output and desired attribute. Monotonicity is defined as a clearly identifiable increasing or decreasing trend over the range of interest. Sensitivity is concerned with an adequately high signal-to-noise ratio. Sensors also must be stable over time to maintain calibration integrity. A sensor-based approach to detection of precursors of quaity requires that the precursors be identified by relationships that meet the same uniqueness, monotonicity, sensitivity, and stability criteria. These criteria form the background for our review. The review has been organized by technology areas. For each technology, a discussion of the physics is followed by a discussion of the applications.

I. Optical Evaluation: Surface Appearance and Internal Attributes

A. Physics

1. Ultraviolet, visible, and near-infrared radiation

Optical methods are based on the transmittance, reflectance, or absorbance of polychromatic or monochromatic radiation in the ultraviolet (UV), visible, and near-infrared (NIR) range of the electromagnetic spectrum. Birth (1976) and Birth and Hecht (1987) summarized the phenomenon, describing the interaction of UV, visible, NIR, and infrared (IR) radiation with food materials using (1) the Fresnel Equation relating to reflection; (2) Snell's Law of refraction; (3) the Beer-Lambert Law describing absorption; and (4) the law of conservation of energy. Transmittance, reflectance, and absorbance of light are generally wavelength dependent.

Basic concepts for using optical radiation to evaluate composition of fresh fruits and vegetables are shown in Fig. 2. In each case, incident radiation at selected wavelengths (or polychromatic or bichromatic wavelengths) is directed toward the product. Incident radiation is reflected by the product based on Fresnel's Equation or transmitted into the product based on Snell's Law.

Optical configuration for measuring direct transmittance or absorbance is shown in Fig. 2A. The inherently simple idea has been extremely difficult to implement because of difficulties in obtaining adequate illumination levels and in detecting the resulting signal. However, lasers and new stronger NIR emitters and detectors capable of measuring at the picowatt level are facilitating many advances (G.G. Dull, personal communication, 1991). Particularly with transmittance and absorbance processes, and to some extent with reflection processes, near infrared wavelengths are useful for detecting soluble solids (Dull *et al.*, 1989a), soluble sugars, and carbohydrates (Birth *et al.*, 1985).

The body transmittance approach shown in Fig. 2B can be used to determine an additional optical property, that of light scattering. Light scattering is caused by reflection, refraction, and diffraction processes that occur at numerous microscopic interfaces within the product (Birth, 1976). Birth (1978,1982) presents a mathematical description of the physics of light scattering in foods. By moving the detectors in Fig. 2B around the fruit or by using more detectors, the exiting energy over a substantial percentage of total product surface area can be mapped. Some investigators hypothesize that the distributors of exiting energy, brought about by light scattering within the product, may be related to texture and firmness as well as composition (G.G. Dull, personal communication, 1991). Optical methods are becoming successful because unique associations with recognized quality attributes,

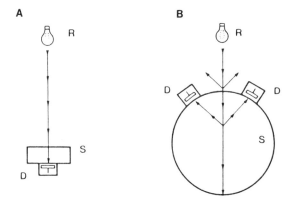

Figure 2 (A) Schematic of direct transmittance system principle. (B) Schematic of body transmittance principle. R, radiation source; S, sample, D, detector. (Courtesy of G. G. Dull.)

for example, sugars in melons, have been found. The relationships are monotonic in many cases. New equipment appears to have the required sensitivity.

Color results from combinations of light in the 400–700-nm range. Color can be described using a spectrophotometer, giving magnitude of intensity at specific wavelengths. Depending on the application, many if not all wavelengths may be used or simplified descriptions involving only selected wavelengths may be used. More concise descriptions of color have been reported using a variety of color coordinate systems. Some popular color coordinate systems in use are the RGB (red, green, blue) systems in color computer monitors, the Hunter Lab system, the CIE (Commission International del'Eclairage) system, and LCH coordinates. The RGB system has not been used extensively to quantify colors numerically. On the other hand, the Lab and LCH systems have seen extensive use for color quantification. Francis and Clydesdale (1975), Billmeyer and Saltzman (1981), and Francis (1980) give excellent discussions of color technology and color expression. Hunt (1987) gives conversion relationships between the Hunter and CIE color coordinate systems. Thai and Shewfelt (1990) concluded that the CIE system better relates to human color perception; therefore, this system is recommended for describing color attributes of fresh market product (see Chapters 5, 8). The CIE system uses the "L" of the Hunter system to signify lightness (0 is black and 100 is white). "H" is the hue, commonly referred to as color. Hue angle can vary from 0° (red) through 90° (yellow), 180° (green), 270° (blue), and so forth. The "C" term is chroma and refers to the vividness of colors (0 is washed out whereas 60 is strong and distinct).

2. Delayed light emission

Delayed light emission (DLE) is a phenomenon whereby an object will reradiate electromagnetic energy after having been subjected to electromagnetic energy in the optical range. The intensity of reradiated energy emitted by the object is usually in the nanowatt range and persists 2–5 sec after excitation is removed. Wavelength and intensity of excitation, dark exposure prior to excitation, sample thickness, area of excitation, ambient temperature, and chlorophyll content affect reradiated energy (Gunasekaran, 1990). DLE phenomena have been studied primarily in the visible light range. Many of these attributes may vary on "good" products. Thus, uniqueness and sensitivity are problematic.

3. Machine vision

Machine vision has been defined by Searcy (1987) as "the use of automatically generated, visual-based knowledge of an object or objects to accomplish assigned tasks in an automated system." This combination of electronic imaging and computer technologies enables machines to perform visual inspection tasks automatically. It may include such components as image capture, image processing, and pattern recognition. Machine vision is not simply image processing, is not interactive, and is always two dimensional or more (Searcy, 1987). Based on a definition of computer vision given by Ballard and Brown (1982), computer vision and machine vision are synonymous.

Perhaps the oldest nondestructive technique is the measurement of size and weight. Is an identifiable fruit dimension or weight (mass) in a high quality or optimally mature fruit significantly different from the same dimension in a lower grade of fruit? The same question applies to shape measurement. Shape is defined as spatial form of a particular item. Shape description is a matter of degree, ranging from one dimension (such as size) to the large number of dimensions possible with machine vision techniques, such as Hough transforms, chain codes, and other machine vision representations of structures (Ballard and Brown, 1982).

B. Applications

1. Ultraviolet, visible, and near-infrared radiation

The composition of fruits and vegetables changes constantly during development (Gortner et al., 1967; Coombe, 1976). In addition to the changes in structural carbohydrates and water content that are related to texture and the changes in pigment concentrations that are related to color, other compounds that increase or decrease in conjunction with maturation of fruits and vegetables include soluble sugars and starch, organic acids, ions, proteins, amino acids, phenolics, lipids, vitamins, and a vast array of volatile aromatic compounds. Although several of these compounds are important maturity indices for many crops, measurement generally has been by a destructive process. However, nondestructive techniques for measurement of these indices have been reported. Soluble sugars and carbohydrates have been defined by Gortner et al. (1967) as maturity indicators, and by numerous investigators (see Gunasekaran et al., 1985) as quality indicators.

Since the chemical compounds just mentioned absorb energy of specific wavelengths, transmittance spectrophotometry identical to that used for internal color measurement has been used, primarily for measuring sugars. Watada and Abbott (1985) reported correlations between optical density measurements at various wavelengths and sensory attributes such as sweetness, acidity, and juiciness, as well as destructive measurements of titratable acidity and soluble solids content in apples. The five best wavelengths correlating with sensory sweetness in 'York Imperial' apples were 690 nm, 710 nm, 930 nm, 940 nm, and 950 nm. The optical density differences at 690 nm and 710 nm were due to chlorophyll differences, whereas the near infrared wavelengths (930 nm, 940 nm, and 950 nm) were related to carbohydrate (sugar) levels. Kawano and Iwamoto (1990) report in a review that NIR has been correlated to the taste of rice and coffee; sugar in tomatoes, peaches, sugars, oranges, and apples; and acids in tomatoes.

Transmittance, absorbance, and reflectance are all exploited in the process shown in Fig. 2B. Birth et al. (1984,1985) and Dull et al. (1989a,b) have implemented this approach, called body transmittance, to characterize nondestructively papaya maturity, onion and potato dry matter, and cantaloupe soluble solids. Fiber optic technology, which has sample precedent in fruit and vegetable handling (McClure et al., 1976), probably will be seen in future applications of the body transmittance concept in handling environments once suitable light sources and detectors are available. Typical energy levels, once light traverses fruit or melons, are in the picowatt range. High correlations exist between NIR outputs and dry matter levels in onions (Birth et al., 1985) and potatoes (Dull et al., 1989a) and soluble solids content of melons (Dull et al., 1989b). For neons, on-line equipment is expected in the near future.

Visible light and NIR transmittance through apples has been used to detect watercore. Birth and Olsen (1964) used a single detector to measure the amount of light energy existing the stem end after illuminating the calyx of the apple. Throop *et al.* (1989) extended this concept by using a machine vision system to collect spatial information as well as intensity. Both techniques have limitations and are subject to effects of temperature and time after harvest.

Surface optical reflectance has been used to detect surface defects on a variety of commodities. To evaluate optical reflectance as a method for detecting surface defects on citrus fruits, spectroscopic curves from blemished and nonblemished fruit were examined to identify wavelengths that showed at least a 15% difference between the two tissue types (Gaffney, 1976a). Gaffney (1976c) showed that a wavelength band around 600 nm could be used to detect various surface defects on citrus fruits. Upchurch *et al.* (1989) reported the need for 12 wavelengths and four classification models to distinguish russet and bruising on 'Empire' apples. This work agreed with that of Rehkugler and Throop (1989), who showed the similarity in optical reflectance for bruises and russet on apples. Other defects such as punctures, cuts, hail damage, bird pecks, and scab could be distinguished from bruises and nonblemished areas (Rehkugler and Throop, 1989). Spectrophotometric studies were conducted to detect 9 different defects on radishes (Gaffney, 1976b). All defects were detected at a wavelength of either 550 or 675 nm. Improved contrast between a blemish and a nonblemish on carrots for a machine vision system made 3 of 4 defects on carrots detectable using wavelengths between 535 and 722 nm (Howarth *et al.*, 1990). Variability in the amount of radiation reflected from similar defects or from nonblemish tissue is expected and makes selection of a wavelength band difficult. This variability must be considered and minimized for an automatic grading system. Burkhardt and Mrozek (1973) and Delwiche *et al.* (1990) reported that optical reflectance was generally higher for defects than for nonblemished areas on prunes; however, the closeness of the reflectance curves would suggest difficulty in separating the samples.

Some diseases or contaminants have fluorescent properties. Detection of aflatoxin in infected pecans using fluorescence was reported by Tyson and Clark (1976). These researchers demonstrated that determining the ratio between two short wavelengths (440/490 nm or 450/490 nm) was a promising technique for detecting aflatoxin on pecan halves. The authors suggested fluorescence as a useful technique for detecting toxins on other agricultural products; however, very little effort has been devoted to exploring this technique.

Much research has emphasized the transmission and reflection of visible light and NIR as a method for detecting bruises. Researchers have reported decreases in reflectivity in the NIR range for bruised apple tissue (Brown *et al.*, 1974; Reid, 1976). Bilanski *et al.* (1984), Upchurch *et al.* (1990), Affeldt and Abbott (1989), and Affeldt *et al.* (1989), extended the basic observations and identified specific wavelengths that distinguished bruised from nonbruised apple tissue. For peeled apples, Bilanski *et al.* (1984) investigated the range 350–750-nm; Upchurch *et al.* (1990) and Affeldt *et al.* (1989) studied 400–1000 nm for unpeeled fruit. Comparing the work by Affeldt *et al.* (1989) and Upchurch *et al.* (1990), the anthocyanin and/or chlorophyll content of the outer skin is a major factor influencing the reflection of NIR radiation and its ability to detect internal bruises on unpeeled fruit.

Potential on-line devices generally sense light at only one or two wavelengths. Usually these wavelengths are chosen to represent relative red and green color of the product. Gaffney (1969) showed that 'Valencia' oranges could be sorted prior to degreening using a single wavelength (600 nm) which essentially corresponds to the amount of chlorophyll in the peel. Hayes et al. (1988) demonstrated that a two-detector on-line device could be fitted with two filters so the b value of the Hunter Lab scale could be calculated. The Hunter b value is used to determine the amount of yellow color on papayas for maturity qualification for the "double-dip" quarantine method of fruit fly disinfestation (Couey and Hayes, 1986).

2. Delayed light emission

Internal color of fruits and vegetables can be measured using a high-intensity spectrophotometer to measure optical density at wavelengths corresponding to the light absorbed by different pigments (Watada and Norris, 1978; Dull et al., 1980; Deshpande et al., 1984). This technique has been used to sort many different fruits, including apples (Yeatman and Norris, 1965; Aulenbach et al., 1972), peaches (Sidwell et al., 1961), tomatoes (Worthington et al., 1976; Nattuvetty and Chen, 1980), papayas (Birth et al., 1984), blueberries (Dekazos and Birth, 1970), and grapes (Ballinger et al., 1978). However, the instruments that measure light trans-mittance or absorbance generally require that the fruit be immobilized and in contact with the detector during measurement, limiting the usefulness of the technique for on-line sorting. DLE appears to be applicable to crops in which chlorophyll and pigmentation near the surface is related to maturity and quality. Forbus and Chan (1990) used a DLE meter for nondestructive evaluation purposes. Thus, measure-ments appears to be unique, but potential for sensitivity difficulties exists because of the extremely low signal strengths.

3. Machine vision

The size of a horticultural commodity often is related closely to its maturity, since many fruits and vegetables are harvested during the growth phase of development or soon after attainment of full size. For example, smaller green tomatoes have higher proportions of immature fruit than do larger tomatoes (Kader and Morris, 1976). Newer methods of sizing use optical dimensional sizing. Single or multiple camera images of each passing item are analyzed for diameter and length using computer software (Ballard and Brown, 1982). Shape or volume classifications may be made based on the dimensions measured. Howarth and Searcy (1990) have developed a machine vision inspection system for carrots based on shape and on defect features.

In some cases, the overall shape of fruits and fruit features is related to consumer acceptance. Fullness of bananas, mangoes, peaches (Sistrunk, 1985), and tomatoes (Stevens, 1985), for example, implies attainment of maturity; more angular or flattened shapes are characteristic of lower grade specimens. Compactness of certain vegetables such as broccoli and cauliflower is a maturity factor since over-mature inflorescences begin to open and spread. Some of these features also may be detectable by optical sorting systems.

Computer vision systems for shape and defect sorting soon will be viable alternatives to visual inspection for some products. The charge couple device (CCD) camera and the lens scan camera are now most widely used in machine vision applications. Machine vision systems that are currently available use polychromatic and monochromatic energy in the visible and ultraviolet spectrum (Muir *et al.*, 1989; Novini, 1990; Shaw, 1990). Bilton (D. E. Bilton, personal communication, 1991) reports that Sunkist Corporation is using machine vision for external appearance evaluation, employing a four-lane machine that processes 480 fruits per minute per lane. Packinghouse applications will, no doubt, increase as research advances.With the ever-declining costs and increasing speed and capabilities of computer systems, it seems most likely that this technology will be incorporated into other applications of nondestructive evaluation in the packinghouse. Promising systems have been described for tomatoes (Sarkar and Wolfe, 1985a,b), apples (Rehkugler and Throop, 1986; Davenel *et al.*, 1988; Throop *et al.*, 1989), citrus (Johnson, 1985; Miller *et al.*, 1988; Miller and Verba, 1987), and peaches (Miller and Delwiche, 1989a,b,1990).

Digital image processing systems (machine vision, computer vision) that include chrominance (color) or luminance (gray level intensity) in the image analysis have been described, for example, for color sorting of peaches (Miller and Delwiche, 1989a) and machine harvest of oranges (Slaughter and Harrell, 1989). In contrast to commercial color sorters, which integrate over the fruit surface to obtain an average color signal, computer vision systems have the potential to measure the spatial distribution of color over the surface of the fruit (Miller and Delwiche, 1989a). Color cameras use a red, green, blue (RBG) color scale. In a peach sorting system similar to that in Fig. 3, peaches were sorted based both on color and on percentage of blushed surface area.

In addition to spectral information, a machine vision system provides spatial information that is useful in distinguishing blemished from nonblemished areas. Delwiche *et al.* (1990) extended the amount of information used to detect defects on prunes by using a machine vision system. Muir *et al.* (1989) used both spectral and spatial information to detect 12 to 15 defects on potato tubers. Before implementing the system on-line, problems such as viewing the whole tuber at each of the 8 wavelengths and pixel resolution to detect the smallest lesion must be resolved. Machine vision techniques with a color camera and a monochromatic camera fitted with a NIR filter were explored for detecting various surface blemishes on peaches (Miller and Delwiche, 1989a,b). Although classification errors were less using the monochromatic camera with a filter, errors with both systems exceeded acceptable levels. Detection of defects on peaches is complicated by the variation of color over the surface of the peach (Miller and Delwiche, 1990). To compensate for the color variation, multiple wavelengths in the NIR region were selected for distinguishing each defect from nonblemished areas on peaches. Additional wavelengths served to reduce the effect of surface curvature of the product. Using spatial information in addition to the difference in reflectance between blemished and nonblemished areas, machine vision techniques have been used to evaluate defects on the blossom and stem end of tomatoes (Sarkar and Wolfe, 1985a,b), as well as bruises on apples (Graf and Rehkugler, 1981; Taylor *et al.*, 1984; Rehkugler and Throop, 1986,1989; Davenel *et al.*, 1988).

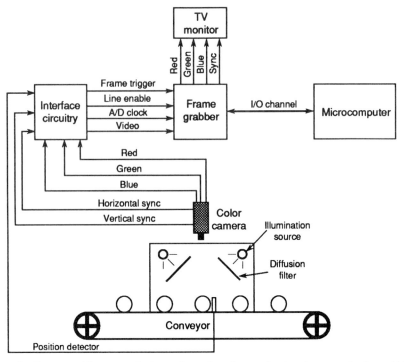

Figure 3 Computer vision system for inspecting and grading peaches based on ground color and blush area. Fruit are presented to the image sensor on a flat belt conveyor. (Reprinted with permission from Miller and Delwiche, 1989.)

II. Acoustic or Ultrasonic Evaluation: Firmness, Texture, and Maturity

A. Physics

Texture or firmness of fruits and vegetables generally is evaluated by the consumer by squeezing the product. Firmness of vegetables is probably primarily a function of tissue turgor and often is described as crispness, whereas firmness of fruits may be due to a combination of turgor and cell wall integrity (Bourne, 1982).

The dynamic elastic characterization of fruits and vegetables frequently is related to product quality, especially textural quality and maturity (Finney and Abbott, 1978). A wide variety of approaches for dynamic elastic property determinations was reviewed by Gaffney (1976a,b,c), Finney and Abbott (1978), and Chen and Sun (1981). Finney and Abbott (1978) identified three categories for measuring textural properties: (1) resonant frequency; (2) vibration transmission; and (3) rebound or resilience techniques. Vibrational testing is the transmission of mechanical waves through a medium, which is acoustics, so vibrational testing will be considered synonymous with acoustics herein.

Direct transmissibility approaches can enable dynamic measurements of the modulus of elasticity (E), which is related to texture and firmness, particularly when the geometry is well defined. For irregularly shaped objects (frequently true of fresh produce), the method is still useful for measurements of damping, which also can be related to texture and firmness. Direct transmissibility tests can be used to measure the dynamic modulus of elasticity, which can be defined as

$$E^* = \frac{\sigma(t)}{\varepsilon(t)} = E^1 + iE^{11} \tag{1}$$

where E^* is the dynamic modulus, E^1 is the real component, frequently called the stor modulus, E^{11} is the imaginary component, called the loss modulus, $\sigma(t)$ is stress as a function of time, and $\varepsilon(t)$ is strain as a function of time.

The shift in phase between peak stress and peak strain as energy passes through the sample can be represented as a loss angle (φ) and is given by the equation

$$\varphi = \tan^{-1} \frac{E^1}{E^{11}} \tag{2}$$

High loss angles correlate with high damping. The damping coefficient is the ratio of peak strain near the opposite boundary of the fruit specimen to peak strain near the point of excitation.

Related to direct transmissibility is the pulse or wave propagation velocity, for which the pulse velocity is measured and related to the elastic modulus (phase shifts not considered) by

$$E^1 = \rho V^2 \tag{3}$$

where V is the pulse velocity and ρ is the mass density (Garrett and Furry, 1972). Pulse velocity concepts coupled with pulse reflection at boundaries forms the basis of sonic tomography, most frequently associated with "ultrasound" in the nondestructive evaluation literature. Sound waves obey Snell's Law of refraction at tissue boundaries. A law similar to Fresnel's Law governs reflection. Both laws depend on acoustic impedances and geometry of the boundary between tissues of different acoustical impedances. The interface could be the boundary between sound flesh and bruised flesh, for example. Acoustic impedance is defined as $(Z_2 - Z_1)/(Z_2 + Z_1)$, where Z_2 and Z_1 are reciprocals of the damping coefficient of the respective media.

Elaborate techniques for using sonic and ultrasound tomography are given by, among others, Goldberg *et al.* (1975), Measurements and Data Corporation (1986), and Thompson and Chimenti (1990). Refer to these references for more complete discussions of sonic tomography.

Although the theory is simple, implementation has been hindered by uniqueness problems and sensitivity issues. For example, acoustic interfaces can be quite different in equally good fruit. Difficulties in coupling the energy to the fruit can cause sensitivity problems.

B. Research Applications

Researchers have used several approaches to evaluate texture or firmness. Non-destructive firmness measurements in the laboratory most commonly involve deformation testing using instruments such as the Instron Universal Testing Machine (Watada *et al.*, 1976) or the Asco Firmness Tester (Sistrunk, 1985). However, deformation testing is simply too slow a procedure to be applied to packingline sorting. An alternative approach is measurement of vibrational force response of the product. The stress exerted by a fruit on a vibrating plate should decrease as the viscoelastic constants decrease, corresponding to softening of the flesh. Such systems have been tested using apples (Finney, 1970,1971a,b), pears (Chen and Fridley, 1972), peaches (Watada *et al.*, 1976; Mahan and Delwiche, 1989), and avocados. A drawback of such a system in on-line testing in packinghouses is sensitivity to outside sources of vibration. The technique also requires somewhat complicated fruit handling or the coupling of surface transducers (Delwiche, 1987).

A fruit or vegetable has been conceptualized as a complex assortment of masses, springs, and dashpots; therefore, mechanical resonance tests have been studied. Resonance characteristics sometimes can be related to quality and maturity indicators. Mechanical resonance tests have been performed by using a vibrating rod driven by an acoustic spectrometer to vibrate apples that were suspended freely by their stems (Abbott *et al.*, 1968a,b). Finney and Norris (1968), using an electromagnetic vibration exciter, measured vibration responses using small accelerometers attached to apples, peaches, and pears. Clark and Shackleford (1973a,b) measured vibration responses of intact peaches by exciting them with a horn-type speaker driver. In each case, the vibration response consists of peak-to-peak displacements, converted to decibels and plotted as a function of frequency. Cooke (1972), following Abbott *et al.* (1968a,b) and Garrett (1970), established theoretical bases for relating the stiffness coefficient, $f^2 m^{2/3}$ (where f represents resonant frequency and m is product mass), to product firmness. Acoustic properties of fruits and vegetables have been correlated with ripeness and firmness (Finney, 1967; Abbott *et al.*, 1968a,b; Finney *et al.*, 1968; Clark and Shackleford, 1973a,b; Yamamoto *et al.*, 1980, 1981).

Impact force response of fruits is related to the elastic properties of the tissue and is correlated with tissue firmness. Diener *et al.* (1971) attempted to build an instrument for measuring firmness. The technique of dropping a fruit onto a load cell or force transducer to obtain impact force with respect to time information has been used to separate tomatoes (Nahir *et al.*, 1986), peaches, and pears (Delwiche, 1987; Delwiche and Sarig, 1989) of different ripeness stages. Meredith *et al.* (1990) used the coefficient of restitution (R), a measure of the relative velocity before and after impact, to separate peaches. The R value was not highly dependent on fruit weight or drop height within the ranges 92–217 gm and 0.5–1.5 cm, respectively, and was correlated well with Instron peak force firmness measurements, as shown in Fig. 4 ($R^2 = 0.87$). Firmness sorting in both cases involved detection of differences in the elastic properties of fruit in different firmness categories. This approach seems most promising for application to on-line sorting.

To achieve a higher resolution, ultrasound in the range of 50 kHz to 5 MHz has

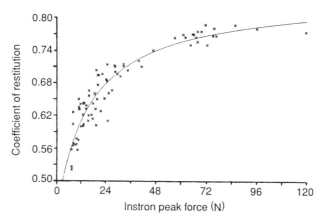

Figure 4 Correlation between nondestructive impact force response (coefficient of restitution, r) and destructive Instron peak force firmness measurements. Data conform to a curve of the form $r = A (f + B)/(f + C)$ where $A = 8508$, $B = 10.56$, $C = 19.45$, and the coefficient of determination for the fit, R^2 is 0.886. (Reprinted with permission from Meredith *et al.*, 1990.)

been applied to detect blemishes and firmness of agricultural products. Ultrasonic detection of bruises on apples was attempted by Upchurch *et al.* (1987); however, several properties of apples prevented the technique from being successful. The high percentage of air in apple fruit and the cuticle contributes to a very high acoustical impedance for apples, causing most of the incident energy to be reflected. Excessive signal attenuation in the porous tissues of fruits and vegetables (Sarkar and Wolfe, 1983) appears to be a limitation in the transmission of ultrasonic energy through intact samples. Use of high power excitation amplitude and lower frequency (50–500 kHz) may help overcome this problem (Mizrach *et al.*, 1989). Ultrasonic measurements have been researched to inspect for surface cracks in tomatoes (Sarkar and Wolfe, 1983). Reflection from a smooth skin was specular with little scattering, whereas signal from cracked skin contained multiple reflections because of the additional surfaces from which the incident energy could reflect. In this case, the differences in interfaces (which ultrasound detects very well) was related strongly to skin smoothness.

Because of irregular shapes of fruits and vegetables, variable acoustical boundaries in bruised and nonbruised tissue, and extraneous vibration in the packinghouse, mechanical methods and ultrasonic techniques have not enjoyed widespread use, except in some notable special cases such as the cranberry bounce test (Franklin, 1917). Issues related to uniqueness, monotonicity, and adequacy of correlation are problematic because of irregular shape and ill-defined boundaries between sound and bruised tissue. Difficulties in coupling the excitation energy source decrease reliability, as does extraneous vibration. The notable exception in the case of ultrasound appears to be the detection of cracks in tomatoes. Most ultrasonic applications in agriculture have been to study animal flesh composition and to characterize wood products. Some success in the postharvest area, in addition to tomato crack detection, was seen in the application of ultrasonic level sensors to potatoes (Studer and Smith, 1984; Thornley, 1984).

III.   X-Ray and Gamma Ray Evaluation: Solids Distribution and Density

---

A.   Physics

Fundamentals of X-ray absorption for some agricultural applications with emphases on low X-ray absorbers are presented by Garrett and Lenker (1976). Beer's Law is used to describe the X-ray absorption process. The absorption coefficient (μ) is a mass average for the solids plus water over the distance traveled (1) by the photons. The absorption coefficient for X rays (and gamma rays) could be expanded as

$$\mu_a = \mu_s f_s + \mu_w f_w \tag{4}$$

where μ_s is the X-ray absorption coefficient for pure solids (liter^{-1}), μ_w is the X-ray absorption coefficient for water (liter^{-1}), f_s is the volume fraction of pure solids, and f_w is the volume fraction of pure water.

Equation 4 can be written as

$$\mu = \mu_s \left(\frac{\rho_{DB}}{\rho_{SG}}\right) + \mu_w \theta_w \tag{5}$$

where ρ_{DB} is the dry bulk density (mg/m^3), ρ_{SG} is the specific density of solid particle (mg/m^3), and θ_w is the volumetric water content.

A substantial body of knowledge exists on processes that affect absorption of monochromatic X rays. X-Ray absorption can be described as

$$\mu = K \lambda^3 Z_n^3 \rho_{SG} \tag{6}$$

where K is the empirical constant of proportionality, λ is the monochromatic X-ray wavelength (m), and Z_n is the effective atomic number (electron density) (Richards *et al.*, 1960).

Because the water component has the dominant atomic number, the water components greatly dominate Eq. 4. Thus, in fresh fruits and vegetables, one is in effect observing differences in volumetric water content (wet based gravimetric water times mass density of wet solids).

X-Ray tomography extends conventional radiography concepts because it allows mapping of the X-ray absorption coefficient μ over user-selected regions in fruits and vegetables, facilitating the visualization of small structural differences on a millimeter. Herman (1980) and Tollner *et al.* (1989) further discuss the theory of X-ray tomography. Gamma radiation generally is discussed in the context of food irradiation and not as a nondestructive testing method; as such, it will not be addressed in detail. Relationships used to characterize gamma radiation are generally similar to those used for X rays.

B.   Applications

Measurement of solidity is quite amenable to newer nondestructive evaluation approaches. Lenker and Adrian (1971) have described the use of X-ray transmission to distinguish among soft, firm, and hard heads of lettuce. Garret and Talley (1969) used gamma ray transmission to select mature heads of lettuce for harvest.

Since X-ray and gamma ray absorbance are correlated with tissue density and water content, these techniques may be applicable to agricultural crops in which these factors change during maturation or internal defect development. Recently, X-ray computed tomography was used to detect the density change in tomato fruit locules, which occurs during the maturation of green fruit, as shown in Fig. 5 (Brecht *et al.*, 1990). Bilton (D. E. Bilton, personal communication, 1991) reports that Sunkist Corporation uses line scan X-ray technology to grade citrus fruit as well as to detect freeze damage (see *Citograph*, 1979). Internal disorders such as split pits in peaches (Bowers *et al.*, 1988) and watercore in apples (Tollner *et al.*, 1989; Upchurch *et al.*, 1989) have been detected using X rays. X-ray computed tomography images of non-watercore- and severely watercore-affected apples are shown in Fig. 6A,B. Ziegler and Morrow (1970) and Diener *et al.* (1970) noted that bruised tissue had different X-ray absorption properties than does nonbruised tissue. Safety concerns must be addressed in any implementation of X-ray technology.

IV. Nuclear Magnetic Resonance Evaluation: Internal Features and Composition

A. Background

Nuclear magnetic resonance (NMR) as a basis for measurement derives from nuclei (usually ^{1}H; other selected nuclei also have magnetic moments) that accept energy when radio frequency (RF), usually very high frequency, is supplied in combination with an external nonvarying magnetic field. Once the energy is received and the

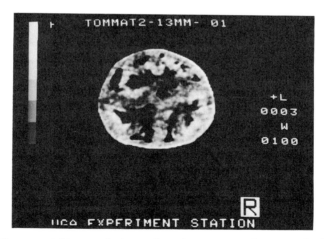

Figure 5   X-Ray computed tomography image of a mature, green, preripe tomato showing the range of signal intensity representing X-ray absorption.

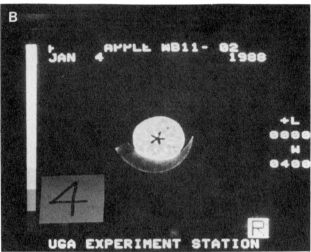

Figure 6 X-Ray computed tomography image of non-watercore-affected apple (A) and severely water-core-affected apple (B).

RF source is turned off, the nuclei release the energy. The magnitude and shape of the return signal contains essential NMR information. The RF energy pulses (called 90° pulses) usually persist on the order of microseconds (depending on the transmitter power and antenna configuration in the detector) whereas the return signal persists on the order of milliseconds. Thus, NMR is a partially reversible absorption process. A schematic of an NMR pulse sequence is shown in Fig. 7.

The relationship between the strength of the external magnetic field and the RF required to absorb the release the RF energy is given by the Larmor Equation which relates resonant frequency to the gyromagnetic coefficient (a function of the element under study) and the external magnetic field strength. Magnetic stability is of utmost

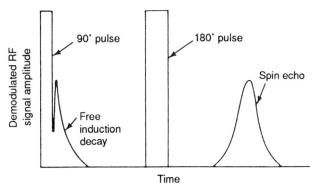

Figure 7 Schematic of low resolution NMR FID and spin echo pulse sequences showing transmitted pulses and received NMR signals.

importance. Temperature variation greatly influences the magnets. Permanent magnets and water-cooled electromagnetics have been used.

The NMR return signal, called the free induction decay (FID), theoretically contains the NMR parameters of interest. Specifically, these parameters are the FID peak magnitude and relaxation time constants, T_1 and T_2. The magnitude of the FID peak is correlated to total hydrogen present (dry matter plus water) and T_2 is a measure of the tenacity with which hydrogen atoms are bound to other hydrogen atoms. Fukushima and Roeder (1981) give an excellent discussion of NMR systems from a technical and practical viewpoint.

Rollwitz (1964) states that the nuclear relaxation times T_1 and T_2 are intrinsic properties of the nuclei and their surroundings. T_1 is the characteristic time for the nuclei to establish thermal equilibrium between their magnetic energy levels and those of their surroundings and is denoted as the spin–lattice relaxation time. Thus, T_1 gives information about the coupling of the nuclei to their surroundings or "lattice." Typical magnitudes reported for T_1 are in the range 10^{-4}–10^{-5} sec. The study of the variation of the spin–lattice relaxation time with temperature can provide accurate and detailed information about (1) molecular motion, (2) variation in coupling to paramagnetic impurities, or (3) any other changes in the lattice surrounding the nuclei. T_2, on the other hand, is the characteristic time of interaction between the nuclei and is denoted as the spin–spin relaxation time. It is the time required for the nuclei to establish thermal equilibrium among themselves. Typically, T_2 is less than (for solids) or equal (for liquid with a low viscosity) to T_1.

There may be multiple T_1 or T_2 values, indicating more than one level of bonding tenacity. This translates to being able to, for example, separate oil-bound hydrogen from water-bound hydrogen. Water also can exist in multiple states (Rollwitz, 1964) within a solid matrix. Rollwitz (1964) made detailed NMR measurements on paper as moisture content varied and showed that up to three T_2 values coexisted. The presence of three T_2 values was related to three separate degrees to which water was bound to the paper. In practical terms, the T_2 values affect the shape of the FID curve. Therefore, the shape of the FID curve should be related to moisture content.

Using more sophisticated NMR approaches (Hahn spin echo) coupled with FID analysis, Paetzold *et al.* (1987) and Tollner and Rollwitz (1988) observed that bound water had a T_2 in the range 0.1–0.5 msec, whereas free water had a T_2 in the range 200–2000 msec. Both free and bound water are very significant characterizations of fruits and vegetables. Product must remain motionless during testing to measure usable signal-to-signal ratios. Sampling time can be in excess of 1 min.

Magnetic resonance imaging (MRI) provides a two- or three-dimensional visualization of NMR parameters. Pulse sequences used with MRI are identical to those used with conventional NMR spectroscopy. MRI is not likely to be useful in a packinghouse environment because of its expense; however, it can be very useful in research that relates quality and maturity to parameters that are more easily measured.

Electron spin resonance (ESR) is similar in concept to NMR except that the ESR signal arises from interactions with unpaired electrons rather than with the nucleus. RF energy is usually in the GHz range. ESR detects unpaired electrons and has had little application in fresh produce evaluation, but may be of interest in operations involving baking, where oxidation reactions bring about radiation with unpaired electrons. Some accounts (e.g., Goodman *et al.*, 1989) also suggest that ESR can detect changes in bones caused by irradiation. NMR theory is highly developed in a chemistry and biochemistry context, for which excellent sample homogeneity is possible; however, most techniques are relevant in a high resolution NMR context when various chemical molecular structures are of interest. Low resolution NMR relationships have a weak theory base, that is, many factors affecting the time constants are not accounted for. Sensitivity is a problem. Lack of uniqueness because of inhomogeneity is a problem. Many unknowns remain.

B. Applications

Nuclear magnetic resonance (NMR) spectroscopy has been used for compounds other than water. NMR is useful for nondestructively measuring changes in various physiologically important compounds in intact plant tissues (Roberts, 1984; Pfeffer and Gerasimowicz, 1989). ^{31}P-NMR spectra allow intracellular measurement of inorganic phosphorus and various phosphorylated compounds, including ATP, ADP, NAD, and UDP-glucose. Inorganic phosphate resonance has been used widely to measure intracellular pH as well. ^{13}C is present in much lower natural abundance than ^{31}P, but can be used to study carbon-containing compounds that are accumulated to relatively high levels. ^{13}C-NMR spectroscopy has been used to measure *in vivo* sugar composition in potatoes (Kainosho and Ajisaka, 1978), grapes (Coombe and Jones, 1983), and carrots (Thomas and Ratliffe, 1985). ^{1}H-NMR spectroscopy can give information concerning tissue water content and mobility, even in waterfree areas.

MRI usually is conducted using ^{1}H as the nucleus examined. Thus, images are primarily a reflection of water (and, in some cases, oil) distribution (Gassner, 1989). McCarthy *et al.* (1989) have demonstrated that unripe tomatoes and avocados can be distinguished easily from ripe fruit using MRI. Ishida *et al.* (1989) similarly have shown differences in water content and mobility in images of unripe and ripe tomatoes. The appearance of the locule tissue in MRI images changed dramatically as green tomatoes progressed from immature through mature stages of development (Fig. 8).

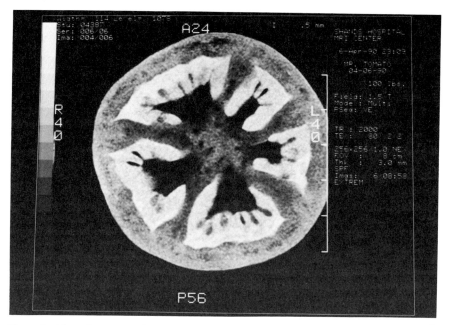

Figure 8 Magnetic resonance image of a mature, green, preripe tomato. Bright image areas (high ^{1}H concentration) correspond to the semiliquid locular gel tissue.

Chen *et al.* (1989) used MRI techniques to obtain images of bruises on apples, peaches, pears, and onions, pits in olives and prunes, and worm damage in pears. Image quality was dependent on experimental parameters such as echo and interpulse delays. Although MRI imaging exhibits great potential, it is costly and time consuming.

V. Electrical Property Evaluation

Electrical properties include the complex impedance properties of fruits and vegetables as functions of excitation voltage and frequency. Impedance is composed of a real component (resistance) and an imaginary component, related to capacitive and/or inductive impedance. Emphasis usually is placed on the dielectric properties (which relate to resistance and capacitance) as functions of frequency; excitation voltage is held constant. Zachariah (1976), in a review of electrical applications, was optimistic about possibilities for applying various electrical approaches to quality evaluation of fruits and vegetables. He gives an excellent discussion of equipment needed to make electrical measurements properly. Resistance and dielectric properties appear to be the most promising. Inductance, the other major contributor to complex impedance, evidently is not changed by most plant materials and therefore is not considered further.

Nelson (1973) identifies Maxwell's Equations as fundamental to evaluating electrical properties of any material. These equations (1) relate electrical field intensity to magnetic field intensity in an electromagnetic wage, (2) relate magnetic flux density to electric flux density in the product, (3) state that magnetic flux is a conserved property, and (4) state that rate of change of electric flux density in a region is equal to the number of charges in the region (charges are conserved). In additional to Maxwell's Equations, Nelson (1973) introduced two constitutive relationships to describe electrical properties. The first relationship relates charge displacement and the dielectric constant. A generalized form of Ohm's Law for alternating currents which is similar to Eq. 1 was introduced also. Nelson (1973) showed how one can relate alternating current conductivity to the complex dielectric constant. Loss modulus was similar to that defined in Eq. 2. Nelson (1973) showed that a strong relationship between moisture and dielectric constant exists for a variety of materials.

Bulk magnetic properties such as magnetic permeability are essentially constant with most plant tissues, including fresh fruits and vegetables. They are generally classed as weak paramagnetic materials. Thus, one can put plant materials in a magnetic resonance imager without altering significantly the bulk magnetic environment of the region. NMR and MRI combine rf waves with a magnetic field and induce changes in magnetization direction with selected elements such as hydrogen. Other electrical properties such as capacitance and dielectric constant are dominated by water, present in virtually all plant tissues destined for fresh-market use. Nelson (1980) could not correlate soluble solids with dielectric properties, but did suggest frequency dependency of dielectric properties for a wide variety of products. Electrical approaches in the fresh fruit and vegetable postharvest environment appear to suffer from uniqueness and low sensitivity difficulties.

Resistance measurements provided a good method for distinguishing internally bruised from nonbruised apple tissue. After bruising, the electrical resistance of the apple tissue dropped and continued to drop with time (Holcomb et al., 1977; Rotz and Mohsenin, 1978). The technique is not practical because (1) insertion of the measurement probes causes damage to the tissues and (2) the inductive coupling required to eliminate the need to insert probes would be difficult to achieve on-line. Blahovec and Janal (1987) attempted to relate direct current conductivity to apple flesh firmness with limited success. In spite of the difficulties of measuring electrical properties, it appears that suggestions for research given in Zachariah (1976) should be reiterated. Advances in modern instrumentation has increased sensitivity and may improve the feasibility of using an electrical approach, particularly one related to dielectric properties.

IV. Far-Infrared Thermography Evaluation: Surface Damage

All media with an absolute temperature above 0 K emit electromagnetic radiation to and absorb electromagnetic radiation from their surroundings in accordance with the Stephan–Boltzman Law. This thermal emission is a result of the accelerations experienced by individual charges in the course of their thermal motions. At typical ambient temperatures in the range of 21°C, maximum intensity occurs near a wavelength of 10,000 nm according to Wein's Displacement Law. This has led to the operation of infrared thermographs at wavelengths near 10,000 nm. The radiation transfer depends strongly on the absorption properties of high- and low-water tissue (Meyers *et al.*, 1979).

Plant tissue moisture also changes with bruises; thus, thermography conceivably could be used to detect bruises in tissue. The major limiting factor in infrared thermography is adequate sensitivity.

Beverly *et al.* (1987) found that 0.5°C sensitivity was not sufficient to detect bruises in broccoli and tomatoes reliably. Danno *et al.* (1978) had success using equipment with 0.2°C sensitivity. Thermography was unable to distinguish between bruised and nonbruised tissue because (1) the reflection coefficient difference between the damaged and undamaged tissues was too small and (2) poor energy propagation into the fruit would result if the surrounding medium had an impedance greatly different from that of the sample.

VII. Fluidized Bed Evaluation: Density

A. Background

The oldest measure of solidity and density is based on dividing weight or mass by volume. Flotation and use of fluidized beds represent actual and potential strategies for nondestructive evaluation in specific cases. Flotation is based on Archimedes' principles and fluidized beds involve classical flotation and drag forces. Both achieve separation by density differentiation between product and surrounding media. Flotation generally uses a fluid with a certain specific gravity as the surrounding media whereas fluidized beds use a solid made to be mobile by drag forces imparted on surrounding solid particles by passing a gas or liquid through the solid, causing flotation. Differences in flotation of the media and the produce become the separation basis. Differences in specific gravity between media and product may be detected to the second or third decimal place.

B. Applications

Solidity and density are important maturity factors for some heading vegetables that become more solid as they grow and mature. Density gradient solutions are used to separate potatoes on the basis of specific gravity, which is related to total solids content (Ryall and Lipton, 1979). Zaltzman *et al.* (1987) developed a fluidized density separator for potatoes which is in field use. The solidity of cabbage heads can be evaluated by use of specific gravity measurements (Pearson, 191). Specific gravity might be expected to change in fruits during maturation as carbohydrates or lipids (e.g., in avocado) accumulate. However, in the case of avocados, variation in specific gravity between individual fruits and the generally narrow range of values seems to preclude its use as a measure of fruit maturity (Lewis, 1978). Sweet cherries are apparently the only fruit for which specific gravity separation is used as a maturity test (Ryall and Pentzer, 1982).

VIII.   "Pseudoproduct" and "Biosensors"

A.   Research

The "pseudoproduct" and "biosensor" approach is fundamentally different from other approaches. Energy propagation mechanisms as described in the previous section are not the primary basis for evaluation. Here the emphasis is on system characterization and the resulting effects on product quality. "Pseudoproduct" is a term believed to have been coined by Rider *et al.* (1973) and O'Brien *et al.* (1973) An object with acceptable surface characteristics is placed in the sample batch of product and various signals are either radiated through the product (e.g., FM telemetry) or stored within the object until retrieval. Tennes *et al.* (1988) reported a device they called an instrumented sphere, which contained a computer and accelerometers. The instrumented sphere or "pseudoproduct" (114-mm cubes and 140-mm diameter spheres have been evaluated) is constructed from epoxy casting resins. Signals from the triaxial accelerometer are filtered, digitized, and stored by the sphere. Given the variety of transducers available, one conceivably could substitute a variety of force, temperature, or "biosensor" transducers, assuming power requirements could be met. The instrumented sphere quantifies impacts, but attempts to correlate impacts with resulting quality have not succeeded.

Using the same principles as the instrumented sphere experiment, the Johns Hopkins Applied Physics Laboratory has reported a "temperature pill" (National Aeronautics and Space Association, 1990). Temperature is sensed in this device using a piezoelectric crystal. After appropriate conditioning, the signal is telemetered to a nearby receiver. These devices now can measure exogenous conditions (such as temperature), satisfying monotonicity, uniqueness, and sensitivity requirements. Continued work is needed to detect internally produced compounds and to make inferences on quality and maturity based on the measurements.

Taste and smell conceivably can be mimicked using "biosensors" (Datta, 1988; Taylor, 1990). The three major types of biosensors are (1) ion selective electrodes, (2) bioaffinity electrodes, and (3) modified metal oxide gas detectors. The ion selective electrode operates similar to the pH electrode except that the H atoms that are sensed are in proportion to enzyme or antibody films placed over the glass electrode. The bioaffinity sensor is an extension of the platinum electrode. Microcapacitors, surface acoustic wave devices, optimal wave guides, and other techniques (R. M. White, personal communication, 1990) are being applied to bioaffinity sensors also. The platinum electrode covered with selective antibody or enzyme film is an amperometric sensor, whereas the coated glass electrode is a potentiometric sensor.

A third type of sensor, the metal oxide gas sensor, also has been evaluated for on-line applications. A typical gas sensor consists of a small tubular ceramic form with a metallization pattern on the outer surface, on which active materials are deposited, usually tin oxide plus catalytic dopants, notably palladium (Datta, 1988). Oxidizing or reducing gases react with the active materials in proportion to concentration at which they are present. Biosensors are subject to fouling, resulting in reduced sensitivity over time. Given the complexity of biochemical processes, it is easy to envision problems with monotonicity and stability in these sensors.

B. Applications

Equipment for temperature measurement over time is available commercially and used routinely by some produce handlers. In a gross sense, this equipment could be termed a "pseudoproduct," although this designation becomes more critical when one wants measures of acceleration over time. Thai (see Chapter 8) has shown that knowledge of time–temperature history can be used to predict color (ripening) of tomatoes and peaches.

The fish freshness sensor is one of the few biosensors developed for specific application to the food industry. Datta (1988) reviewed an enzyme electrode that determined sucrose concentrations directly in the juice of sugar beets and in samples of instant cocoa. These and similar methods are currently off-line, but have potential for on-line use. Taylor (1990) and Datta (1988) each give excellent reviews of the application of biosensors in the laboratory. The use of biosensors to detect precursors to damage require that precursors be identified and that sensor research toward that end be initiated.

IX. Conclusions

The purpose of the postharvest handling system is to deliver nutritious appealing food to the consumer in an economical manner. All disciplines must realize that this goal requires that the product be handled in batches, subject to possible environmental shocks of various kinds, and subjected to a wide variety of unit operations. Nondestructive testing and the use of sensor platforms can aid immensely in the understanding of how systems affect individual fruits.

The goal of early detection is to reduce the time required between damage inception and detection, so steps can be taken with increased confidence both to correct various deficiencies in the postharvest system (feedback information) and to handle product through the remainder of the system to the consumer more appropriately (feedforward information). In Fig. 1, the ultimate goal is to move the detection measure as close as possible to the damage inception point. Feedforward information then would be most useful because of maximized opportunity time. Also feedback information then would be most immediate because of reduced lag time between detection and adjustment.

Application of nondestructive testing methods will be concerned increasingly with product effects on various energy fields. Nondestructiveness implies that the mechanical, electrical, electromagnetic, or magnetic field has absolutely no effect on the product attribute being measured. Research is needed to insure that other product attributes also are unchanged when various energy fields are applied. The controversy over irradiated foods demonstrates that gamma irradiation is perceived by some consumers to affect foods in a negative way.

Products flowing through the postharvest system are notoriously variable. Quality is psychological as well as physical and chemical. For example, many shapes are acceptable; many shapes are not acceptable. Thus, the uniqueness issue arises. As more precise quality attributes, such as the presence of certain biochemical compounds, are defined, the research issue becomes one of detecting the attributes with adequate signal-to-noise ratios. Stability over time becomes an increasingly important issue because of possibilities of drift. Monotonicity and stability questions must be resolved to develop a viable sensor technology.

For purposes of this review, sensor technologies were categorized according to the problem area most often addressed by past applications of the technology. Sensor physics, however, is universal. Particularly with the new nondestructive electrical approaches, the technology may have been researched in only one area, but there may be additional applications. Additional research is needed, particularly with X rays, NMR, NIR, and machine vision approaches. For example, machine vision approaches may be useful for studying computed tomography images. It is well known that the laws of optics are not confined to the NIR and visual portions of the electromagnetic spectrum. Much research remains to be done in this area.

A wide variety of sensing technologies was evaluated with respect to uniqueness, monotonicity, sensitivity, and stability criteria. Technologies meeting these criteria are moving into use in postharvest systems. A motivating factor in the trend toward nondestructive testing techniques and, to some extent, toward new sensor platforms is the desire to increase uniqueness. The counterbalance appears to be that of adequate sensitivity. For example, in the case of machine vision and in the case of melon quality detection with NIR, increased uniqueness was obtained; new developments in technology in the last decade have enabled the sensitivity issue to be addressed satisfactorily (or nearly so). Other techniques reviewed face a variety of uniqueness and sensitivity uses that must be worked out.

Issues of monotonicity and stability over time also affect the movement of a sensing technology from the laboratory to the postharvest system. Monotonicity and stability appear to be formidable issues for biosensors. Stability and sensitivity issues are still important in the successful development of sensor platforms to be used in various postharvest systems or unit operations. Continued advances in biochemistry and microelectronics may overcome these limitations.

Nondestructive sensing approaches for individual products are moving from the laboratory to various points in the postharvest handling systems. Reflectance color sorters and optical size and external blemish sorters (computer vision) for sorting fresh fruits and vegetables are commercially available for an increasingly wide variety of applications. Mechanical sizing and density-based approaches are so commonplace that they are almost overlooked. If sizing or density is an objective, economical nondestructive techniques are available. Commercial X-ray sorters for freeze damage and hollow heart detection are available. NIR approaches for testing soluble solids and possibly texture in melons and fruits appear to be on the verge of commercial success. Many other techniques, such as delayed light emission, nuclear magnetic resonance, magnetic resonance imaging, X-ray tomography, and ultrasound, are available for research purposes. These tools can be useful in answering specific questions in specific instances. Sensor platforms for detecting temperature history through the system are becoming widely used in the industry. Platforms for measuring shock and vibration through postharvest systems are now commercially available. Perhaps biosensors will be available soon for incorporation into sensor platforms, enabling better correlation of problems with problem location. Biosensors appear to be on the verge of practicality in the processed food industry, but not in the postharvest system.

Bibliography

Abbott, J. A., Bachman, G. S., Childers, N. F., Fitzgerald, L. V., and Matusik, F. J. (1968a). Sonic techniques for measuring texture of fruits and vegetables. *Food Technol.* **22(5)**, 101.

Abbott, J. A., Childers, N. F., Bachman, G. S., Fitzgerald, J. V., and Matusik, F. J. (1968b). Acoustic vibration for detecting textural quality of apples. *Proc. Amer. Soc. Hort. Sci.* **93**, 725–737.

Affeldt, H. A., and Abbott, J. A. (1989a). Apple firmness and sensory quality using contact acoustic transmission. *In* "Proceedings of the Eleventh International Congress of Agricultural Engineers" (V. A. Dodd and P. M. Grace, eds.), pp. 2037–2045.

Affeldt, H. A., Upchurch, B. L., Norris, K. H., and Throop, J. A. (1989). "Bruise detection on 'Golden Delicious' Apples." ASAE Tech. Paper No. 89-3011. American Society of Agricultural Engineers, St. Joseph, Michigan.

Aulenbach, B. B., Yeatman, J. N., and Worthington, J. T. (1972). "Quality Sorting of Red Delicious Apples by Light Transmission." USDA Marketing Research Report No. 936. U.S. Govt. Printing Office, Washington, D.C.

Ballard, D. H., and Brown, C. M. (1982). "Computer Vision." Prentice-Hall, Englewood Cliffs, New Jersey.

Ballinger, W. E., McClure, W. F., Nesbitt, W. B., and Maness, E. P. (1978). Light-sorting Muscadine grapes (*Vitis rotundifolia* Michx.) for ripeness. *J. Amer. Soc. Hort. Sci.* **103**, 629–634.

Beverly, R., Hung, Y.-C., Prussia, S. E., Schewfelt, R. L., Tollner, E. W., Garner, J. C., and Robinson, D. (1987). Thermography as a nondestructive method to detect invisible quality damage in fruits and vegetables. *HortScience* **22(5)**, 1056.

Bilanski, W. K., Pen, C. L., and Fuzzen, D. R. (1984). Apple bruise detection using optical reflectance parameters. *Can. Agric. Eng.* **26(2)**, 111–114.

Billmeyer, F. S., Jr. and Saltzman, M. (1981). "Principles of Color Technology." John Wiley & Sons, New York.

Birth, G. S. (1976). How light interacts with foods. *In* "Quality Detection in Foods" (J. J. Gaffney, ed.), pp. 6–11. American Society of Agricultural Engineers, St. Joseph, Michigan.

Birth, G. S. (1978). The light scattering properties of foods. *J. Food Sci.* **43**, 916–925.

Birth, G. S. (1982). Diffuse thickness as a measure of light scattering. *Appl. Spectrosc.* **36(6)**, 675–681.

Birth, G. S., and Hecht, G. H. (1987). The physics of near infrared reflectance. "Near-Infrared Technology in the Agricultural and Food Industries" (P. C. Williams and K. H. Norris, eds.). American Association of Cereal Chemists. pp. 1–15. St. Paul, Minnesota.

Birth, G. S., and Olsen, K. L. (1964). Nondestructive detection of water core in Delicious apples. *J. Amer. Soc. Hort. Sci.* **85**, 74–84.

Birth, G. S., Dull, G. G., Magee, J. B., Chan, H. T., and Cavaletto, C. G. (1984). An optical method for estimating papaya maturity. *J. Amer. Soc. Hort. Sci.* **109(1)**, 62–66.

Birth, G. S., Dull, G. G., Renfro, W. P., and Kays, S. J. (1985). Nondestructive spectrophotometric determination of dry matter in onions. *J. Amer. Soc. Hort. Sci.* **110(2)**, 297–303.

Blahovec, J., and Janal, R. (1987). Firmness and electrical conductivity of apple flesh. *Sbornik Uvitz, Zahradnictvi* **14(4)**, 252–258.

Bourne, M. C. (1982). Physical properties and structure of horticultural crops. *In* "Food Texture and Viscosity" (M. C. Bourne, ed.), Academic Press, New York, pp. 207–228.

Bowers, S. V., Dodd, R. B., and Han, Y. J. (1988). Nondestructive testing to determine internal quality of fruit. St. Joseph, Michigan. American Society of Agricultural Engineers, ASAE Techn. Paper No. 88-6569.

Brecht, J. K., Shewfelt, R. L., Garner, J. C., and Tollner, E. W. (1990). Using X-ray computed tomography (X-ray CT) to nondestructively determine maturity of green tomatoes. *HortScience* **26(1)**, 45–47.

Brown, G. K., Segerlind, L. J., and Summitt, R. (1974). Near-infrared reflectance of bruised apples. *Trans. ASAE* **17(1)**, 17–19.

Burkhardt, T. H., and Mrozek, R. F. (1973). Light reflectance as a criterion for sorting dried prunes. *Trans. ASAE* **16(4)**, 683–685.

Chen, P., and Fridley, R. B. (1972). Analytical method of determining viscoelastic constants for agricultural materials. *Trans. ASAE* **15**, 1103–1106.

Chen, P., and Sun, Z. (1991). A review of nondestructive methods for quality evaluation and sorting of agricultural products. *J. Agric. Res.* **49**, 85–98.

Chen, P., McCarthy, M. J., and Kauten, R. (1989). NMR for internal quality evaluation of fruits and vegetables. *Trans. ASAE* **32(5)**, 1747–1753.

Citrograph (1979). The EMG: An electronic grader, fruit separator. *Citrograph* **64(4)**, 77–80.

Clark, R. L., and Shackleford, P. S. (1973a). Resonance and optical properties of peaches as related to flesh firmness. *Trans. ASAE* **16**, 1140–1142.

Clark, R. L., and Shackleford, P. S., Jr. (1973b). Resonance and optical properties of peaches as related to firmness. *In* "Quality Detection in Foods" (J. Gaffney, ed.), pp. 143–145. American Society of Agricultural Engineers, St. Joseph, Michigan.

Cooke, J. R. (1972). An interpretation of the resonant behavior of intact fruits and vegetables. *Trans. ASAE* **15(6)**, 1075–1080.

Coombe, B. G. (1976). The development of fleshy fruits. *Ann. Rev. Plant Physiol.* **27**, 507–528.

Coombe, B. G., and Jones, G. P. (1983). Measurement of the changes in the composition of developing undetached grape berries by using ^{13}C NMR techniques, *Phytochem.* **22**, 2185–2187.

Couey, H. M., and Hayes, C. F. (1986). Quarantine procedure for Hawaiian papaya using fruit selection and a two-stage hot water immersion. *J. Econ. Entomol.* **79**, 1307–1314.

Danno, A., Miyazato, M., and Ishiguro, E. (1978). Quality evaluation of agricultural products by infrared imaging method. I. Grading of fruits for bruise and other surface defects. *Memoirs Fac. Agric.* **14(23)**, 123–138.

Datta, A. K. (1988). Novel chemical and biological sensors for monitoring and control of food processing operations. *In* "Proceedings of the SME Conference on Intelligent Sensors for Automated and Process Control in Food Processing, October 25–26, 1988," pp. 1–18. Society of manufacturing engineers, Dearborn, Michigan.

Davenel, A., Guizard, C., Labarre, T., and Sevila, F. (1988). Automatic detection of surface defects on fruit by using a vision system. *J. Agr. Eng. Res.* **41**, 1–9.

Dekazos, E. D., and Birth, G. S. (1970). A maturity index for blueberries using light transmittance. *J. Amer. Soc. Hort. Soc.* **95**, 610–614.

Delwiche, M. J. (1987). Theory of fruit firmness sorting by impact forces. *Trans. ASAE* **30**, 1160–1166, 1171.

Delwiche, M. J., and Sarig, Y (1989). "A Probe Impact Sensor for Fruit Texture Measurement." ASAE Tech. Paper No. 89-6609. American Society of Agricultural Engineers, St. Joseph, Michigan.

Delwiche, M. J., Tang, S., and Thompson, J. F. (1990). Prune defect detection by line-scan imaging. *Trans. ASAE* **33(3)**, 950–954.

Deshpande, S. E., Cheryan, M., Gunasekaran, S., Paulsen, M. R., Salunkhe, D. K., and Chancellor, V. (1984). Nondestructive optical methods of food quality evaluation. *CRC Crit. Rev. Food Sci. Nutr.* **21(4)**, 323–379.

Diener, R. G., Mitchell, J. P., and Rhoten, M. L. (1970). "Using an X-Ray Image Scan to Sort Bruised Apples." ASAE Rep. J-380, pp. 115–117. American Society of Agricultural Engineers, St. Joseph, Michigan.

Diener, R. G., Sobotka, F. E., Watada, A. E. (1971). An accurate, low cost Firmness Measuring Instrument. *J. Text. Stud.* **2**, 373–384.

Dull, G. G. (1986). Nondestructive evaluation of quality of stored fruits and vegetables. *Food Technol.*, **40(5)**, 110–116.

Dull, G. G., Birth, G. S., and Magee, J. B. (1980). Nondestructive evaluation of internal quality. *HortScience* **15(1)**, 60–63.

Dull, G. G., Birth, G. S., and Leffler, R. G. (1989). Use of near infrared analysis for the nondestructive measurement of dry matter in potatoes. *Amer. Potato J.* **66**, 215–225.

Dull, G. G., Birth, G. S., Smittle, D. A., and Leffler, R. G. (1989b). Near-infrared analysis of soluble solids in intact cantaloupe. *J. Food Sci.* **54(2)**, 393–395.

Finney, E. E (1967). Dynamic elastic properties of some fruits during growth and development. *J. Agric. Eng. Res.* **14**, 249–255.

Finney, E. E. (1970). Mechanical resonance within Red Delicious apples and its relation to fruit texture. *Trans. ASAE* **13(2)**, 177.

Finney, E. E. (1971a). Random vibration techniques for nondestructive evaluation of peach firmness. *J. Agric. Engin. Res.* **16(1)**, 81–87.

Finney, E. E. (1971b). Dynamic elastic properties and sensory quality of apple fruit. *J. Text. Stud.* **2**, 62–74.

Finney, E. E., and Abbott, J. A. (1978). Methods for testing dynamic mechanical response of solid foods. *J. Food Qual.* **2**, 55–74.

Finney, E. E., and Norris, K. H. (1968). Instrumentation for investigating dynamic mechanical properties of fruits and vegetables. *Trans. ASAE* **11(1)**, 94–97.

Finney, E. E., Ben-Gara, I., and Massie, D. R. (1968). An objective evaluation of changes in firmness of ripening bananas using a sonic technique. *J. Food Sci.* **32**, 642–646.

Forbus, W. R., and Chan, J. (1990). The effects of quarantine heat treatments in delayed light emission measurements for papaya maturity. *J. Food Sci.* (In Press).

Francis, F. J. (1980). Color quality evaluation of horticultural crops. *HortScience* **15**, 58–59.

Francis, F. J., and Clydesdale, F.M. (1975). "Food Colorimetry: Theory and Applications." AVI, Westport, Connecticut.

Franklin, H. J. (1917). "Report of the Cranberry Substation for 1916." Massachusetts Agricultural Experiment Station Bulletin No. 180. Amherst.

Fukushima, E., and Roeder, S. B. W. (1981). Experimental Probe NMR: A Nuts and Bolts Approach." Addison-Wesley, Reading, Massachusetts.

Gaffney, J. J. (1969). Reflectance properties of citrus fruit. *Trans. ASAE* **16**, 310–314.

Gaffney, J. J. (ed) (1976a). "Quality Detection in Foods." American Society of Agricultural Engineers, St. Joseph, Michigan.

Gaffney, J. J. (1976b). Reflectance properties of citrus fruits. *In* "Quality Detection in Foods," pp. 60–64. American Society of Agricultural Engineers, St. Joseph, Michigan.

Gaffney, J. J. (1976c). Light reflectance of radishes as a basis for automatic grading. *In* "Quality Detection in Foods," pp. 75–79. American Society of Agricultural Engineers, St. Joseph, Michigan.

Garrett, R. E. (1970). Velocity of propagation of mechanical disturbances in apples. Ph.D. Thesis, Cornell University, Ithaca, New York.

Garrett, R. E., and Furry, R. B. (1972). Velocity of sonic pulses in apples. *Trans. ASAE* **15(4)**, 770–774.

Garrett, R., and Lenker, D. H. (1976). Selecting and sensing X and gamma rays. *In* "Quality Detection of Foods," pp. 107–115. American Society of Agricultural Engineers, St. Joseph, Michigan.

Garrett, R. E., and Talley, W. K. (1969). "Use of Gamma Ray Transmission in Selecting Lettuce for Harvest." ASAE Techn. Paper No. 69-310. American Society of Agricultural Engineers, St. Joseph, Michigan.

Gassner, G. (1989). Magnetic resonance imaging in agricultural research. *In* "Nuclear Magnetic Resonance in Agriculture" (P. E. Pfeffer and W. V. Gerasimowicz, eds.), pp. 405–428. CRC Press, Boca Raton, Florida.

Goldberg, B. B., Kotler, M. N., Ziskin, M. C., and Waxham, R. D. (1975). "Diagnostic Uses of Ultrasound." Grune and Stratton, New York.

Goodman, B. A., McPhail, D. B., and Duthie, D. M. L. (1989). Electron spin resonance spectroscopy of some irradiated foodstuffs. *J.Sci. Food Agric.* **47**, 101–111.

Gortner, W.A., Dull, G. G., and Krauss, B. H. (1967). Fruit development, maturation, ripening and senescence: A biochemical basis for horticultural terminology. *HortScience* **2(4)**, 1–4.

Graf, G. L., and Rehkugler, G. E. (1981). "Automatic Detection of Surface Flaws on Apples Using

Digital Image Processing." AsAE Tech. Paper No. 81-3537. American Society of Agricultural Engineers, St. Joseph, Michigan.

Gunasekaran, S. (1990). Delayed light emission as a means of quality evaluation of fruits and vegetables. *CRC Crit. Rev. Food Sci. Nutr.* **29(1),** 19–34.

Gunasekaran, S., Paulsen, M. R., and Shore, G. C. (1985). Optical methods for nondestructive quality evaluation of agricultural and biological materials. *J. Agric. Eng. Res.* **32(3),** 209–241.

Hayes, C. F., Chingon, H. T. G., and Young, H. G. C. (1988). Hunter *b* color measurements of papaya using a two-filter system. *HortScience* **23,** 399.

Herman, G. T. (1980). "Image Reconstruction from Projections: The Fundamentals of Computed Tomography." Academic Press, New York.

Holcomb, D. P., Cooke, J. R., and Hartman, P. L. (1977). "A Study of Electrical Thermal and Mechanical Properties of Apples in Relation to Bruise Detection." ASAE Tech. Paper. No. 77-3512. American Society of Agricultural Engineers, St. Joseph, Michigan.

Howarth, M. S., and Searcy, S. W. (1990). Fresh market carrot inspection by machine vision. *SPIE-Optics Agric.* **1379,** 141–150.

Howarth, M. S., Searcy, S. W., and Birth, G. S. (1990). Reflectance characteristics of freshmarket carrots. *Trans. ASAE* **33(3),** 961–964.

Hunt, R. W. G. (1987). "Measuring Color." Ellis Howard, New York.

Ishida, N., Kobayashi, T., Koizumi, M., and Kano, H. (1989). ^{1}H-NMR imaging of tomato fruits. *Agric. Biol. Chem.* **53,** 2363–2367.

Johnson, M. (1985). Automation in citrus sorting and packing. *In* "Proceedings of Agrimation I Congress and Exposition, Chicago, Illinois," pp. 63–68.

Kader, A. A., and Morris, L. L. (1976). Appearance factors other than color and their contribution to quality. *In* "Proceedings of the 2nd Tomato Quality Workshop, University of California, Davis," pp. 8–14.

Kainosho, M., and Ajisaka, K. (1978). Carbon-13 NMR spectra of gross plant tissues containing starch. *Tetr. Lett.* **18,** 1563–1566.

Kawano, S., and Iwamoto, M. (1990). Advances in R&D on near infrared spectroscopy in Japan. *SPIE-Optics Agric.* **1379,** 2–9.

Lenker, D. H., and Adrian, P. A. (1971). Use of X-ray for selecting mature lettuce heads. *Trans. ASAE,* **14,** 894–898.

Lewis, C. E. (1978). The maturity of avocados—A general review. *J. Sci. Food Agric.* **29,** 857–866.

McCarthy, M. J., Chen, P., and Kauten, R. (1989). "Maturity Evaluation of Avocados by NMR Methods." ASAE Tech. Paper No. 89-3548. American Society of Agricultural Engineers, St. Joseph, Michigan.

McClure, W. F., Rohrback, R.P., Kushman, L. J., and Ballinger, W. E. (1976). Design of a high-speed fiber optic blueberry sorter. *In* "Quality Detection in Foods" (J. J. Gaffney, ed.), pp. 189–192. American Society of Agricultural Engineers, St. Joseph, Michigan.

Mahan, S. A., and Delwiche, M. J. (1989). "Fruit Firmness Detection by Vibrational Force Response." ASAE Tech. Paper No. 89-3010. American Society of Agricultural Engineers, St. Joseph, Michigan.

Measurements and Data Corporation (1986). "Ultrasonics for Medical Diagnostics: A Professional Course." Measurements and Data Corporation, Pittsburgh.

Meredith, F. I., Leffler, R. G., and Lyon, C. E. (1990). Detection of firmness in peaches by impact force response. *Trans. ASAE* **33,** 186–188.

Meyers, P. C., Sodowsky, N. L., and Barrett, A. H. (1979). Microwave thermography: Principles, methods and clinical applications. *J. Microwave Power* **14(2),** 105–113.

Miller, B.K., and Delwiche, M. J. (1989a). A color vision system for peach grading. *Trans. ASAE* **32,** 1484–1490.

Miller, B. K., and Delwiche, M. J. (1989b). "Peach Defect Detection with Machine Vision." ASAE Tech. Paper No. 89-6019. American Society of Agricultural Engineers, St. Joseph, Michigan.

Miller, B. K., and Delwiche, M. J. (1990). "Spectral Analysis of Peach Surface Defects." ASAE Tech. Paper No. 90-6040. American Society of Agricultural Engineers, St. Joseph, Michigan.

Miller, W.M., and Verba, W. L. (1987). Automated density measurements for quality sorting of 'Marsh' grapefruit. ASAE Tech Paper No. 87-6518, Am Soc. Agr Eng., St. Joseph, Michigan.

Miller, W. M., Peleg, K., and Briggs, P. (1988). Automated density separation for freeze-damaged citrus. *Appl. Eng. Agric.* **4(4),** 344–348.

Mizrach, A., Galili, N., and Rosenhouse, G. (1989). Determination of fruit and vegetable properties by ultrasonic excitation. *Trans. ASAE* **32**, 2053–2058.

Muir, A. Y., Shirlaw, I. D. G., and McRae, D. C. (1989). Machine vision using spectral imaging techniques. *Agric. Eng.* **44(3)**, 79–81.

Nahir, D., Schmilovitch, Z., and Ronen, B. (1986). "Tomato Grading by Impact Force Response." ASAE Tech. Paper No. 86-3028. American Society of Agricultural Engineers, St. Joseph, Michigan.

National Aeronautics and Space Association (1990). "Temperature Pill." NASA Technical Briefs, Houston.

Nattuvetty, V. R., and Chen, P. (1980). Maturity sorting of green tomatoes based on light transmittance through regions of the fruit. *Trans. ASAE* **23**, 515–518.

Nelson, S. O. (1973). Electrical properties of agricultural products—A critical review. *Trans. ASAE* **16(3)**, 384–400.

Nelson, S. O. (1980). Microwave dielectric properties of fresh fruit and vegetables. *Trans. ASAE* **23(5)**, 1314–1317.

Novini, H. (1990). Fundamentals of machine vision component selection. *In* "Food Processing Automation: Proceedings of the 1990 Conference," pp. 60–71. American Society of Agricultural Engineers, St. Joseph. Michigan.

O'Brien, M., Fridley, R. B., Goss, J. R., and Schubert, J. F. (1973). Telemetry for investigating forces on fruits during handling. *Trans. ASAE* **16(2)**, 245–247.

Paetzold, R. F., de los Santos, A., and Matzkanin, G. A. (1987). Pulsed nuclear resonance instrument for soil-water content measurement: Sensor configuration. *SSSA* **51(2)**, 287–290.

Pearson, O. H. (1931). Methods for determining the solidity of cabbage heads. *Hilgardia* **5**, 383–393.

Pfeffer, P. E., and Gerasimowicz, W. V. (1989). Introduction to high-resolution NMR spectroscopy and its application to *in vivo* studies of agricultural systems. *In* "Nuclear Magnetic Resonance in Agriculture" (P. E. Pfeffer and W. V. Gerasimowicz, eds.), pp. 3–70. CRC Press, Boca Raton, Florida.

Powers, J. B., Gunn, J. T., and Jacob, F. C. (1953). Electronic color sorting of fruits and vegetables. *Agric. Eng.* **34**, 149.

Rehkugler, G. E., and Throop, J. A. (1986). Apple sorting with machine vision. *Trans. ASAE* **29(5)**, 1388–1397.

Rehkugler, G. E., and Throop, J. A. (1989). Image processing algorithm for apple defect detection. *Trans. ASAE* **32(1)**, 267–272.

Reid, W. S. (1976). Optical detection of apple skin, bruise, flesh, stem, and calyx. *J. Agric. Eng. Res.* **21(3)**, 291–295.

Richards, J. W., Sears, F. W., Wehr, M. R., and Zemansky, M. W. (1960). "Modern University Physics." Addison-Wesley, New York.

Rider, R. C., Fridley, R. B., and O'Brien, M. (1973). Elastic behavior of a pseudo fruit for determining bruise damage to fruit during mechanized handling. *Trans. ASAE* **16(2)**, 241–244.

Robert, J. K. M. (1984). Study of plant metabolism *in vivo* using NMR spectroscopy. *Annu. Rev. Plant Physiol.* **35**, 375–386.

Rollwitz, W. L. (1964). "A Study of Moisture in Paper by Spin Echo Nuclear Magnetic Resonance." Internal Research Report. Southwest Research Institute, San Antonio, Texas.

Rotz, C. A., and Mohsenin, N. N. (1978). A note of potential applications of physical properties of bruised tissue of apples for automatic sorting. *Trans. ASAE* **21(4)**, 790–792.

Ryall, A. L., and Lipton, W. J. (1979). "Handling, Transportation, and Storage of Fruits and Vegetables," Vol. 1. AVI, Westport, Connecticut.

Ryall, A. L., and Pentzer, W. T (1982). "Handling, Transportation, and Storage of Fruits and Vegetables," Vol. 2. AVI, Westport, Connecticut.

Sarkar, N., and Wolfe, R. R. (1983). Potential of ultrasonic measurements in food quality evaluation. *Trans. ASAE,* **26**, 624–629.

Sarkar, N., and Wolfe, R. R. (1985a). Feature extraction techniques for sorting tomatoes by computer vision. *Trans. ASAE* **28**, 970–974.

Sarkar, N., and Wolfe, R. R. (1985b). Computer vision based system for quality separation of fresh market tomatoes. *Trans. ASAE* **28(5)**, 1714–1718.

Searcy, S. W. (1987). "Machine Vision Workshop." Presented at ASAE Summer Meeting, June 27–28, Baltimore, Maryland.

Shaw, W. E. (1990). Machine vision for detecting defects in fruits and vegetables. *In* "Food Processing Automation: Proceedings of the 1990 Conference," pp. 50–59. American Society of Agricultural Engineers, St. Joseph, Michigan.

Sidwell, A. P., Birth, G. S., Ernest, J. V., and Golumbic, C. (1961). The use of light transmittance techniques to estimate the chlorophyll content and stage of maturation of Elberta peaches. *Food Technol.* **19,** 123.

Sistrunk, W. A. (1985). Peach quality assessment: fresh and processed. "Evaluation of Quality of Fruits and Vegetables" (H. E. Pattee, ed.), pp. 1–46. AVI, Westport, Connecticut.

Slaughter, D. C., and Harrell, R. C. (1989). Discriminating fruit for robotic harvest using color in natural outdoor scenes. *Trans. ASAE* **32,** 757–763.

Stevens, M. A. (1985). Tomato flavor: Effect of genotype cultural practice and maturity at picking. *In* "Evaluation of Quality of Fruits and Vegetables." (H. E. Pattee, ed.), pp. 367–386. AVI, Westport, Connecticut.

Studer, H. E., and Smith, N. E. (1984). "Bulk Handling Equipment for High Moisture Raisins." ASAE Tech. Paper No. 84-1571. American Society of Agricultural Engineers, St. Joseph, Michigan.

Taylor, R. F. (1990). Applications of biosensors in the food processing industry. *In* "Food Processing Automation: Proceedings of the 1990 Conference," pp. 156–166. American Society of Agricultural Engineers, St. Joseph, Michigan.

Taylor, R. W., Rehkugler, G. E., and Throop, J. A. (1984). Apple bruise detection. *In* "Agricultural Electronics—1983 and Beyond," Vol. II, pp. 652–662. American Society of Agricultural Engineers, St. Joseph, Michigan.

Tennes, B. R., Zap, H. R., Brown, G. K., and Ehlert, S. H. (1988). Self contained impact detection device. *Trans. ASAE* **31(6),** 1869–1874.

Thai, C. N., and Shewfelt, R. L. (1990). "Modeling Sensory Color Quality: Neural Network vs. Statistical Regression." ASAE Tech. Paper No. 906038. American Society of Agricultural Engineers, St. Joseph, Michigan.

Thomas, T. H., and Ratliffe, R. G. (1985). ^{13}C NMR determination of sugar levels in storage roots of carrot (*Daucus carota*). *Physiol. Plant.* **63,** 284.

Thompson, D. O., and Chimenti, D. E. (1990). "Review of Progress in Quantitative Nondestructive Evaluation," Vols. 9A,B. Plenum Press, New York.

Thornley, W. R. (1984). Ultrasonic sensors applied to potato piling. *In* "Agricultural Electronics—1983 and Beyond," Vol. II, pp. 754–747. American Society of Agricultural Engineers, St. Joseph, Michigan.

Throop, J. A., Rehkugler, G. E., and Upchurch, B. L. (1989). Application of computer vision for detecting watercore in apples. *Trans. ASAE* **32(6),** 2087–2092.

Tollner, E. W., and Rollwitz, W. L. (1988). Magnetic resonance for moisture analysis of meals and soils. *Trans. ASAE* **31(5),** 1608–1615.

Tollner, E. W., Davis, J. W., and Verma, B. P. (1989). Managing errors with x-ray computed tomography (x-ray CT) when measuring physical properties. *Trans. ASAE* **32(3),** 1090–1096.

Tyson, T. W., and Clark, R. L. (1976). An investigation of the properties of aflatoxin infected pecans. *In* "Quality Detection of Foods" (J. J. Gaffney, ed.), pp. 53–56. American Society of Agricultural Engineers, St. Joseph, Michigan.

Upchurch, B. L., Miles, G. E., Stroshine, R. L., Furgason, E. S., and Emerson, F. H. (1987). Ultrasonic measurement for detecting apple bruises. *Trans. ASAE* **30(3),** 803–809.

Upchurch, B. L., Affeldt, H. A., Hruschka, W. R., and Throop, J. A. (1989). Optical detection of bruised and early frost damage on apples. ASAE Tech. Paper. No. 89-3013. American Society of Agricultural Engineers, St. Joseph, Michigan.

Upchurch, B. L., Affeldt, H. A., Hruschka, W. R., Norris, K. H., and Throop, J. A. (1990). Spectrophotometric study of bruises on whole 'Red Delicious' apples. *Trans. ASAE* **33(2),** 585–589.

Watada, A. E., and Abbott, J. A. (1985). Apple quality: Influences of pre- and postharvest factors and estimation by objective methods. *In* "Evaluation of Quality of Fruits and Vegetables" (H. E. Pattee, ed.), pp. 63–81. AVI, Westport, Connecticut.

Watada, A. E., and Norris, K. H. (1978). Quality of fresh commodities estimated by spectrophotometric technique. *In* "Encyclopedia of Food Technology and Food Science Series" (M. Peterson and A. J. Johnson, eds) Vol. 3, pp. 648–653. AVI, Westport, Connecticut.

Watada, A. E., Abbott, J. A., and Finney, E. E. (1976). Firmness of peaches measured nondestructively. *J. Amer. Soc. Hort. Sci.* **101,** 404–406.

Worthington, J. T., Massie, D.R., and Norris, K. H. (1976). Light transmission technique for predicting ripening time for intact green tomatoes. *In* "Quality Detection in Foods" (J. J. Gaffney, ed.), pp. 46–49. American Society of Agricultural Engineers, St. Joseph, Michigan.

Yamamoto, H., Iwamoto, M., and Haginuma, S. (1980). Acoustic impulse response method for measuring natural frequency of intact fruits and preliminary applications to internal quaity evaluations of apples and watermelons. *J. Text. Stud.* **11,** 117–136.

Yamamoto, H., Iwamoto, M., and Haginuma, S. (1981). Nondestructive acoustic impulse response method for measuring internal quality of apples and watermelons. *J. Jap. Soc. Hort. Sci.* **50,** 247–261.

Yeatman, J. N., and Norris, K. H. (1965). Evaluating internal quality of apples with new automatic fruit sorter. *Food Technol.* **19,** 123.

Zachariah, G. (1976). Electrical properties of fruits and vegetables for quality evaluation. *In* "Quality Detection in Foods" (J. J. Gaffney, ed.), pp. 98–101. American Society of Agricultural Engineers, St. Joseph, Michigan.

Zaltzman, A., Verma, B. P., and Schmilovitch, Z. (1987). Potentials of quality sorting with a fluidized bed medium. *Trans. ASAE* **30(3),** 823–831.

Ziegler, G. E., and Morrow, C. T. (1970). "Radiographic Bruise Detection." ASAE Tech. Paper No. 70-553. American Society of Agricultural Engineers, St. Joseph, Michigan.

CHAPTER 12

STRESS PHYSIOLOGY: A CELLULAR APPROACH TO QUALITY

Robert L. Shewfelt

Living cells constitute fresh and minimally processed products. Storage and handling techniques are designed to keep these cells alive in a high-quality state. Postharvest physiologists seek a better understanding of response of harvested plant products to handling conditions, presuming that basic knowledge will provide answers to practical problems. Unfortunately, few direct links between physiological processes and quality maintenance have been established clearly. Thus, most handling and storage studies remain empirical in nature; commercial quality programs are limited to managing conditions to retard quality losses (see Chapter 13). True quality assurance of fresh and minimally processed products requires an understanding of the basis of quality at the cellular and molecular levels.

One area that shows promise in relating cell response to product quality is stress physiology. Stress is defined as "one or a combination of environmental and biological factors" resulting in "an aberrant change in physiological processes" (Hale and Orcutt, 1987). Stress conditions lead to disorders in susceptible tissue that result in clear, visible symptoms that directly affect quality. Differences in susceptibility among and within species provide geneticists and physiologists with models to study the cellular and molecular basis of tissue response. In this chapter, a current understanding of stress physiology provides an example of the potential use of cell biology as a basis for quality management. Current limitations in knowledge prevent its practical application.

I. Types of Postharvest Stress

Stress usually is categorized by environmental or biological factors that induce injury and quality degradation. Whenever possible, physiologists seek to relate specific responses of plant tissue to specific levels of stress. Unfortunately, many of the symptoms of injury are broad based and defy strict quantification. Thus, much of the literature tends to be qualitative rather than quantitative.

Fruits and vegetables are exposed to several types of stress, some intentional and some unintentional, as they pass through the handling system from farm to consumer. Examples of intentional stresses include low temperatures and controlled atmospheres. Unintentional stresses include impacts of fruit on hard surfaces or other fruit and exposure to high temperatures during harvest operations. Postharvest stress is induced by exposure to extremes in temperature or relative humidity, microbial invasion, deviation in atmospheric gas composition from that of air, and mechanical impact.

A. Temperature Extremes

Among the most common type of postharvest stresses, temperature extremes can be classified further into low- and high-temperature stress. Low temperatures can induce freezing and chilling injury whereas high-temperature injury occurs at temperatures above 30°C.

Low-temperature storage is the most effective tool used to maintain quality and extend shelf life. As the temperature of storage decreases and approaches the freezing point, enzymatic reactions slow, including those responsible for respiration and senescence. Low temperatures also slow the rate of growth of decay micro-organisms. Optimal storage temperatures for numerous commodities have been studied and compiled [American Society of Heating, Refrigeration, and Air Conditioning Engineers (ASHRAE), 1986; Hardenburg *et al.*, 1986]. Low temperatures only become a problem in the postharvest system when they result in injury.

Freezing injury is the direct result of unintentional freezing of fresh tissue. This response should not be confused with frozen foods, which are typically blanched first to inactivate degradative enzymes, then frozen under controlled conditions. Mere exposure of living tissue to temperatures below the freezing point does not necessarily result in injury. Supercooling of tissue up to $-10°C$ has been observed without ice formation (Lindow, 1983). Agitation of a supercooled item, however, can result in ice formation, which is a primary cause of freezing injury. Prevention of freezing injury is best achieved by maintaining storage temperatures above the freezing point of the commodity in question. When temperatures have cycled below the freezing temperature, the most prudent action is to bring the tissue temperature above freezing slowly before surveying the possible damage. When establishing the lowest temperature for commercial storage that will not result in freezing, the normal range of temperature fluctuation must be considered. As refrigeration systems become more precise, storage temperatures can be lowered. Such regimes would be more beneficial to items with longer projected storage lives, for example, apples, for which the difference of $1-3°C$ over months of storage is more critical than for highly perishable items such as strawberries which will be marketed within days of harvest.

Chilling injury is "physiological damage that occurs to many tropical and subtropical plants when exposed to low but nonfreezing temperatures in the chilling range (generally 15°C down to 0°C)" (Raison and Lyons, 1986). Chilling injury is a disorder of both living plants and harvested organs exposed to chilling temperatures. Some generalizations of chilling injury follow.

1. Some species are susceptible whereas others are not.
2. Within a susceptible species, some genotypes are more susceptible than others.
3. Within a genotype, most, if not all, organs are susceptible but some organs are more susceptible than others.
4. Within an organ, response does not tend to be uniform and often is localized.
5. Damage incurred at low temperatures may not become evident until the plant or organ is exposed to nonchilling temperatures.

(For review, see Lyons, 1973; Wolfe, 1978; Berry and Bjorkman, 1980; Steponkus, 1981; Graham and Patterson, 1982; Oquist, 1983; Markhart, 1986; Lyons and Briedenbach, 1987; Parkin *et al.*, 1989; Wang, 1989). The variation in susceptibility of organs within a plant and how this susceptibility interacts with the variation of susceptibility among genotypes has been demonstrated (Bodner and Larcher, 1978). The variation of sensitivity of membrane systems within tomato pericarp cells has been described (Cheng and Shewfelt, 1988).

Chilling injury is one of the most intensely studied postharvest disorders. It affects many high-value crops such as tomatoes, avocados, bananas, and other tropical fruits. The most obvious way to prevent chilling injury is to store susceptible commodities above the temperature at which injury is incurred. In commercial practice, however, a wide range of commodities is stored at one or two temperatures. Frequently, a compromise temperature of 4–8°C is selected, which serves to reduce chilling injury in susceptible commodities and shorten shelf life of nonsusceptible items. Rapid turnover of stock tends to minimize losses.

Symptoms of chilling are diverse, ranging from surface pitting and water-soaked surface lesions in cucumbers, to internal discoloration in avocados, to failure to ripen and off flavor development in tomatoes, to scald and abnormal ripening in apples and pears (see Table I). Descriptions of symptoms as they are associated with specific commodities are available (Lyons and Briedenbach, 1987; Jackman *et al.,* 1988; Wang, 1989; Wills *et al.,* 1989; Bramlage and Meir, 1990; Paull, 1990). Despite the wide range of chilling symptoms, it generally is assumed that the disorder proceeds by a single mechanism. Lyons (1973) proposed the phase transition hypothesis, which, despite extensive modification, remains the most viable working model for most physiologists in the field. Reviews on the mechanism of chilling injury (Parkin *et al.,* 1989; Wang, 1989, 1990) provide an in-depth treatment of current knowledge. A brief description of the role of membranes in the etiology of the disorder follows in Section III,A.

High-temperature injury can result from chronic exposure for days or weeks to temperatures above typical room temperatures (ranging from 30 to 40°C) and acute exposures of short duration to higher temperatures (Steponkus, 1981). Much less is known about high-temperature injury than about chilling injury since the former generally can be prevented with normal handling practices. Short-term injury can be induced during harvesting operations in warm climates and during storage at ambient temperatures in developing nations in the tropics. Bananas and pears are among the crops shown to be susceptible to high temperatures (Shewfelt, 1986). The most extensively studied susceptible fruit is the tomato, which fails to ripen

Table I

Secondary Responses Observed in Chilling Injury of Fruits and Vegetables

---

Solute leakage
Respiratory burst
Toxin accumulation
Symptom development
Metabolic imbalances
Loss of compartmentation
Loss of membrane integrity
Increased C_2H_4 production
Ultrastructural disorganization
Decreased rate of photosynthesis
Cessation of protoplasmic streaming
Reduced supply and use of energy
Decreased rate of mitochondrial oxidative activity
Increased activation energy of membrane-bound enzymes

---

properly above 30°C. Abnormal ripening becomes apparent by a lack of lycopene development (Duggar, 1913; Tomes, 1963; Buescher, 1979) and results from a decline in the expression of genes signaling ethylene accumulation (Biggs *et al.*, 1988; Picton and Grierson, 1988) as well as decreased sensitivity to exogenous ethylene (Yang *et al.*, 1990). High temperature response has been studied also in apples (Lurie *et al.*, 1991) and papayas (An and Paull, 1990). Since electrolyte leakage is observed on exposure above 40°C (Inaba and Crandall, 1988), it has been suggested that the injury is mediated through membrane lipids. High-temperature injury, from chronic exposure as described here, appears to be a different response than heat shock, which induces development of heat-shock proteins in response to high temperatures as described in Section IV,D.

B. Relative Humidity

Stress responses in harvested plant tissue are also induced by extremes of relative humidity (RH). Optimal conditions have been compiled for most fresh commodities (Hardenburg *et al.*, 1986; ASHRAE, 1986). When RH is too low, transpiration is enhanced, resulting in loss of moisture. The primary factor controlling the rate of moisture loss is actually the water vapor pressure deficit (VPD), which reflects the difference between the humidity in the tissue and the humidity of the air in the storage room or container (Kays, 1991). As water evaporates from the tissue, turgor pressure decreases and the cells begin to shrink and collapse. Although some damage can be reversed by returning the items to high RH by icing or misting, prolonged stress conditions lead to irreversible injury. Leafy tissue such as lettuce and turnip greens wilts, whereas stems such as broccoli and celery become limp. Moisture barriers, such as the exocarp of tomatoes and cucumbers as well as epicuticular waxes on the surface of peppers, protect the product. Porous exocarp, such as that found in citrus fruits, generally benefits from exogenous waxes to prevent the fruit from drying out. Use of packaging film such as high density polyethylene can create microenvironments of higher RH than the surroundings. The primary explanation offered to explain amelioration of chilling injury by seal packaging is the prevention of water loss (Ben-Yehoshua *et al.*, 1987).

Storage at excessive RH also can result in adverse effects. Free water on the plant tissue provides an excellent environment for decay microorganisms. Although healthy plant tissue is resistant to decay, cuts, bruises, or additional stress can weaken the endogenous resistance of the tissue. Rapid changes of temperature and RH conditions can lead to condensation (Wills *et al.*, 1989). High RH storage combined with low temperatures has been suggested as a means of maintaining quality of bulk products for extended periods of time. Crops amenable to such treatment are carrots (Van den Berg and Lentz, 1966) and cherries (Sharkey and Peggie, 1984). Peach storage life and quality decreases as a result of high RH storage.

C. Microbial Invasion

Another inducer of stress response in plant tissue is invasion by microorganisms. Harvested products are equipped with physical and chemical barriers to microbes (see Chapter 6). Physical barriers can be compromised by cuts or abrasions; removal of trichomes ("fuzz") from peaches during grading and sorting operations (Ben-Yehoshua, 1987); minimal processing such as cutting, slicing, or dicing (Brackett, 1987); or exposure to other types of stress such as chilling (Lyons and Briedenbach, 1987). The second line of defense against microbes is chemical. Phenolics and benzoic acid are examples of endogenous compounds that protect the tissue. Phytoalexins are compounds produced by plant tissue as a defense mechanism in response to microbial attack (Rizk and Wood, 1980). Although primarily antifungal in nature, phytoalexins can be induced by other types of stress such as exposure to ultraviolet light, freezing, heating, or wounding (Darvill and Albersheim, 1984). Some concern has been raised about the safety of stress-induced phytoalexins since they are toxic to microbes. Evidence of human toxicity of potato alkaloids has been suggested (Mundt, 1977).

D. Atmospheric Stress

Atmospheric stress can be induced by high carbon dioxide (CO_2), low oxygen (O_2), and ethylene (C_2H_4). Controlled and modified atmospheres are designed to slow respiration and, thus, senescence by reducing O_2 or increasing CO_2 concentrations. Much early research effort was directed at empirically determining optimal conditions for specific commodities (Kader, 1985,1986a). More recent efforts have been directed at modeling optimal environments using computer-simulation techniques (Chinnan, 1989). The true potential of this research has been limited, however, since a general model to explain response of harvested plant tissue to modified atmospheres has not been formulated (Kader, 1986a). An integrated model would assist the effective design of modified atmosphere packaging greatly (Kader *et al.*, 1989; Labuza and Breene, 1989), since species, tissue, maturity of harvest, and external handling temperatures all affect product performance.

High CO_2 environments may result in CO_2-injury symptoms in many products. Symptoms range from surface pitting and increased decay in brussels sprouts and cucumbers to off odors and flavors in broccoli and sweet corn (Herner, 1987). The most sensitive crops include lettuce, green tomatoes, carrots, and radishes, which cannot tolerate CO_2 levels above 3%. Moderate sensitivity (damage induced by 5–10% CO_2) is observed in beets, cauliflower, bell peppers, and potatoes. Least sensitive to CO_2 are celery, mushrooms, muskmelon, and sweet corn, which do not evidence injury until levels of 20% CO_2 or higher. Usually, low O_2 levels accompany high CO_2 levels but the stress responses are not the same. Low O_2 can induce anaerobiosis, which results in off flavors and odors. Although ethanol production is a conclusive sign of anaerobic degradation, off flavors can develop in broccoli and cauliflower before significant ethanol formation (Weichmann, 1987).

Ethylene is generated by climacteric fruits such as apples, bananas, and tomatoes, which should not be stored or transported with C_2H_4-sensitive items such as lettuce. Controlled-atmosphere storage limits russet spotting of lettuce (apparently by inhibition of C_2H_4 by low 0_2; Ke and Saltviet, 1989), yellowing of broccoli, softening of watermelons, lignin formation in asparagus, and leaf abscission (Kays, 1991). Ethylene is also a component of diesel exhaust. Thus, electric forklift trucks should be used in fresh produce operations and the diesel exhaust pipe of tractors should not be located upwind of the air suction intakes of refrigerated trailers (Reid, 1985).

E. Mechanical Damage

Bruises, cuts, and abrasions are examples of tissue response to mechanical stress. A bruise is initiated by the breakage of cell membranes, allowing cytosolic enzymes to act on previously sequestered substrates. The resultant browning is caused by the action of polyphenol oxidase on phenolic substrates such as tyrosine, caffeic acid, and chlorogenic acid (Richardson and Hyslop, 1985). In addition, phenylalanine ammonia-lyase (PAL) is stimulated on wounding and can lead to lignification; this may represent an early step in enzymatic browning (Jones, 1984). Softening is the result of the degradation of cell wall polysaccharides by hydrolytic enzymes such as polygalacturonase, pectin methylesterase, and cellulase (Huber, 1983) as well as the physical rupture of structural components. In addition, wounding stimulates respiration and production of ethylene, which in turn can trigger degradative processes in the cell. The best protection against bruising is careful handling by reducing drop heights, padding handling equipment where possible, reducing fruit-to-fruit impact, and minimizing vibration during transport. Cuts and abrasions remove the first line of defense of the harvested product, permitting microbial invasion.

F. Other Stresses

Additional stresses are incurred from light, irradiation, chemical additives, and interaction of any of the already mentioned stresses. The most important stress response to light stress is greening of potatoes, in which chlorophyll synthesis is accompanied by the formation of solanine and other toxic alkaloids (Mundt, 1977). Light stress also can induce photooxidation of chlorophyll in photosynthetic tissue such as broccoli, spinach, and other leafy vegetables, which can lead to premature yellowing. Irradiation is used as a postharvest technique to extend shelf life. It is an effective technique for strawberries and some tropical fruits such as mangoes, papayas, and guavas. Low-dose irradiation also can result in injury of some crops, as evidenced by surface blemishes in citrus fruits, accelerated yellowing of cucumbers and peppers, and altered ripening patterns in mangoes (Kader, 1986b). In many of these commodities, physiological damage is induced at levels lower than those at which any beneficial effects can be observed. The greatest promise of irradiation lies in combination with other techniques such as packaging. Chemical additives such as auxins, cytokinins, gibberellins, and ethylene-regulating compounds are growth regulators applied in preharvest sprays or postharvest dips to delay senescence, improve quality, or extend shelf life. Fungicides are added to control decay microorganisms. Waxes are applied to slow water loss due to transpiration (Shewfelt, 1986). Consumer safety concerns about all chemicals have stimulated intensive research efforts to find alternatives to chemical preservation (see Chapter 15).

II.   Factors in Stress Response

As suggested earlier, the response of plant tissue to stress is highly variable. Factors in this variability include genetics, growth conditions, endogenous defense mechanisms such as the nature of the outer surface of the tissue, maturity at harvest, and severity of exposure.

A. Genetics

Genetic factors in stress response are identifiable at both species and genotypic levels. Certain species are susceptible to chilling injury whereas others are not. Closely related tomato species that vary in chilling susceptibility are *Lycopersicon esculentum,* which is susceptible to chilling injury, and *Lycopersicon hirsutum,* which has similar fruits, but neither the plant nor the fruit is susceptible (Patterson and Reid, 1990). Lettuce, carrots, and radish are sensitive to CO_2 levels at or below 3% whereas cantaloupes, okra, and sweet corn can withstand levels of CO_2 of 20% or greater (Herner, 1987). Within a susceptible species, genotypic variation is seen, for example, among *Saintpaulia* and *Coffea* genotypes in response to chilling (Bodner and Larcher, 1987) and among apple cultivars for disease susceptibility (Pierson *et al.,* 1971). Certain fruits and vegetables provide natural protective barriers such as specialized structures in the exocarp that render them less susceptible to stress. The tomato peel does not have stomata and, as such, greatly restricts CO_2 and O_2 diffusion while permitting some water vapor diffusion (Floros and Chinnan, 1990). Thus, the tomato is much less susceptible to shrivelling than are citrus fruits and leafy vegetables, which lose water rapidly. Application of waxes to items such as apples, cucumbers, or oranges reduces water loss and overcomes genetic deficiencies or losses of endogenous protectants incurred during normal postharvest operations.

B. Growth Conditions and Cultural Practices

Production phase conditions can predispose tissue to postharvest damage. Apples grown in warm climates tend not to display chilling symptoms during postharvest handling. Preharvest nutritional deficiencies can lead to increased susceptibility to decay for many fruits during postharvest handling, including apples, oranges, and pears. Excess water during growth and development can affect quality of cherries and tangerines adversely (Ben-Arie and Lurie, 1986).

C. Maturity at Harvest

Tomatoes are most susceptible to chilling injury at the mature-green and pink stages, suggesting that two distinct mechanisms may be involved (Autio and Bramlage, 1986). In other stress responses, mature-green tomatoes tend to be more susceptible to CO_2 injury than pink fruit (Herner, 1987). Tree fruit are generally more susceptible to postharvest decay later in the harvesting season (Rosenberger, 1990). It is widely believed in the peach industry that bruise susceptibility increases with maturity at harvest; this belief frequently justifies harvesting immature fruit. Hung and Prussia (1989) have demonstrated that fruits harvested at optimal maturity (color chips 3 and 4) have no greater susceptibility to bruising than immature fruits (chips 1 and 2).

D. Severity of Exposure

The severity of the injury tends to be a function of the level and duration of the stress. In most cases, plant response to stress is reversible to a point. Prior to the point of irreversibility, symptoms usually do not appear; those that do tend to be superficial. Once irreversibility is reached, symptoms will develop. The severity of response often is linked to the severity of exposure. A graphic demonstration of such responses with respect to chilling of papaya is given in Fig. 1 (Paull, 1990). As the temperature is lowered, the duration of exposure required to induce a symptom decreases. At a given temperature, a pattern of response severity develops as the duration of exposure increases. Similar effects are observed in other responses to stress, including increased bruise volume of peaches exposed to higher levels of input energy (Hung and Prussia, 1989); internal browning or breakdown of cabbage,

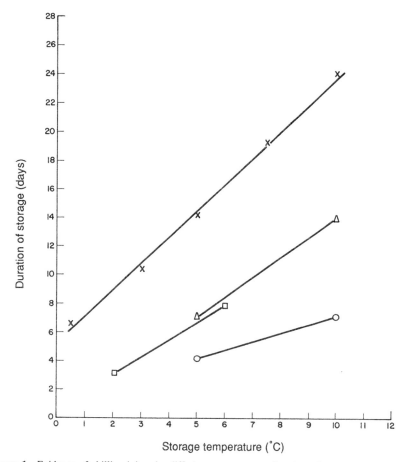

Figure 1 Evidence of chilling injury by differing measures as a function of storage temperature and duration of storage. Visible symptoms, x; electrolyte leakage, △ *Alternia* rot, □; C_2H_4 content, ○. [Reprinted with permission from (Paull, 1990). Chilling injury of crops of tropical and subtropical origin. *In* "Chilling Injury of Horticultural Crops (C. Y. Wang, ed.) pp. 17–36. CRC Press, Boca Raton, Florida.]

onions, and peppers due to CO_2 injury (Herner, 1987); and increased desiccation of spinach and citrus fruits due to transpiration (Ben-Yehoshua, 1987).

Postharvest technologists are developing some techniques that reduce susceptibility to injury. Temperature conditioning is used to decrease susceptibility to chilling injury. Storage of grapefruit at 16°C for up to 7 days prior to storage at chilling temperatures tends to reduce the development of symptoms (Hatton and Cubbedge, 1981). Apples held at 38°C for 4 days showed less superficial scald, a symptom of chilling injury, during subsequent storage at 0°C than those not exposed to high temperatures (Lurie *et al.*, 1991). Curing at high temperatures (32°C) lowers chilling sensitivity of sweet potatoes (Patterson and Reid, 1990). Likewise, a combination of individual seal packaging and curing for 3 days at 36°C reduces decay and water loss of citrus fruits (Ben-Yehoshua *et al.*, 1987). Intermittent warming, the cycling of temperatures above and below the critical chilling temperature, has been shown to be effective for chilling-sensitive crops such as apples, cucumbers, peaches, sweet peppers, and tomatoes (Wang, 1989). Unfortunately, the technique has not been adopted within the industry, probably because of the lack of flexibility of temperature conditions at wholesale warehouses (see Chapter 2).

III. Site of Stress Response

To understand the mechanism of stress response, the site of that response must be known. Although the actual site has not been established definitively, several lines of evidence indicate cellular membranes. Stress-induced changes in lipid composition could, in turn, influence the function of membrane-bound proteins. A second area of research investigates the levels of free ions in the cytosol, which can serve as second messengers in the cell-signaling response. Cell walls have been proposed as the source of elicitors formed in response to microorganisms. Gene expression is a fourth line of investigation.

A. Membranes

Since the phase transition hypothesis was proposed (Lyons, 1973), numerous researchers have linked preharvest and postharvest disorders to alterations in membrane structure and function. Such disorders include chilling injury, desiccation (Stewart and Bewley, 1980), loss of viability of seeds (Senaratna and McKersie, 1986), and freezing injury (Kendall and McKersie, 1989). Similar changes in composition are observed during senescence and aging (McKersie *et al.*, 1988; Thompson, 1988). It has been suggested that responses to stress may share pathways with senescence (Parkin *et al.*, 1989).

Lyons' original hypothesis proposed that chilling-sensitive tissue is unable to modify fatty acid composition to maintain membrane lipids in a liquid crystalline phase. Thus, it was postulated, as storage temperature is lowered below the threshold, membrane lipids undergo a transition from a liquid-crystalline to a solid-gel phase. This phase transition leads to decreased activities of mitochondrial enzymes and, in turn, to metabolic imbalances (Lyons, 1973). This simplistic view largely has been discredited, but the basic premise that membrane lipid composition provides the key to changes in the physical properties of membranes, leading to dysfunction of membrane-bound proteins, provides the basis for most studies on membrane-associated disorders (Shewfelt, 1992).

Phase transitions are not as sharp in plant membranes, which are complex mixtures of proteins and numerous molecular species, as in liposomes of defined lipid composition. Wolfe (1978) suggests that the fluidity of lipids rather than a phase transition mediates the activity of membrane-bound proteins. Fatty acid saturation (Vigh *et al.*, 1985), hydrolysis, peroxidation (Thompson, 1988), retailoring of molecular species (Thompson, 1989), and sterol accumulation are processes that affect the fluidity by modifying membrane lipids. Other investigators provide evidence that microdomains are formed within the membrane that have different fluidity than the membrane as a whole (Caldwell and Whitman, 1987; Platt-Aloia and Thomson, 1987). These microdomains could influence membrane-bound enzymes. An alternative view suggests that phase transitions can be induced by major modifications of the molecular species of a relatively minor lipid component such as phosphatidylglycerol in thylakoids (Murata and Nishida, 1990). Research to date has established that plant membranes undergo changes in lipid composition and biophysical properties in response to stress, but have not elucidated the mechanism of change or demonstrated a direct link to plant dysfunction (Shewfelt, 1991).

B.   Cytosol

Calcium has special importance within the plant cell. Many specific physiological disorders are associated with calcium deficiency. Calcium has been shown to stabilize cell walls (Glenn *et al.*, 1988) and membranes (Nur *et al.*, 1986) as well as regulate protein phosphorylation (Poovaiah, 1988). More importantly, calcium serves as a secondary messenger in cells; the concentration of free cytosolic Ca^{2+} is associated with the regulation of many cellular processes, including protoplasmic streaming, phloem transport, exocytosis, and membrane turnover (Minorsky, 1989). Ca^{2+} concentration in the cytosol is controlled by a series of calcium-binding proteins, including calmodulin. It has been suggested that calcium binding to calmodulin triggers a cascade of reactions through a series of phosphoinositol intermediates. Although Ca^{2+} signaling mechanisms have been suggested as primary events for senescence (Leshem, 1988) and chilling injury (Minorsky, 1985), conclusive evidence that phosphoinositols play a major role in plant membrane function is still not available (Morré, 1990).

C.   Cell Walls

Degradation of cell walls plays an obvious role in the development of the visible symptoms of stress response. Localized softening associated with fruit bruising is the most obvious example of enzymatic breakdown of cell walls. Examples of cell-wall breakdown during chilling injury include the development of pitting in cucumbers, melons, and tomatoes (Jackman *et al.*, 1988), subcellular changes in cucumbers (Fukushima, 1978), and wooliness in peaches (Ben-Arie and Lavee, 1971; von Mollendorff and deVilliers, 1988). Enzymatic degradation also has been implicated in the development of watercore during storage of Japanese pears (Yamaki and Kajiura, 1983).

Plant cells deposit callose, a polysaccharide containing a high proportion of 1,3-β-linked glucose molecules, in cell walls in response to wounding (Dekazos and Worley, 1967; Tighe and Heath, 1982) and fungal invasion (Hinch and Clarke, 1982). Callose formation also can be elicited in cell culture by the addition of chitosan (Kohle *et al.*, 1985). The enzyme 1,3-β-D glucan glucosyltransferase, responsible for callose synthesis, is located in the plasma membrane (Delmer, 1990) and has been characterized extensively in red beet tissue (Frost *et al.*, 1990). The results of Bonhoff *et al.* (1987) suggest that callose formation is triggered by cytoplasmic calcium concentration via a mechanism not mediated by calmodulin. Callose appears to rigidify and toughen bruised tissue in red cherries (Dekazos and Worley, 1967). It also may serve to protect tissue from fungal invasion, since Hinch and Clarke (1982) observe that roots from corn (a plant resistant to fungal invasion) produce callose in response to *Phytophthora cinnamomi* whereas lupines (a susceptible species) do not produce callose.

Phytoalexins have been associated with resistance to microbes, particularly fungi. Evidence suggests that oligosaccharides are cleared from the complex carbohydrates composing the host plant cell wall that are involved in cell-to-cell signaling of microbial invasion and elicit phytoalexin synthesis (Darvill and Albersheim, 1984). Other potential elicitors include compounds in fungal cell walls and microbial enzymes. The host–pathogen relationship is very complex, however, since certain pathogens have response mechanisms to detoxify phytoalexins synthesized by the host (Van Etten *et al.*, 1989).

D. Gene Expression

A dramatic change in the level of a specific protein in response to stress suggests a change in gene expression. A dramatic plant-stress response directly tied to changes in expression is heat shock. Heat shock occurs during short-term exposure (4–8 hr) of plants to temperatures of 40–50°C. It is characterized by the production of distinct patterns of mRNA and heat-shock proteins (more appropriately, polypeptides) (Schoffl *et al.*, 1986). High-temperature injury from chronic exposure (described in Section I,A) is not as well characterized and develops during the course of exposure to lower temperatures (30–40°C) over a period of days or weeks, depending on the species. Heat-shock proteins have been suggested to confer protection against high-temperature damage, but the mechanism of protection has not been elucidated. The heat-shock response has been observed in fungi, insects, mammals, and plants. Most of the work with heat-shock proteins in plant tissue has been done with seedlings (Key *et al.*, 1987), but some reports have established their presence in carrot cells (Hwang and Zimmerman, 1989), papaya fruit (Paull and Chen, 1990), and tomato fruit (Duck *et al.*, 1989). A similar line of research has identified cold-shock proteins in barley (Marmiroli *et al.*, 1986) and maize

(Yacoob and Filion, 1987), but work is still too preliminary to draw firm conclusions (Minorsky, 1989).

IV.   Molecular Biology of Harvested Tissue

Molecular biology now plays an important role in postharvest research. Using genetic or molecular approaches, specific traits such as tomato firmness (see Fig. 2) can be traced from quality characteristic back to the expression of mRNA, and to the DNA in the gene comprising the code for that trait. Through a myriad of techniques including RFLP (restriction fragment length polymorphism) mapping, monoclonal and polyclonal antibody development, Western, Northern, and Southern blotting, development and incorporation of antisense genes, and cell culturing, molecular biologists provide fresh insight into the inner workings of the cell (Hille *et al.*, 1989; Wasserman, 1990). Of particular interest in postharvest physiology are the molecular biology of fruit ripening (Tucker and Grierson, 1987), tomato softening (Grierson *et al.*, 1986; Giovannoni *et al.*, 1989), and flower senescence (Lawton *et al.*, 1989).

The primary limitation to the practical application of molecular biology to postharvest systems lies in the inability to identify molecular targets for modification. This problem results from a lack of clear specifications of desirable quality attributes, the inability to link quality attributes to specific aspects of cell physiology, and the lack of understanding of basic cell physiology. In designing a plant to be stress resistant or to produce harvested tissue that is stress resistant, one must know the characteristics that are most limiting in response to stress, the physiological processes that are responsible for these characteristics, and the genes that encode the proteins responsible for the process.

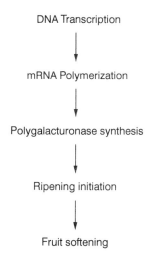

DNA Transcription

mRNA Polymerization

Polygalacturonase synthesis

Ripening initiation

Fruit softening

Figure 2   Proposed sequence of events leading to softening of tomato fruits.

A. Molecular Approach to Stress Response

The molecular approach to stress response has been limited compared with other areas of postharvest physiology. Response to temperature stress has been limited primarily to the isolation and identification of heat-shock proteins (described in Section IV, D). Other promising progress is being made in the areas of callose synthase activity in response to wounding (Frost *et al.,* 1990) and phytoalexin production in response to microbial invasion.

The molecular approach to gaining a better understanding of membrane-associated disorders such as chilling injury has been limited by a lack of gene targets. The inability to identify a primary event clearly has prevented focusing the research effort. Many studies suggest that initial changes related to the response are in the lipid component of the membrane, which is much less amenable to molecular techniques than the protein component. Membrane-bound enzymes are less tractable to analysis than soluble enzymes, although molecular approaches have been developed, as pictured in Fig. 3. Also, the response may be caused primarily by the physical properties of membrane lipids rather than by their chemical properties, another area not readily exploitable with current molecular techniques. Despite these problems, it is clear that genetic factors are involved in the stress response (described in Section II.A), but the pure genetic differences in responses are mitigated by acclimation and conditioning. A better understanding of the mechanism of cellular response to chilling or another membrane-associated stress could provide the needed gene targets (Wasserman, 1990). Promising targets include enzymes responsible for lipid degradation (Paliyath *et al.,* 1987), for compositional modification, for example, the desaturases (Vigh *et al.,* 1985) and acyltransferases (Thompson, 1989), and for sterol biosynthesis (Guye, 1987). If Ca^{2+} regulation is the primary event in chilling injury, then the associated calcium-regulating enzymes provide primary targets.

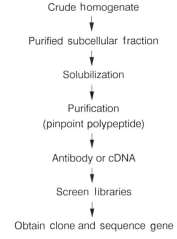

Figure 3 Classical molecular approach to characterize membrane-bound enzymes. (Courtesy of B. P. Wasserman.)

B. Molecular Appoach to Quality

The molecular approach has shown promise for commercial benefit from quality enhancement in postharvest technology. Although senescence is technically not a stress response, it is the ultimate membrane-associated disorder in terms of quality degradation. The pattern of degradation observed in chilling injury and in other membrane-associated disorders has been described as "senescence-like" (Thompson, 1988; Parkin *et al.*, 1989). An understanding of the molecular biology of senescence should lead to genetic modifications that enhance quality and extend shelf life. Gene expression has been characterized during carnation senescence (Lawton *et al.*, 1989) and tomato ripening (Picton and Grierson, 1988). Premature softening has been identified as the major limiting quality characteristic of tomatoes, so the antisense gene for polygalacturonase has been inserted into tomato plants to maintain firmness relative to color and flavor development (Smith *et al.*, 1988). At this point, however, it is not clear that polygacturonase is the primary enzyme responsible for softening, since modification of the gene does not produce the expected effect on softening (Giovannoni *et al.*, 1989).

V. Implications for Quality Management

The establishment of quality specifications and maintenance of the product within specifications during handling and storage are the keys to quality management of fresh products (see Chapters 5, 13). These specifications must be related to consumer acceptance and be readily quantifiable throughout the complete handling system.

Stress is of commercial consequence when it results in a detectable loss of quality. The symptoms of stress do not affect all quality attributes equally, but are apparent by selective degradation of specific attributes. A hazard of postharvest management programs is the sacrifice of consumption quality to maintain purchase quality (Chapter 5).

Exposure of fruits and vegetables to shorter durations or lower levels of stress can result in a reversible response. At some point, however, the response becomes irreversible. A readily detectable zone of reversibility suggests its potential exploitation to extend shelf life without adversely affecting quality. Short-term exposure of tomatoes to either high or low temperatures slows ripening sufficiently at a critical juncture in handling to prolong shelf life without degrading quality (Shewfelt *et al.*, 1989).

Management of quality does not always require biotechnological solutions. Some simple management techniques have potential benefit. When a product is received that is suspected of exposure to a particular stress, the quality characteristic(s) most likely to be affected by that specific stress should be monitored carefully. In many cases, the symptoms of injury have been characterized for specific commodities (Hardenburg *et al.*, 1986; Weichmann, 1987). Storage of a subsample at higher than normal temperatures (accelerated storage tests) may lead to a more rapid development of symptoms. Care must be exercised, however, to prevent substituting one type of stress for another. For example, lettuce suspected of exposure to C_2H_4 would develop symptoms more rapidly at 20°C than at 5°C. Tomatoes suspected of chilling exposure would develop symptoms much more rapidly at 25°C than at 15°C, but storage above 30°C would induce high-temperature injury.

A major limitation encountered in quality management of fresh products is sample-to-sample variation within a lot. Predicting the behavior of the "average" item provides poor correlations and gives little information about the range of item variability. If a predictable path of quality degradation occurs under given conditions and attributes can be determined by nondestructive evaluation, individual items can be modeled (see Chapter 8). Such models can be used to provide a much more accurate prediction of the average item and of the range of variability within a lot, as well as to provide information on the time of optimal marketability for that lot.

By assuming that changes in quality of whole fruits and vegetables are manifestations of activities at the cellular and molecular levels, development of cellular and molecular indices of quality provides a promising opportunity. Such indices should (1) have relevance to the quality specifications, (2) be readily measured throughout the handling system, and (3) provide sufficient predictability of product behavior to permit management decisions. Cellular or molecular indices that can distinguish between reversible and irreversible damage caused by short-term stress could be used to manipulate storage conditions to prolong shelf life without sacrificing quality.

Goals of quality management include the delivery of fresh products at an optimal level of quality while minimizing waste. A better understanding of stress response at the cellular and molecular levels can provide critical information needed to minimize losses due to stress, to develop indices for predicting quality changes, and to use stress conditions to extend shelf life while maintaining quality.

Bibliography

An, J.-F., and Paull, R. E. (1990). Storage temperature and ethylene influence on ripening of papaya fruit. *J. Amer. Soc. Hort. Sci.* **115**, 949–953.

American Society of Heating, Refrigeration, and Air Conditioning Engineers (1986). "ASHRAE Handbook—Refrigeration Systems and Applications." American Society of Heating, Refrigeration, and Air Conditioning Engineers, Atlanta, Georgia.

Autio, W. R., and Bramlage, W. J. (1986). Chilling sensitivity of tomato fruit in relation to ripening and senescence. *J. Amer. Soc. Hort. Sci.* **111**, 201–204.

Ben-Arie, R., and Lavee, S. (1971). Pectic changes occurring in Elberta peaches suffering from wooly breakdown. *Phytochemistry* **10**, 531–538.

Ben-Arie, R., and Lurie, S. (1986). Prolongation of fruit life after harvest. *In* "CRC Handbook of Fruit Set and Development" (S. Monselise, ed.), pp. 493–520. CRC Press, Boca Raton, Florida.

Ben-Yehoshua, S. (1987). Transpiration, water stress, and gas exchange. *In* "Postharvest Physiology of Vegetables" (J. Weichmann, ed.), pp. 113–170. Marcel Dekker, New York.

Ben-Yehoshua, S., Shapiro, B., and Moran, R. (1987). Individual seal-packaging enables the use of curing at high temperatures to reduce decay and heal injury of citrus fruits. *HortScience* **22**, 777–783.

Berry, J., and Bjorkman, O. (1980). Photosynthetic response and adaptation to temperature in higher plants. *Ann. Rev. Plant Physiol.* **31**, 491–543.

Biggs, M. S., Woodson, W. R., and Handa, A. K. (1988). Biochemical basis of high-temperature inhibition of ethylene biosynthesis in ripening tomato fruits. *Physiol. Plant* **72**, 572–578.

Bodner, M., and Larcher, W. (1987). Chilling susceptibility of different organs and tissues of *Saintpaulia ionantha* and *Coffea arabica*. *Agnew. Botanik* **61**, 225–242.

Bonhoff, A., Rieth, B., Goleck, J., and Grisebach, H. (1987). Race cultivar-specific differences in callose deposition in soybean roots following infection with *Phytophtora megasperma* F. sp. *glycinea*. *Planta* **172**, 101–105.

Brackett, R. E. (1987). Microbiological consequences of minimally processed fruits and vegetables. *J. Food Qual.* **10**, 195–206.

Bramlage, W. J., and Meir S. (1990). Chilling injury of crops of temperate origin. *In* "Chilling Injury of Horticultural Crops" (C. Y. Wang, ed.), pp. 37–49. CRC Press, Boca Raton, Florida.

Buescher, R. W. (1979). Influence of high temperature on physiological and compositional characteristics of tomato fruits. *Lebensm. Wiss. Technol.* **12**, 162–164.

Caldwell, C. R., and Whitman, C. E. (1987). Temperature-induced protein conformational changes in barley root plasma membrane-enriched microsomes. I. Effect of temperature on membrane protein and lipid mobility. *Plant Physiol.* **84**, 918–923.

Cheng, T.-S., and Shewfelt, R. L. (1988). Effect of chilling exposure of tomatoes during subsequent ripening. *J. Food Sci.* **53**, 1160–1162.

Chinnan, M. S. (1989). Modeling gaseous environment and physiochemical changes of fresh fruits and vegetables in modified atmospheric storage. *In* "Quality Factors of Fruits and Vegetables" (J. J. Jen, ed.), pp. 189–202. American Chemical Society, Washington, D.C.

Darvill, A. G., and Albersheim, P. (1984). Phytoalexins and their elicitors—A defense against microbial infection in plants. *Ann. Rev. Plant Physiol.* **35**, 243–275.

Dekazos, E. D., and Worley, J. F. (1967). Induction of callose formation by bruising and aging of red tart cherries. *J. Food Sci.* **32**, 287–289.

Delmer, D. P. (1990). Role of the plasma membrane in cellulose synthesis. *In* "The Plant Plasma Membrane: Structure, Function and Molecular Biology" (C. Larsson and I. M. Moller, eds.), pp. 256–268. Springer-Verlag, New York.

Duck, N., McCormick, S., and Winter, J. (1989). Heat shock protein hsp 70 cognate gene expression in vegetative and reproductive organs of *Lycopersicon esculentum*. *Proc. Natl. Acad. Sci. U.S.A.* **86**, 3674–3678.

Duggar, B. M. (1913). Lycopersicon, the red pigment of the tomato, and the effect of conditions upon the development. *Washington Univ. Studies* **1**, 22–45.

Floros, J. D., and Chinnan, M. S. (1990). Effect of film perforation on the quality of individually seal packaged tomatoes. *J. Food Qual.* **13**, 317–329.

Frost, D. J., Read, S. M., Drake, R. R., Haley, B. E., and Wasserman, B. P. (1990). Identification of the UDP-glucose-binding polypeptide of callose synthase from *Beta vulgaris* L. by photoaffinity labeling with 5-azido-UDP-glucose. *J. Biol. Chem.* **265**, 2162–2167.

Fukushima, T. (1978). Chilling-injury in cucumber fruits. VI. The mechanism of pectin de-methylation. *Sci. Hort.* **9**, 215–226.

Giovannoni, J. J., Della Penna, D., Bennett, A. B., and Fischer, R. L. (1989). Expression of a chimeric polygalacturonase gene in transgenic *rin* (ripening inhibitor) tomato fruit results in polyuronide degradation but not fruit softening. *Plant Cell* **1**, 53–63.

Glenn, G. M., Reddy, A. S. M., and Poovaiah, B. W. (1988). Effect of calcium on cell wall structure, protein phosphorylation, and protein profile in senescing apples. *Plant Cell Physiol.* **29**, 565–572.

Graham, D., and Patterson, B. D. (1982). Responses of plants to low, nonfreezing temperatures: Proteins, metabolism, and acclimation. *Ann. Rev. Plant Physiol.* **33**, 347–372.

Grierson, D., Maunders, M. J., Slater, A., Ray, J., Bird, C. R., Schuch, W., Holdsworth, M. J., Tucker, G. A., and Knapp, J. E. (1986). Gene expression during tomato ripening. *Phil. Trans. R. Soc. Lond. B* **314**, 399–410.

Guye, M. G. (1987). Chilling and age related changes in the free sterol composition of *Phaseolus vulgaris* L. primary leaves. *Plant Science* **53**, 209–213.

Hale, M. G., and Orcutt, D. (1987). "The Physiology of Plants under Stress." John Wiley & Sons, New York.

Hardenburg, R. E., Watada, A. E., and Wang, C. Y. (1986). "The Commercial Storage of Fruits, Vegetables, and Florist and Nursery Stocks." USDA Agriculture Handbook No. 66. U.S. Govt. Printing Office, Washington, D.C.

Hatton, T. T., and Cubbedge, R. H. (1981). Effects of ethylene on chilling injury and subsequent decay of conditioned early 'Marsh' grapefruit during low-temperature storage. *HortScience* **16**, 783–784.

Herner, R. C. (1987). High CO_2 effects on plant organs. *In* "Postharvest Physiology of Vegetables" (J. Weichmann, ed.), pp. 239–253. Marcel Dekker, New York.

Hille, J., Koornneef, M., Ramanna, M. S., and Zabel, P. (1989). Tomato: A crop species amenable to improvement by cellular and molecular methods. *Euphytica* **42**, 1–23.

Hinch, J. M., and Clarke, A. E. (1982). Callose formation in *Zea mays* as a response to infection with *Phytophthora cinnamomi*. *Physiol. Plant Pathol.* **21**, 113–124.

Huber, D. J. (1983). The role of cell wall hydrolases in fruit softening. *Hort. Rev.* **5**, 169–219.

Hung, Y.-C., and Prussia, S. E. (1989). Effect of maturity and storage time on bruise susceptibility of peaches (cv. 'Red Globe'). *Trans. ASAE* **32**, 1377–1382.

Hwang, C. H., and Zimmerman, J. L. (1989). The heat shock response of carrot cells. *Plant Physiol.* **91**, 552–558.

Inaba, M., and Crandall, P. G. (1988). Electrolyte leakage as an indicator of high-temperature injury to harvested mature green tomatoes. *J. Amer. Soc. Hort. Sci.* **113**, 96–99.

Jackman, R. L., Yada, R. Y., Marangoni, A., Parkin, K. L., and Stanley, D. W. (1988). Chilling injury, a review of quality aspects. *J. Food Qual.* **11**, 253–278.

Jones, D. H. (1984). Phenylalanine ammonia-lyase: Regulation of its induction and its role in plant development. *Phytochemistry* **23**, 1349–1359.

Kader, A. A. (1985). "Postharvest Technology of Horticultural Crops." Agriculture and Natural Resources Publications, University of California, Berkeley.

Kader, A. A. (1986a). Biochemical and physiological basis for effects of controlled and modified atmospheres on fruits and vegetables. *Food Technol.* **40(5)**, 99–104.

Kader, A. A. (1986b). Potential applications of ionizing radiation in postharvest handling of fresh fruits and vegetables. *Food Technol.* **40-(5)**, 117–121.

Kader, A. A., Zagory, D., and Kerbel, E. L. (1989). Modified atmosphere packaging of fruit and vegetables. *Crit. Rev. Food Sci. Nutr.* **28**, 1–30.

Kays, S. J. (1991). "Postharvest Physiology and Handling of Perishable Plant Products." Van Nostrand-Reinhold, New York.

Ke, D., and Saltviet, M. E. (1989). Regulation of russet spotting, phenolic metabolism, and IAA oxidase by low oxygen in iceberg lettuce. *J. Amer. Soc. Hort. Sci.* **114**, 638–642.

Kendall, E. J., and McKersie, B. D. (1989). Free radical and freezing injury to cell membranes of winter wheat. *Physiol. Plant* **76**, 86–94.

Key, J. L., Kimpel, J., and Nagao, R. T. (1987). Heat shock gene families of soybean and the regulation of their expression. *In* "Plant Gene Systems and Their Biology" (J. L. Key and L. McIntosh, eds.), pp. 87–97. Liss, New York.

Kohle, H., Jeblick, W., Poten, F., Blaschek, W., and Kauss, H. (1985). Chitosan-elicited callose synthesis in soybean cells as a Ca^{2+}-dependent process. *Plant Physiol.* **77**, 544–551.

Labuza, T. P., and Breene, W. M. (1989). Applications of "active packaging" for improvement of shelf-life and nutritional quality of fresh and extended shelf-life foods. *J. Food Proc. Pres.* **13**, 1–69.

Lawton, K. A., Huang, B., Goldsbrough, P. B., and Woodson, W. R. (1989). Molecular cloning and characterization of senescence-related genes from carnation flower petals. *Plant Physiol.* **90**, 690–696.

Leshem, Y. Y. (1988). Plant senescence processes and free radicals. *Free Rad. Biol. Med.* **5**, 39–49.

Lindow, S. E. (1983). Methods of preventing frost injury caused by epiphytic ice-nucleation-active bacteria. *Plant Dis.* **67**, 327–333.

Lurie, S., Klein, J. D., and Ben Arie, R. (1991). Prestorage heat treatment delays development of superficial scald on 'Granny Smith' apples. *HortScience* **26**, 166–167.

Lyons, J. M. (1973). Chilling injury in plants. *Ann. Rev. Plant Physiol.* 24, 445–466.

Lyons, J. M., and Briedenbach, R. W. (1987). Chilling injury. *In* "Postharvest Physiology of Vegetables" (J. Weichmann, ed.), pp. 305–326. Marcel Dekker, New York.

McKersie, B. D., Senaratra, T., Walker, M. A., Kendall, E. J., and Hetherington, P. R. (1988). Deterioration of membranes during aging in plants: Evidence for free radicals. *In* "Sesescence and Aging in Plants" (L. D. Nooden and A. C. Leopold, eds.), pp. 441–464. Academic Press, New York.

Markhart, A. H. (1986). Chilling injury: A review of possible causes. *HortScience* **21**, 1329–1333.

Marmiroli, N., Terzi, V., Odoardi Stanca, M., Lorenzoni, C., and Stanca, A. M. (1986). Protein synthesis during cold shock in barley tissues: Comparison of two genotypes with winter and spring growth habit. *Theor. Appl. Genetics* **73**, 190–196.

Minorsky, P. V. (1985). An heuristic hypothesis of chilling injury in plants: A role for calcium as the primary physiological transducer. *Plant Cell Environ.* **8**, 75–94.

Minorsky, P. V. (1989). Temperature sensing by plants: A review and hypothesis. *Plant Cell Environ.* **12**, 199–135.

Morré, D. J. (1990). Comparison of plant and animal signal transducing systems. *In* "Signal, Perception and Transduction in Higher Plants" (R. Ranjeva and A. M. Boudet, eds.), pp. 307–322. Springer-Verlag, Amsterdam.

Mundt, J. O. (1977). Fungi in the spoilage of vegetables. *In* "Food and Beverage Mycology" (L. R. Beuchat, ed.), pp. 110–128. AVI/Van Nostrand Reinhold, New York.

Murata, N., and Nishida, I. (1990). Lipids in relation to chilling sensitivity in plants. *In* "Chilling Injury of Horticultural Crops" (C. Y. Wang, ed.), pp. 181–199. CRC Press, Boca Raton, Florida.

Nur, T., Ben Arie, R., Lurie, S., and Altman, A. (1986). Involvement of divalent cations in maintaining cell membrane integrity in stressed apple fruit tissue. *J. Plant Physiol.* **83**, 63–68.

Oquist, G. (1983). Effects of low temperature on photosynthesis. *Plant Cell Environ.* **6**, 281–300.

Paliyath, G., Lynch, D. V., and Thompson, J. E. (1987). Regulation of membrane phospholipid catabolism in senescing carnation flowers. *Physiol. Plant.* **71**, 503–511.

Parkin, K. L., Marangoni, A., Jackman, R. L., Yada, R. Y., and Stanley, D. W. (1989). Chilling injury. A review of possible mechanisms. *J. Food Biochem.* **13**, 127–153.

Patterson, B. D., and Reid, M. S. (1990). Genetics and environmental influences on the expression of chilling injury. *In* "Chilling Injury of Horticultural Crops" (C. Y. Wang, ed.), pp. 87–112. CRC Press, Boca Raton, Florida.

Paull, R. E. (1990). Chilling injury of crops of tropical and subtropical origin. *In* "Chilling Injury of Horticultural Crops" (C. Y. Wang, ed.), pp. 17–36. CRC Press, Boca Raton, Florida.

Paull, R. E., and Chen, N. J. (1990). Heat shock response in field-grown ripening papaya fruit. *J. Amer. Soc. Hort. Sci.* **115**, 623–631.

Picton, S., and Grierson, D. (1988). Inhibition of expression of tomato-ripening genes at high temperature. *Plant. Cell Environ.* **11**, 265–272.

Pierson, C. F., Ceponis, M. J., and McColloch, L. P. (1971). "Market Diseases of Apples, Pears, and Quinces." USDA ARS Agricultural Handbook No. 376. U.S. Govt. Printing Office, Washington, D.C.

Platt-Aloia, K. A., and Thomson, W. W. (1987). Freeze-fracture evidence for lateral phase separations in the plasmalemma of chilling-injured avocado fruit. *Protoplasma* **136**, 71–80.

Poovaiah, B. W. (1988). Molecular and cellular aspects of calcium action in plants. *HortScience* **23**, 267–271.

Raison, J. K., and Lyons, J. M. (1986). Chilling injury: A plea for uniform terminology. *Plant Cell Environ.* **9**, 685–686.

Reid, M. S. (1985). Ethylene in postharvest technology. *In* "Postharvest Technology of Horticultural Crops" (A. A. Kader, ed.), pp. 68–74. Agriculture and Natural Resources Publications, University of California, Berkeley.

Richardson, T. and Hyslop D. B. (1985). Enzymes. *In* "Food Chemistry" (O. R. Fennema, ed.) pp. 371–476. Marcel Dekker, New York.

Rizk, A.-F., and Wood, G. E. (1980). Phytoalexins of leguminous plants. *Crit. Rev. Food Sci. Technol.* **13**, 245–295.

Rosenberger, D. A. (1990). Postharvest integrated pest management. *Tree Fruit Postharvest J.* **1(1)**, 8–12.

Schoffl, F., Bauman, G., Raschke, E., and Bevan, M. (1986). The expression of heat-shock genes in higher plants. *Phil. Trans. Royal Soc. Lond. B* **314**, 453–468.

Senaratna, T., and McKersie, B. D. (1986). Loss of desiccation tolerance during seed germination: A free radical mechanism of injury. *In* "Membranes, Metabolism and Dry Organisms" (C.A. Leopold, ed.), pp. 85–101. Comstock, Ithaca, New York.

Sharkey, P. J., and Peggie, I. D. (1984). Effects of high-humidity storage on quality, decay, and storage life of cherry, lemon, and peach fruits. *Sci. Hort.* **23**, 181–190.

Shewfelt, R. L. (1986). Postharvest treatment for extending the shelf life of fruits and vegetables. *Food Technol.* **40(5),** 70–89.

Shewfelt, R. L. (1992). Response of plant membranes to chilling and freezing. *In* "The Plant Membrane: A Biophysical Approach." pp. 192–219. Kluwer, Amsterdam.

Shewfelt, R. L., Brecht, J. K., Beverly, R. B., and Garner, J. C. (1989). Modification of conditions at the wholesale warehouse to improve quality of fresh-market tomatoes. *J. Food Qual.* **11,** 397–409.

Smith, C. J. S., Watson, C. F., Ray, J., Bird, C. R., Morris, P. C., Schuch, W., and Grierson, D. (1988). Antisense RNA inhibition of polygalacturonase gene expression in transgenic tomatoes. *Nature (London)* **334,** 724–726.

Steponkus, P. L. (1981). Responses to extreme temperatures—Cellular and subcellular bases. *In* "Encyclopedia of Plant Physiology" (O. L. Lange, P. S. Nobel, C. B. Osmid, and H. Ziegler, eds.), Vol. 12A, pp. 71–102. Springer-Verlag, New York.

Stewart, R. R. C., and Bewley, J. D. (1980). Lipid peroxidation associated with accelerated aging of soybean axes. *Plant Physiol.* **65,** 245–248.

Thompson, G. A. (1989). Lipid molecular species retailoring and membrane fluidity. *Biochem. Soc. Trans.* **17,** 286–289.

Thompson, J. E. (1988). The molecular basis for membrane deterioration during senescence. *In* "Senescence and Aging in Plants" (L. D. Nodden and A. C. Leopold, eds.), pp. 51–83. Academic Press, New York.

Tighe, D. M., and Heath, M. C. (1982). Callose induction in cowpea by uridine diphosphate glucose and calcium phosphate-boric acid treatments. *Plant Physiol.* **69,** 366–370.

Tomes, M. L. (1963). Temperature inhibition of carotene synthesis in tomato. *Bot. Gaz.* **124,** 180–185.

Tucker, G. A., and Grierson, D. (1987). Fruit ripening. *Biochem. Plants* **12,** 265–318.

Van den Berg, L., and Lentz, C. P. (1966). Effect of temperature, relative humidity, and atmospheric composition on changes in quality of carrots during storage. *Food Technol.* **20,** 954–957.

Van Etten, H. D., Matthews, D. E., and Matthews, P. S. (1989). Phytoalexin detoxification: Importance for pathogenicity and practical implications. *Ann. Rev. Phytopathol.* **27,** 143–164.

Vigh, L., Gombos, Z., and Joo, F. (1985). Selective modification of cytoplasmic membrane fluidity by catalytic hydrogenation provides evidence of its primary role in chilling susceptibility of the blue-green alga *Anacystis nidulans. FEBS Lett.* **191,** 200–204.

von Mollendorff, L. J., and deVilliers, O. T. (1988). Role of pectolytic enzymes in the development of woolliness in peaches. *J. Hort. Sci.* **63,** 53–58.

Wang, C. Y. (1989). Chilling injury of fruits and vegetables. *Food Rev. Int.* **5,** 209–236.

Wang, C. Y. (1990). "Chilling Injury of Horticultural Crops." CRC Press, Boca Raton, Florida.

Wasserman, B. P. (1990). Expectations and role of biotechnology in improving fruit and vegetable quality. *Food Technol.* **44(2),** 68–71.

Weichmann, J. (1987). "Postharvest Physiology of Vegetables." Marcel Dekker, New York.

Wills, R. H. H., McGlasson, W. B., Graham, D., Lee, T. H., and Hall, F. G. (1989). "Postharvest: An Introduction to the Physiology and Handling of Fruits and Vegetables." AVI/Van Nostrand, New York.

Wolfe, J. (1978). Chilling injury in plants—The role of the membrane. *Plant Cell Environ.* **1,** 241–247.

Yacoob, R. K., and Filion, W. G. (1987). The effects of cold-temperature stress on gene expression in maize. *Biochem. Cell Biol.* **65,** 112–119.

Yamaki, S., and Kajiura, I. (1983). Changes in the polysaccharides of cell wall, their constituent monosaccharides and some wall-degrading enzyme activities in the watercore fruit of Japanese pear (*Pyrus serotina* Rehder var. *culta* Rehder). *J. Jap. Soc. Hort.* **52,** 250–255.

Yang, R. F., Cheng, T.-S., and Shewfelt, R. L. (1990). The effect of high temperature and ethylene treatment on the ripening of tomatoes. *J. Plant Physiol.* **136,** 368–372.

CHAPTER 13

QUALITY MANAGEMENT: AN INDUSTRIAL APPROACH TO PRODUCE HANDLING

Amos Lidror and Stanley E. Prussia

I. Quality Management

A. Approach

The fresh fruit and vegetable industry has the potential to benefit greatly from recent advances in quality management techniques that have revitalized the electronics, automotive, food processing, and other major industries. Major improvements in profits and quality have resulted by progressing from inspection for finding and removing defects, to quality control systems for reducing defect production, to quality assurance programs and total quality management approaches that involve the complete business cycle.

Most businesses in the fresh produce postharvest system have only an inspection function as an integral part of their operations. However, quality control and other managerial tools for monitoring and controlling fresh produce quality now are being developed to overcome difficulties unique to this industry. The most basic difficulty is the inherent lack of control during the growing process. Extreme variation in quality exists even under the best known cultural conditions including soil preparation, certified seed, plant spacing, fertilizing, irrigation, and other care (as discussed in Chapter 4). Improvements in the quality of fresh produce available to consumers at reasonable prices will require that production and delivery businesses in the postharvest system implement appropriate quality management systems for delivering consistent quality.

Agricultural production and postharvest systems for produce consists of a series of components, equipment, and human resources that are complexly independent. Achieving high quality produce should be the aim and responsibility of each member in the production and distribution system. Inferior quality reflects on the entire postharvest system.

In the modern world, retail stores and consumers require produce with consistent quality. The most reliable path to success in the fresh produce industry is to offer products with superior and consistent properties. In the future, as consumers expect and demand higher quality products, producers must respond to maintain market share. Producers must be sensitive to the fluctuating and dynamic requirements of consumers. Quality of harvested crops must be maintained between field and consumer. The packinghouse serves as the point of integration in the handling system by converting bulk loads of variable quality into a uniform pack with consistent quality. Hence, quality management strategies must be developed to decrease variability, enhance quality, and maintain stability within production and postharvest systems.

B. Produce Quality

Buyers perceive that the products of certain suppliers are significantly higher in quality than those of their competition and they buy accordingly (Feigenbaum, 1983). Quality and excellence are related to consumer perception of the product and to safety. The lack of quality as related to safety, health, and wholesomeness can result in personal injury, sickness, or even death (Tybor *et al.*, 1988). Consequently, quality has become the single most important force leading to business success in markets.

An agricultural production organization cannot function well as a system without a well defined goal. Thus, the meaning of quality must be defined, and the definition with the same interpretation by most individuals in the system. It is necessary to insure as much agreement as possible among the trade interests, the official inspection control, and the producers.

Many people have favorite definitions for quality. In production, it is not simply a degree of excellence, peculiar character, or distinguishing attribute, as defined in dictionaries. Quality as a goal should be formulated from the perspective of the consumer; thus, the expression "quality" requires a comparison of the criteria with reality (see Chapter 7). Experts have defined quality briefly: "conformance to requirements" (Crosby, 1979), "fitness for purpose" (Juran, 1988), "the sum of the attributes or properties that describe the product" (McDermott and Cound, 1971), "conformance to a customer's price-limited need" (Groocock, 1986), "conformance to a customer's price-limited anticipated needs" (Lidror and Prussia, 1990), and other analogous definitions. A common definition was formulated by the International Organization for Standardization (ISO) in Europe: "Quality is the degree in which the whole of characteristics of a product meets the requirements that spring from the goal of use."

Attributes should be expressed quantitatively in terms that can be detected and evaluated objectively. Quality of design or development is the accumulative of the product characteristics or a measure of how well the product achieves its expected purpose. Quality of conformance is a degree of realizing the quality of design.

C. Quality Inspection, Control, and Assurance

Grade inspection of agricultural products is the function of comparing products at the end of a production process with accepted specifications or other recognized requirements. Specifications can be used to motivate producers to supply products of better quality, but principally inspection is no more than a postmortem operation performed after the product has been prepared for shipment completely. The inspection activity adds nothing to the value of the shipment. The purpose of grade inspection is to certify that lots have been separated into different grades or to differentiate between acceptable and unacceptable lots. The lots rated as conforming are accepted, and others are rejected.

Quality control (QC) is more than inspection. QC consists of systematized activities. Before QC was used in industry, quality production was achieved by inspection to remove defective items. Effective control prevents dispensable and wrong actions, from field work to packing and delivery of produce. QC also keeps operations from deviating from consumer expectations. QC must demonstrate continuous progress by decreasing the rejections according to inspection results. To achieve quality of conformance, it is suggested to follow a systematic approach and control the main handling stages, for example, crop production, harvesting, transporting, reception at the packinghouse, sorting, packing, and delivery of the final product.

Specifications should consist of quality factors or characteristic attributes, criteria with definitions and descriptions, using understandable terminology, must be delineated. Specifications must be understood easily, and be simple, precise, and practicable using measuring techniques available to the producer. High inspection reliability can be achieved best by clarifying the quality criteria.

To promote international trade in fruits and vegetables, the European Organization for Economic Cooperation and Development (OECD, 1983) established and applied international standards. More than 20 countries were members of the scheme for the standards, including some who trade from outside Europe: Canada, New Zealand, the United States, and Israel. To make the specifications clear, explanatory brochures provide well-designed common interpretations for the various provisions contained in the standards through the use of clear terminology, illustrations, and color photographs. A study carried out under the scheme of the OECD shows much influence of the standardization of the marketing of fruits and vegetables. The producer must realize that the standardization of products begins when they are still on the tree or in the field, and must endeavor constantly to adapt to market requirements. Market organizations were influenced, as were trade organizations, quality inspection procedures, costs, and prices.

Total quality management (TQM) is a structured and organized management approach, with the ultimate goal of meeting customer expectations. Its main application is on the management process, which is responsible for planning, controlling, and creating quality culture and continuous improvement to function effectively and ultimately to reach levels of high quality. Quality assurance (QA) involves all processes related to maintaining and improving quality conformance, including the systems approach for quality management, TQM and QC facilities, essential feedback information, and quality pricing for producers (Gudnason, 1982). The first generation for quality maintenance in industry involved only quality inspection at the end of a production process. The second generation involved QC during the execution of a process. The third generation is now developing: QA facilities and techniques are an integral part of a total management system. QA methods used successfully by manufacturing industries are not directly transferable to packinghouse operations, but newly developed and adapted QA programs for agricultural fresh produce should improve consistency and reduce losses in harvesting, transportation, packing, and all other handling operations.

II. Outlines of Quality Control Techniques for Agricultural Products

A. Data Collection

Data should be recorded for analysis by using appropriate statistical methods, including basic statistics, frequency distributions, histograms, and graphs, as described in various basic publications for statistics. Data collection forms must be made as simple as possible, saving effort required to fill in forms, and have clear instructions that do not require the control charts typical of manufacturing applications. (An example is given in Fig. 1.)

Statistically valid sampling at every change-over point to determine if processes are under control and to identify corrective actions is the main means of controlling the quality of agricultural products as they flow from the field through the marketing and distribution system (Lidror and Prussia, 1990).

Peach Quality Control—on receiving.

Date _____   *Specifications*: Size (diameter) ["]

                                                             Ground Color (chip #) [   ]

Crew (opt.) _____   Serious defects _____

Sample size = *50*   Warning limit = *30* defectives, for under size only.
                                                   (sign W for warning, near to findings #).

Code (crew)	Arrival time	Bins		Findings			Marks		Calculated bonus (%)
				Under size #	Under color #	Serious defects #	Under Quality	Quality value	
			1						
			2						
			3						
			4						
			5						
			6						
			7						
			8						
			9						
			10						

Under quality $= \dfrac{\text{\% under size}}{100} + \dfrac{\text{\% under color} + \text{\% serious defects}}{200}$

Quality value $= 1.00 -$ under quality

Calculated bonus $= 20\% \times$ quality value $(= \%$ of additional payment)

Figure 1 Inspection form for peaches at receiving.

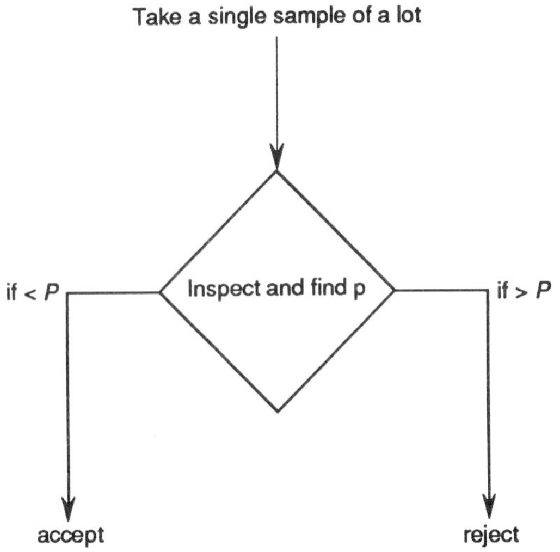

Take a single sample of a lot

if < P ——— Inspect and find p ——— if > P

accept   reject

Figure 2 Single sampling plan for produce attributes. *P*, Predetermined proportion number; p, inspected proportion (percentage defective).

B. Sampling Inspection

In acceptance sampling inspection, the examination is used to determine if the lot of fresh produce meets specifications. Supervisors are concerned about the disposition of the lot. Diverse sampling plans are discussed in basic publications and manuals for statistical QC methods (McDermott and Cound, 1971; Shainin, 1971; Brumbaugh, 1982; Feigenbaum, 1983; Messina, 1987).

Any sampling plan is dependent on lot size, planned sample size, and the anticipated acceptance number of percentage defective. When using a single sampling plan (Fig. 2), rejection of a lot on the basis of substandard quality is not an excellent test of the quality of the lot. Results may be used better as feedback information or for pricing according to quality evaluation.

By using a multistage sequential sampling plan, decision making becomes more efficient and rejection decisions are more reliable at a cost of increased complexity in the sampling procedure (Fig. 3). Any sampling plan has a certain element of risk, represented by its operating characteristic (OC) curve, which is discussed more in basic publications and manuals for statistical QC methods. Deriving the OC curve for some situations may clarify matters and lessen the perceived problems of involved mathematics needed for statistical sampling. The most suitable sampling plan and sample sizes for each purpose, for example, single, double, or multistage plans, must be identified.

C. From Acceptance Sampling Inspection to Quality Control

Traditionally, inspection data have been used to classify produce as acceptable or rejectable. Quality may be improved if supervisors are concerned about the pro-

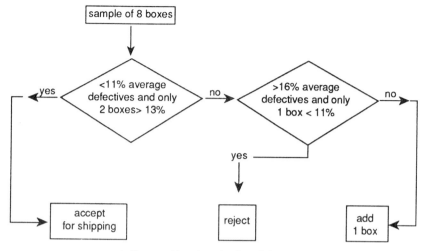

Figure 3 Example of part of a sequential sampling plan (for quality control of produce attributes at the shipping point). Decisions for a lot, at the second stage, after taking 8 boxes. Similar considerations will be made after the 9th box, and so on. (Lidror *et al.*, 1992.)

duction process. Better quality can be achieved by quality monitoring and control, checking if the production process is in statistical control, or using corrective actions and quality pricing. If a process is working property and has not been altered improperly, it will produce an acceptable product.

Ordinary and advanced QC techniques, generally used in manufacturing, are Pareto analysis, cause and effect analysis, scatter diagrams (regressions), and statistical process control (SPC) methods and charts. These are discussed in basic publications and manuals for statistical QC methods as well. The SPC charts consist of two basic types; attributes and variables, each with its own techniques for analysis. For agricultural fresh produce, it is more convenient to use attribute control charts than variable control charts to analyze go–no–go data. Feedback and results may be used for troubleshooting, improving processes, corrective actions, or pricing decisions according to quality evaluation.

The QC methods are used to identify and measure the assignable and common causes of variation so the appropriate strategy can be developed to reduce the degree of variation. These methods may provide immediate reliable feedback on process performance, prevent problems that occur on-line, and determine when to adjust the process and how effective the process is. The P charts for attribute inspection often are used to determine if the percentage defective from time to time is larger than is reasonable to expect under specific conditions. (An example of a P chart is provided in Fig. 4.) Flow chart techniques are very useful for selecting the best effective sampling points in the flow of fresh fruits and vegetables to market. (An example is provided in Fig. 5.)

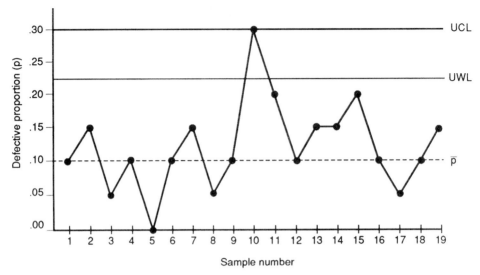

Figure 4 P chart for the fruit-sorting process (for quality control of produce attributes at sorting belts). $\overline{p}$, Expected defective proportion; $\overline{STD}$, expected standard deviation; UCL, upper control line, $\overline{p}$ + 3 $\overline{STD}$; UWL, upper warning line, $\overline{p}$ + 2 $\overline{STD}$.

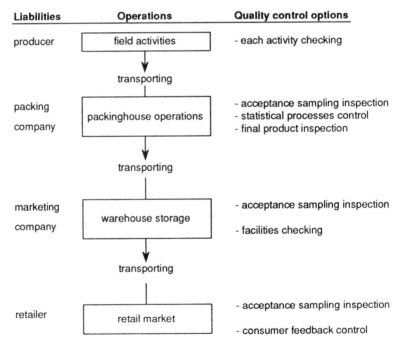

Figure 5 Typical flow chart for fresh produce with selected sampling and process control points.

III. Quality Assurance Techniques for Further Improving Produce Quality

A. Improving Quality by Total Quality Management

A systems approach to TQM considers all the interactions necessary among the various elements, and facilitates an integrated awareness of the importance of quality throughout the production process (Badiru, 1990). Communication, cooperation, and coordination functions are carried out to facilitate TQM. Production management must play an active role in implementing a systems approach, not only by proclaiming the need for better quality, but also by committing the necessary resources to its attainment.

Quality definitions must be discussed and acknowledged by everyone in the production chain. Clear specifications are needed when controlling a process such as harvesting or sorting. For example, the instructions given to the harvesters must be known to the inspector who examines the received produce. The inspector should, in turn, notify the harvester about defects that should have been prevented.

A quality assurance manual should be published by the operation, describing the quality assurance program, to present all the written descriptions of operating procedures and to supply details of quality assurance policy, plans, and procedures to workers, supervisors, managers, and all others involved. In the manual, it is advantageous to clarify the responsibilities of each group or individual and to identify the exchanges of information necessary between various sectors, producers, companies, or organizations. The concepts must be followed throughout the production process, considering it a complete system, and be accepted by most of the producers, public and private organizations, and individuals (Lidror *et al.*, 1986). Coordination and information flow among all the production, control and marketing bodies involved must be improved and control work should be simplified.

B. Quality Policy

Everyone who participates in and has responsibilities for any fraction of the agricultural production chain, for example, growers, packinghouse companies, contractors, and marketing and exporting organizations, must make a policy statement. For example, a policy may be to implement a quality assurance program as a management tool for improving product quality consistency and insuring that each shipment of crops meets or exceeds quality specifications. Policy objectives are to develop a description of actions for fulfilling the quality policy by incorporating it into a manual, making immediate improvements whenever needed, and evaluating this program after each season for any required changes.

C. Corrective Action Investigations

If product quality deviates from expectations, a case study or a corrective action, such as a problem-solving step, must be completed. Its purpose is to find the cause of any reduction in quality level and to make the changes necessary to raise the quality to expected levels or to adjust the processing variables to maintain the output at a desired set point. An investigation must be conducted when customer inspection records show unexplained changes in quality, when customer complaints indicate a change in quality, or when a request is made by a supervisor of the producer or the packinghouse. Several production organizations appoint corrective action teams whenever necessary, manned by high-level personnel, who perform this function.

IV. Systems Used for Quality Assurance

A. Systems Control for Field Activities

The grower or field production manager is responsible for growing, harvesting, and delivering the crop to the postharvest operation. Each field activity, for example, cultivation, fertilization, irrigation, pest control, harvesting, and transporting, should be specified and planned before it is conducted. To maintain the desired end-product quality, every activity must be examined on completion and performance level should be recorded. The quality inspector could check performance of all relevant activities. Consequently, produce arriving at the packinghouse or factory should meet the previously agreed on quality specifications.

Statistically analyzed data provide valuable feedback information about field activities, growing methods, areas or blocks, and fields or orchards. The data can be used for corrective actions, pricing to growers, incentives, or rewards, and as a guide on how to manage subsequent operations efficiently.

B. Sampling at Packinghouse Entrance

When products move from the field to the packinghouse, the responsibility for quality usually changes. Transport systems for fresh fruit and vegetable crops conventionally accumulate the product by lots. A quality sampling inspection technique is suitable for monitoring each lot at the change-over point. On reception of the produce at the packinghouse, inspectors take samples, inspect the product, and record their findings.

"Acceptance inspections" at the reception points are traditionally the most critical part of the quality control system for the food industry (Kramer, 1973). Most contracts to growers, packinghouses, and processors contain some incentive clauses for quality attributes. Each packinghouse manager intends to include some terms that best suit his requirements. Cooperatives sometimes institute quality inspections at the entrance to the packinghouse owned by the growers themselves. Effective control systems are necessary to disqualify low quality fruit and to motivate the growers, through financial incentives, to produce high quality produce since quality factors influence the marketplace.

Most fruits and vegetables in the United States are subject to inspection tests for minimum quality before they can be used at a processing plant or shipped from a packinghouse. Additional grade standards apply to fresh fruit.

A suitable example is a controlled production and quality system that was developed for exported citrus fruit and put into partial operation on a national scale in Israel for citrus exports (Lidror *et al.,* 1986). To determine quality, every lot of citrus fruit for export was sampled at the packinghouse entrance. Fruit samples were collected from every truck or trailer as it reached the packinghouse gate. Samples were examined immediately and the results indicated the interim storage treatment to be applied, or when and to which production line to feed the fruit, according to its quality on arrival from the orchard. This insured uniform quality on each line for better productivity and quality. Growers can use the supplied feedback information to improve their harvesting operations.

C. Process Control in Packinghouse Operations

When products flow in a continuous process, such as on a sorting or packaging line in a packinghouse, a control inspection (using statistical sampling or charts techniques) is suitable for monitoring production. Inspection should be conducted near the most quality-affecting activities to provide prompt feedback. Quality factors should be evaluated immediately to permit rapid adjustments to the process or the initiation of corrective actions.

Statistical analysis of data gathered from samples collected frequently from lines following the sorting operation provides valuable information about grading and sorting, different growers or fields, and processing methods. Data can be used for

- corrective actions and appropriate changes
- keeping sorting and other operations under control
- providing warnings and incentives to sorting groups
- planning efficient execution of subsequent treatments

Some packinghouses check samples on-line systematically at the feeding belts. The results may be used both as regular indications for further processing operations and for cross-checking the entrance inspection, thereby providing growers with continuous feedback about various quality parameters of their own produce.

An experiment was conducted to establish a process control technique similar to SPC for the official quality inspection at the end of the grade sorting operation in some citrus packinghouses in Israel. A similar experiment was performed later in some peach packinghouses in the United States. Inspectors were asked to draw samples of 20 fruits at constant short intervals of time and to evaluate selected factors. Easy-to-use control forms were provided, marked with upper control limits (UCL) for the defective proportion. (An example for peach sorting is provided in Fig. 6.) Instructions were given to inspectors about what actions to take if the defective proportion was above the predetermined limits, either as a warning to the foreman or, in the case of critical data, to make a process adjustment or undertake a critical sampling procedure.

D. Final Product Quality Examination

Statistical sampling is the principal method for monitoring final product quality. Food processing plants and some packinghouses use quality control laboratories to

Peach Quality Control—on line, grade A (U.S. No. 1).

Sample size = *20*   Upper limit = *3* defectives

Date	Time		Defectives		* W, C	Specify major defects
			No. 2	under grade		
		1				
		2				
		3				
		4				
		5				
		6				
		7				
		8				
		9				
		10				
		11				
		12				
		13				
		14				
		15				
		•				
		•				
		•				
		•				
		•				

* sign W for Warning if = 3 defectives (No. 2 + undergrades)
  sign C for Corrective action if >3 defectives.

Figure 6 Control form for peaches at sorting belt.

monitor final product quality, as demanded by some customers. Various packing-houses use simple sampling techniques for final product quality examination at the end of the packingline. The results are used to monitor the product and take corrective actions at processing operations, for example, supervising errors made by the sorters and packers and recording information to compare quality data from marketing company with quality data from packinghouse.

Agricultural fresh produce in most of the world markets must be certified officially through inspection of grading. Inspectors are present to take samples of final products for almost every import and export delivery, and to certify produce conformance to grade standards. Shipments of poor quality may be rejected or downgraded at a delivery point, packinghouse, storehouse, export terminal, warehouse, or market by authorized inspectors.

E.   Sampling at Export Terminals

An acceptance sampling inspection method ordinarily is used at the entrance to export terminals by official inspectors. A statistical sequential sampling procedure has been developed for citrus delivery inspection using a microcomputer and portable terminals (Lidror *et al.*, 1992). Its aim is to identify deliveries with a high probability of substandard quality (Fig. 7). The number of packs that are to be sampled has been reduced over that in previous seasons, due to statistical considerations, whereas reliability has been improved, thereby making it possible to focus on the deliveries most likely to be of unacceptable quality. The results show much improvement: a reduction in the time needed to inspect trucks and an increase in the number of deliveries that are permitted to unload directly for export. The sampling procedure is immediate and more efficient, reliable, and objective than the usual procedure for inspection control.

To integrate the conclusions of diverse inspectors and unify the significance of decisions by inspectors at the export terminals, a computerized system based on internal control was developed and examined (Lidror and Prussia, 1989). The internal control was operated continuously to improve the official inspection service based on systematic feedback of quality data. It became clear that quality criteria needed to be defined more precisely, as did the sampling procedure. Greater

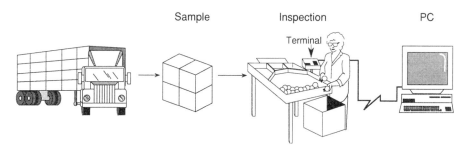

Sample   Inspection   PC

Terminal

Figure 7   Computerized citrus quality control at shipping point.

inspection reliability can be reached by continuous control with feedback of correct and incorrect decisions. This system helped enhance inspector ability, develop uniform criteria for decision making, and achieve higher reliability in decision making. Differences among inspectors were reduced when they became more conscious of certain relevant criteria.

A quality control system for fresh fruit and vegetables was developed and instituted by Agrexco (Agricultural Export Company, Israel) and has been operating on a national scale. It included sampling and inspection of products, computerized statistical data processing, quality improvements through incentive payments awarded to the numerous producers, and immediate information feedback to the producers and to the export authorities (Lidror and Kissos, 1986). The quality control center comprised sampling teams stationed near the air and sea export terminals, sampling by specially designed procedures, inspecting the incoming products from the deliveries, and submitting the results to statistical processing and analysis by a central computer system (Lidror and Silberstein, 1986).

This information reached the producers before the day is over, so they can improve the quality of their products further for the following day. A significant improvement was noted in the quality level of all the products supplied for export and controlled by the system. Mechanical damage was reduced to less than half, rot was reduced to a minimum, and sizing and package marking were improved in most of the products.

The Citrus Marketing Board of Israel sampled the loaded trucks coming from packinghouses at the entrance to ports of export (Davidson *et al.*, 1977). The samples were inspected carefully for grade, size, and packaging at a special station by quality checking staff. The collected data were computer processed and the output on waste and substandard inspection factors was used for information and monitoring purposes. A major portion of this sample was stored under predicted conditions for a simulated delay. After 10 days under controlled conditions, blemishes that have developed were inspected at the checking station for decay, and data were computer processed to determine packinghouse quality rewards and debits.

F. Sampling at Markets

An unusual and noteworthy company, with effective performance for many years, is the Outspan Company (South African Citrus Board). They maintain a waste sample examination line at a European warehouse to check the quality of citrus fruit exported from South Africa to Europe, to gather data used to make incentive payments and awards to the packinghouses in South Africa. Supermarket chain companies in Europe sample almost every dispatch at delivery centers and inspect the quality carefully for suitability to the local quality requirements.

Most countries conduct official acceptance sampling inspections for imported agricultural produce on entry into the country. Large marketing companies carry out acceptance sampling inspections for agricultural products at the entrance to the market. The collected quality data can be used to monitor incoming quality for acceptance or rejection of deliveries, pricing to trade companies, choosing markets, and further treatments needed by the produce.

Bibliography

Badiru, A. B. (1990). A systems approach to total quality management. *Indus. Eng.* **22(3)**, 33–36.

Brumbaugh, P. S. (1982). Quality control. *In* "Handbook of Industrial Engineering" (G. Salvendy, ed.), pp. 8.3.1–8.3.23. John Wiley & Sons, New York.

Crosby, P. B. (1979). "Quality Is Free." McGraw-Hill, New York.

Davidson, M., Peres, H., and Shalit, D. (1977). Quality control system for Israel citrus export. *Proc. Int. Soc. Citriculture* **1**:317–319.

Feigenbaum, A. V. (1983). "Total Quality Control," 3d Ed. McGraw-Hill, New York.

Groocock, J. M. (1986). "The Chain of Quality, Market Dominance through Product Superiority." John Wiley & Sons, New York.

Gudnason, C. H. (1982). The quality assurance system. *In* "Handbook of Indusrial Engineering" (G. Salvendy, ed.), pp. 8.1.1–8.1.17. John Wiley & Sons, New York.

Juran, J. M. (ed.) (1988). "Quality Control Handbook," rev. McGraw-Hill, New York.

Kramer, A. (1973). Fruits and vegetables. *In* "Quality Control for the Food Industry" (A. Kramer and B. A. Twigg, eds.), Vol. 2, pp. 157-228. AVI, Westport, Connecticut.

Lidror, A., Alper, Y., and Prigojin, I. (1986). Controlled production and quality systems for citrus fruit. *In* "Proceedings of the 4th Meeting of Industrial Engineers in Israel." (M. J. Rosenblatt, ed.) pp. 1-3. Ortra Ltd, Tel Aviv, Israel.

Lidror, A., and Kissos, P. (1986). Quality assurance of vegetables exported from Israel. *In* "Proceedings of the Sixth International Conference of the Society for Quality Assurance," (Z. Bluvband, ed.) pp. 6.3.1, 1-4. Ortra Ltd., Tel Aviv, Israel.

Lidror, A., Prigojin, I., and Pasternak, H. (1992). Quality inspection of fruit and vegetables with computer support. ASAE Paper 92-6001, American Society for Agricultural Engineering, St. Joseph, Michigan.

Lidror, A., and Prussia, S. E. (1989). Human factors principals for agricultural quality control. ASAE Paper 89-1116. American Society for Agricultural Engineering, St. Joseph, Michigan.

Lidror, A., and Prussia, S. E. (1990). Applications of quality assurance techniques to production and handling agricultural crops. *J. Food Qual.* **13(3)**:171-184.

Lidror, A., and Silberstein, B. A. (1986). Quality control of growing house tomatoes for export. *Acta Hort.* **191**, 381-385.

McDermott, T. C., and Cound, D. M. (1971). Inspection and quality control. *In* "Handbook of Industrial Engineering and Management" (W. G. Ireson and E. L. Grant, eds.), pp. 703-751. Prentice-Hall, Englewood Cliffs, New Jersey.

Messina, W. S. (1987). "Statistical Quality Control for Manufacturing Managers." John Wiley & Sons, New York.

Organization for Economic Co-Operation and Development (1983). "The OECD Scheme for the Application of International Standards for Fruit and Vegetables." Organization for Economic Co-Operation and Development, Publications Office, Paris, France.

Shainnin, D. (1971). Quality control. *In* "Industrial Engineering Handbook" (H. B. Maynard, ed.), pp. 8.1.19–8.1.39. McGraw-Hill, New York.

Tybor, P. T., Hurst, W. C., Reynolds, A. E., and Schuler, G. A. (1988). "Quality Control: A Model Program for the Food Industry." Cooperative Extension Service Bulletin No. 997. University of Georgia, Athens.

CHAPTER 14

PRODUCE MARKETING: NEW TECHNIQUES AT THE SUPERMARKET

Stanley M. Fletcher

I. Introduction

Before the 1980s, mass marketing strategies for food were the norm. There was less variety in the food products offered in terms of number, form, and quality than there is today. Target marketing strategies have replaced the mass marketing approach of the 1980s. New food product development has increased at record rates, as discussed in Chapter 7. Micromerchandising strategy is being now used to tune the segmentation strategies even more finely. However, none of these marketing developments could have been possible without the advent of computer technology.

United States fresh produce consumption has expanded over 23% since 1978. Produce sales in the supermarket have exceeded $27 billion, representing approximately 10% of supermarket sales. Further, supermarket produce sales have grown between 8 and 10% in the last few years in spite of significant price increases. To put this performance in perspective, red meat has increased in price but has had an overall decline in sales of 0.9%.

The marketing responsibilities of the produce manager in a supermarket have increased dramatically since the 1970s. In 1975, an average produce department handled approximately 65 items. Today, the number of items exceeds 260 and can exceed 300 items during the summer months. Between 1988 and 1989, the number of items carried by a produce department increased about 50-fold. Further, according to *The Packer's* 1992 "Fresh Trends Consumer Profile Study" (*The Packer,* 1992), approximately 60% of consumers agree that the selection of a supermarket at which to shop is influenced more by the quality of the produce department than by any other single department.

The gradual adoption of scanner and related technologies by the food industry is probably the greatest significant event to transform the food industry. This technology adoption has allowed the retailer to move from mass merchandising to target merchandising to micromerchandising. This chapter addresses scanner technology and its current and potential impact on produce marketing in the supermarket.

II. Historical Overview of Scanning

Testing of automated checkstands started as early as the 1950s. By the late 1960s, the commercial availability of the laser beam had made scanning possible. During this period, a study by Bouma (1968) on the criteria for developing the automated checkout led to United States Department of Agriculture (USDA) specifications for optical systems. Meanwhile, food industry trade groups and equipment manufacturers started expressing concerns about different code structures. In August, 1970, the Grocery Industry Ad Hoc Committee on Universal Product Coding was formed. The following year, Ricker and Krueckeberg (1971) completed a study on computerized checkout systems for retail food stores. In April, 1973, the Grocery

Industry Ad Hoc Committee (sometimes known as the Universal Code Council) announced the selection of the now familiar universal product code (UPC). Approximately a year later (June, 1974), Marsh's supermarket in Troy, Ohio, installed one of the first scanners capable of reading the UPC symbol.

In scanning, a checker moves the UPC-coded item across the path of an optical laser beam that interprets the symbol. The code is transmitted to a computer processor that identifies the product associated with that code. The computer processor returns the price, item description, and other related information such as tax status to the electronic cash register. The item price and description are shown on a terminal display, a customer display, and a receipt printer. All these actions occur instantaneously and simultaneously. In addition, the transaction information is kept in computer memory to be added to other transactions. Thus, detailed records are stored for future use in the generation of reports and analyses.

III. Hard and Soft Benefits of Scanning

Scanning savings usually are classified into hard and soft savings. Hard savings are those that result from the improved speed and accuracy of scanner ability to identify a universal product code symbol and retrieve the product price from a computer file. Examples of hard savings include improved checker productivity, greater checker accuracy, improved accounting methods, item price removal, and computerized ordering (Fletcher *et al.*, 1986). Soft savings relate more to improved management information and control, and generally accrue over time as a result of using scanner-generated information that is processed through a computer. In fact, many retailers are using desktop personal computers to achieve soft savings.

The soft savings from scanning are not quantified easily but include many that can be beneficial in making management and marketing decisions. These include

- measuring price elasticity (or price sensitivity)—a good measure of customer reaction to price changes
- monitoring direct-delivery items and measuring their sales
- measuring the effectiveness of in-store promotions as well as coupon item movements and comparing daily redemption of coupons
- evaluating new products and customer acceptance of new product introductions
- keeping track of the movement of date-coded perishables and shelf life
- making store-to-store comparisons of item movement among stores and demographic areas
- keeping track of item movement by product group for stocking and merchandising decisions
- keeping gross margin information as a basis for resetting commodities or departments as well as eliminating individual items

- controlling shrinkage by matching delivery to the store with actual sales
- more accurate and timely price changes
- monitoring the movement of advertised product to appraise the effectiveness of media use and the effect of sales over time and on "affected" items.

These soft benefits are not all conclusive. However, this area of savings has not been captured as well as the hard savings have been, an issue that has been addressed in several articles (McLaughlin and Pierson, 1984; Capps, 1986, 1989; Fletcher, 1987).

IV. Bar Codes

The UPC code found on food products is a rectangle of thick and thin lines, also known as a bar code. There are two groups of these lines, each of which represents a different number. The first set represents the product manufacturer (e.g., Hunt-Wesson, Inc.) and the second set identifies the product (e.g., 18-oz jar of creamy peanut butter). Thus, each food product found in a supermarket potentially could have a UPC code. However, a set of UPC codes for fresh produce was not developed. This deficiency removed the potential benefits of the scanner technology from the fresh produce departments.

Enabling the produce department to attain the benefits that other supermarket departments incur has not been without problems. For example, of the numerous shippers of produce items, a supermarket may purchase produce from one shipper one week and from a different shipper the following week. Because of this situation, retailers are not able to keep entering new UPC codes each time they order produce from a different shipper. If they could, they would run out of computer system memory. Thus, the retailers take a different approach for their produce department. Instead of using UPC codes, the retailers assign their own generic Price Look Up (PLU) number to each produce item. This generic PLU number is usually a two-, three-, or four-digit number that the checker enters on the register keypad. For example, 'Red Delicious' apples may have a PLU number of 16, independent of the shipper. One problem with this approach is that each retailer has its own set of PLUs. Thus, one cannot aggregate the data across retailers to obtain local or regional market quantities. Another concern is that a retailer could have, on the average, 250–300 PLUs and each cashier must be able to identify each produce item.

To address these issues, the Produce Electronic Identification Board (PEIB) was established to develop a standardized set of UPCs and PLUs for produce. After many years, in 1990 the PEIB created a set of codes for the produce industry. The list includes UPCs for produce sold in fixed-weight form as well as a set of UPCs and PLUs for bulk fresh produce. The standard codes for bulk random-weight produce differ from the traditional UPCs. The bar code still has two groups of lines

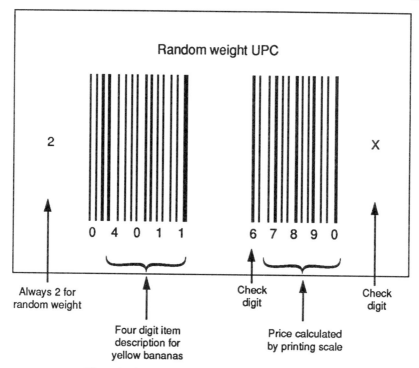

Figure 1 Standard codes for bulk random weight produce.

(Fig. 1). The first group provides a four digit number that identifies the produce (e.g., 4011 for yellow bananas). The second group provides the price of the item. This set of standardized codes will lay the groundwork for the greater sales tracking and market analysis capabilities that other food industries currently have.

V. Marketing Opportunities

Historically, the produce department has exerted the least amount of effort in terms of marketing. The only data available to them were quarterly warehouse withdrawal data that arrived three months after the end of the reported quarter. Thus, the produce manager was not able to perform many marketing functions. He or she basically was able to calculate the gross margin of the produce department.

The produce department has great marketing potential with the advent of the computer technology. Produce is second to meat in terms of average value per item, and has one of the highest gross profit margins. The three general areas of opportunity for the produce department are (1) space management, (2) computer-assisted ordering, and (3) electronic marketing.

A. Space Management

When the topic of space management is discussed, the concept of DPP/DPC (direct product profit/direct product cost) is raised. The DPP/DPC concept has emerged as a retail tool with the growth of computer technology. This concept is designed to aid decision makers in managing space to maximize sales (Stoops and Pearson, 1988). Most of the attention has focused on DPP, which is the difference between the net price paid by the customer for a product and the cost of the product to the retailer. However, DPC studies, total cost of the product to the retailer, may provide the greatest long-term benefits for the produce industry.

The DPC tool provides a means to evaluate the detailed costs of handling a particular produce item in the store. Once the direct product costs are ascertained, the produce manager can model changes in the handling to determine the impact on the DPC of the item. Thus, the DPC model can be used to examine sales benefits and cost savings due to various merchandising, packaging, and handling methods. This model is not limited to retail store operations, but can include producers, wholesalers, and shippers. For example, the method of handling and packaging a particular produce item could be studied from the producer all the way to the retailer. DPC could be used as the basis to measure the cost at each level of the distribution chain. If a modified handling and packaging approach is found to be feasible, the possibility exists that the cost to one segment of the distribution chain could increase but there may be increased profits at the next level. The challenge is for the segments to join and share the profits. Otherwise, any potential improvement will fall by the wayside.

Using DPC and scanner-generated sales information for each produce item provides the opportunity for the produce manager to alter the inventory of an item without adversely affecting customer service. In addition, this set of information can be used to tailor shelf space allocation for each item.

Although DPP/DPC can provide information that can be used by a retailer in shelf space allocation of individual produce items, caution must be exercised. As any produce manager can attest, every produce operation is different. A standard approach does not exist among stores. Further, the status of the produce department in the retail store requires the inclusion of many new, exotic, and highly perishable products that would be removed if the usual DPP/DPC criterion were followed. In addition, given that the produce department is a primary draw, competition is an overriding factor in produce decision making. One must not forget that DPP/DPC and shelf space management are merely one set of tools among many. We must resist the temptation to allow DPP/DPC numbers to dominate the decision-making process. Variety and quality cannot be sacrificed for enhanced numbers. The human and intuitive component must be merged with the numbers in the analysis.

B. Computer-Assisted Ordering

Produce ordering is a time-consuming function that traditionally has been done over the telephone. However, computer technology and scanning have given produce managers some options in ordering that improve accuracy and save time. Further, such options could improve customer service levels, reduce labor, and reduce dependence on trained personnel with unique ordering skills.

Accurate data are critical for computer-assisted ordering. Movement data would be captured from the scanning system. Hand-held computer devices can be carried by the produce manager as he or she moves through the department and storage areas taking inventory. The computer would then create a single report of the needs of the produce department based on a set of decision rules that is incorporated into the computer as an expert system program. Using these concepts, Muttiah *et al.* (1988) developed an expert system for lettuce handling at a supermarket. They found that interfacing with such an expert system could reduce the experience needed by a produce manager to make complex decisions in managing the produce department. Further, such a system also could provide updated information on produce availability.

The benefits of such a reorder system do not stop at the retail store. Sales information could be passed to other segments of the produce distribution chain. These data could help reduce fluctuations in orders. For example, this information could help the producers decide how much produce to plant as well as set harvesting schedules. In addition, packers could adjust their labor requirements and produce needs, which could reduce the amount of inventory required for their operations.

C. Electronic Marketing

The greatest potential opportunity for the use of computer-generated information in the produce department will come from electronic marketing. This new technology addresses the retail movement from mass marketing to targeted marketing to micromerchandising. This trend is facilitated through the use of customer transaction data gathered by the scanning system. This trend is enhanced further if the supermarket has equipped their customers with "frequent shopper" cards. Thus, specific customer sales information can be distinguished from other transactions in the marketing analysis.

A few retailers equipped with electronic marketing systems are offering promotions to a selected group of customers. Thus, their promotional efforts are focused on a group of customers for whom they expect the largest impact. These retailers will be able to follow-up their promotional efforts using customer transaction data to examine the benefits of the marketing effort.

A retailer could target an entire class of customers. For example, the class of customers could be those whose produce purchases are at least 25% less than the weekly average at the store. This class of customer could be given special discounts on all their produce purchases. This targeted marketing has been tried by some retailers and has been very successful, since increased sales covered the cost of the promotion.

Another area of marketing is in-store pricing control. The use of customer transaction data will help the produce manager determine consumer price perception thresholds. This information will help the produce manager determine the appropriate price markdown for each produce item when using that marketing program.

The sales during the markdown program can be captured and analyzed for effectiveness of that marketing strategy.

Retailing experts envision that, in the 1990s, produce will continue to play a critical role in establishing the character of the supermarket. Supermarkets need to use the new technology and the associated techniques fully to enhance their operations. Retailers have been enhancing their grocery departments. It is now time to enhance their produce department.

Bibliography

Bouma, J. C. (1968). Some criteria for developing the automated checkout. *Food Dist. Res. Soc. Proc.* pp. 229–237.

Capps, O., Jr. (1986). The revolutionary and evolutionary universal product code: The intangible benefits. *J. Food Dist. Res.* **17(1)**, 21–28.

Capps, O., Jr. (1989). Uses of scanner information for food industry executives. *J. Food Dist. Res.* **20(1)**, 22–40.

Fletcher, S. M. (1987). Scan data research: The status. *J. Food Dist. Res.* **18(1)**, 41–45.

Fletcher, S. M., Trieb, S. E., and Edwards, D. (1986). "An analysis of scanning computer checkout systems to determine the feasibility of scanning systems for mid- and low-sales-volume food stores." Georgia Agricultural Experiment Station Research Bulletin No.344, University of Georgia, Griffin.

McLaughlin, E. W., and Pierson, T. R. (1984). The fresh fruit and vegetable marketing system: Toward improved coordination. *J. Food Dist. Res.*, **15(1)**, 31–50.

Muttiah, R. S., Thai, C. N., Prussia, S. E., Shewfelt, R. L., and Jordan, J. L. (1988). An expert system for lettuce handling at a retail store. *Trans. ASAE*, **31(2)**, 622–628.

The Packer (1992). Fresh trends. Vance Publishing Corporation, Overland Park, Kansas.

Ricker, H. S., and Krueckeberg, H. F. (1971). "Computerized Checkout Systems for Retail Food Stores." Management Information Bulletin No. 3, Indiana State University. Terre Haute, Indiana.

Stoops, G. T., and Pearson, M. M. (1988). Direct product profit: A view from the supermarket industry. *J. Food Dist. Res.* **19(2)**, 10–14.

CHAPTER 15

FOOD SAFETY: CRITICAL POINTS WITHIN THE PRODUCTION AND DISTRIBUTION SYSTEM

Robert E. Brackett, David M. Smallwood,

Stanley M. Fletcher, and Dan L. Horton

When food is in short supply, the quality and even the safety of food becomes secondary to its availability. In contrast, people tend to take their bountiful food supply for granted and become more concerned about quality and wholesomeness of food when food is abundant.

Consumers in developed countries have come to expect and demand safe foods. Ironically, as the general level of safety of a food supply increases, the public tends to become more concerned about safety. In industrialized nations, a demand for absolute safety by consumer advocates coupled with an inability of governmental agencies to meet these demands has led to feelings of mistrust and concern.

Although the term "food safety" seems self-explanatory, there is much confusion about the hazards that are included. Moreover, opinions differ on the relative importance of food safety hazards. In general, there are several broad groups of food safety hazards, including presence of pesticides or other agricultural chemicals, naturally occurring toxicants, tampering, food additives, pathogenic microorganisms, and nutritional defects. This chapter will address only the issues of agricultural chemicals and pathogenic microorganisms. These issues represent what the public perceives and the scientific community recognizes as the most serious safety hazards associated with fresh fruits and vegetables.

I. Consumer Perceptions

Consumer perceptions of food safety are a major factor shaping the marketplace for agricultural produce. Other major factors traditionally found to be important include product prices, quality, convenience, value-added processing, availability of substitute products, and consumer buying power. These attributes can be detected visually by consumers before purchase or experienced readily through consumption after purchase. Food safety differs from many other product quality attributes such as size, color, appearance, and taste. Safety attributes are much more difficult to assess because they usually are undetectable by the senses, and adverse effects may be delayed or difficult to link to particular foods or eating occasions.

Economists refer to food safety as a credence good, meaning that consumers must rely heavily on the credibility of the marketing system to insure product quality. The difficulty that consumers have in assessing food safety attributes makes it a volatile issue for producers and marketers. New information about particular food safety risks, whether true or not, can shake consumer confidence and cause substantial economic losses to the food system.

Market disruptions are likely to be larger when consumers have substitutes from which to choose available and the product in question is perishable. With the wide variety of alternative high-quality fresh produce available in supermarkets, safety scares can have a devastating effect on suppliers, as was evidenced in the 1989 Alar® scare in red apples and in the Chilean grape crisis.

Tracking of consumer confidence during the Alar® crisis by the Food Marketing Institute (FMI)—a nonprofit research and lobbying organization of over 1500 food retailers and wholesalers—revealed that confidence dropped as low as 65% before rebounding to 79%, slightly less than the level before the scares. Consumer confidence in food safety clearly was shaken; numbers of individuals reporting "complete confidence" in 1990 had dropped 8% from a year earlier. Industry estimates of the impact of the Alar® scare range as high as $140 million in lost sales for the Washington State apple industry (Fitch, 1989).

A.   Consumer Food Safety Perceptions

The consumer of today is more aware of safety issues, more knowledgeable, and more apt to look for assurances that his or her food is safe. A study by the FMI (1990) reported that 91% of shoppers ranked food safety as "very important" or "somewhat important" in food selection; 71% percent ranked food safety as "very important." Except for the crisis of confidence that occurred during the Alar® scare, annual FMI surveys consistently report that 4 of 5 shoppers are satisfied completely or mostly with the safety of food in the supermarket. Some 71% expressed a high degree of confidence in the safety of the food supply and only 2% expressed "very doubtful" as a response. However, although the vast majority express confidence in the food supply, more than 73% continue to harbor significant food safety concerns (Table I). In 1990, 80% of the FMI respondents considered pesticides a serious food safety hazard. Thus, the food industry continues to be at risk from rapid changes in consumer perceptions of food safety.

Consumers need and are more likely to seek food safety information during and soon after a crisis. According to FMI, 37% of their survey respondents reported seeking out food safety information before the Alar® scare and Chilean grape crisis, compared with 46% after the scare. The survey also found that consumers were more likely to seek information from government sources than from consumer organizations or magazines. Almost 28% reported that they would look to government sources for food safety information, compared with 21% who would rely on consumer organizations and 15% who turned to magazines. Only 14% would look to newspapers for more information and less than 5% would look to friends or associates. However, consumer groups and the press probably are more active in

Table I

Particular Food Safety Concerns Reported as Serious Hazards

Safety concern	Shoppers reporting (%)						
	1984	1985	1986	1987	1988	1989	1990
Residues of pesticides and herbicides	77	73	75	76	75	82	80
Antibiotics and hormones in poultry and livestock	—	—	—	61	61	61	56
Nitrites in food	—	—	—	38	44	44	37
Irradiated foods	—	—	37	43	36	42	42
Additives and preservatives	32	36	33	36	29	30	26
Artificial coloring	26	28	26	24	21	28	21

Source: Food Marketing Institute (1990).

communicating their messages than are government sources; thus, their message is heard more frequently.

Most foods have both positive and negative food safety attributes. Thus, consumers are forced to decide based on risk-decreasing and risk-increasing factors. According to "Fresh Trends 1991" (*The Packer Focus,* 1991), an annual survey of consumers conducted by *The Packer,* some 56% of consumers reported diet and health concerns as a reason for increasing consumption of fresh fruits and vegetables. More than half agree with the statement that "the potential health benefits of eating fresh fruits and vegetables outweigh the potential risks from possible residues." The National Academy of Sciences report "Diet and Health" (National Academy of Sciences, 1989) concludes that the health-enhancing aspects of fresh fruits and vegetables outweigh any potential health risks associated with pesticide residues. The survey found that several diet and health related concerns are responsible for increased fruit and vegetable consumption, including calorie control, control of fat intake, and disease prevention. The survey also found that although consumers generally understand why farmers use pesticides, they believe that chemical use could be reduced greatly without reducing product quality. Only 1% of respondents strongly disagreed with the statement that farmers could reduce chemical usage greatly.

Consumers continue to look for assurances of food safety. According to "Fresh Trends 1991" (*The Packer Focus,* 1991), consumers expressed a strong interest in labeling of fresh produce as "certified safe by residue (pesticide) testing." More than 9 of every 10 respondents expressed interest in this type of labeling. Some food retailers have been quick to provide this assurance. The number of food retailers offering testing and certification services grew from a single chain in 1987 to 14 retailers in 1989 operating more than 740 stores (Kaufman and Newton, 1990).

Organic foods fill a growing, but small, market niche in the market for fresh produce. Some consumers and special interest groups urge consumption of organic produce as an alternative to traditionally grown produce based primarily on reduced consumer risks and environmental damage from pesticides. Legislation in the 1990 farm bill established a national certification program for organic produce that could help shape consumer perceptions by establishing a clear national standard of what constitutes an organic product. However, many consumers are not willing to pay the higher prices or accept cosmetic defects that typically are associated with organic produce. In a 1988 survey of consumers in the Atlanta area, Ott (1989) found that 34% were not willing to pay a higher price for pesticide-free fresh produce. He also found that 62% would not accept cosmetic damage, and 88% would not accept insect damage. Thus, several obstacles must be overcome before organic foods become a major factor in produce marketing.

Economists try to quantify consumer perceptions and concerns about food safety risks by measuring their willingness to pay to reduce risks. Conceptually, willingness to pay is related directly to the perceived risk and the implicit value placed on the life of the individual. A study by Hammitt (1986) examined this issue by measuring the price premium consumers pay for organic compared with traditionally grown produce. The study found that the median cost premium for organic produce is about $1 per pound per part per million of pesticide residue. Given estimates of the actual risk reduction associated with consuming organic produce, the study yielded an implied value of life exceeding $180 million. The high estimate of the value of life suggested that risks were perceived to be much higher than suggested by scientific evidence. In focus group discussions, Hammitt found that those who purchased organic produce perceived a very high level of risk from pesticide residues compared with those who consumed only conventional produce.

B.   Factors Shaping Food Safety Perceptions

Social scientists have identified several factors that affect consumer perceptions of risks. Scientists tend to measure risks in term of deaths and other adverse outcomes. In contrast, consumers often include many other psychological dimensions that

Table II
Factors Affecting Consumer Perception and Acceptance of Risk

Risk factor	Description	Example
Voluntariness	Voluntarily assumed risks are perceived as less risk and more acceptable	Microbial contamination from home canning of food and pesticide use in home gardens are likely to be more acceptable risks than unseen microbial and chemical residues in the market place
Control	The more control a consumer perceives over a risk, the more acceptable the risk	Consumers can exert some control over microbial contamination by proper storage, sanitation, and cooking
Fairness	Perceptions that the risk is borne more by some groups than by others due to political factors or consumer income will decrease risk acceptance	Consumers would be more outraged (lower acceptance) about a pesticide risk if they thought a politician was manipulating the regulatory system to protect a special interest group
Process	Making consumers part of the decision making process that affects risk lowers risk perceptions	Excluding consumers from risk management decisions and public comment can decrease consumer acceptance of the risk
Morality	Risks perceived as morally unjust are perceived as greater	If consumers believe that farmers use chemicals only to make a profit, perceived risks will be greater
Familiarity	New high-technology risks that are not well understood by the consumer are perceived to be greater	Consumer perception of food irradiation as a technique to control microbial contamination; synthetic chemicals are likely to generate more public concern than natural plant constituent chemicals
Memorability	Personal experience or other memorable event can heighten risk perception	A consumer who has experienced an episode with a food hazard directly or knows someone who has is likely to be more concerned with the risks
Dread	Some illnesses or consequences are more feared than others and tend to heighten risk perceptions	People sometimes fear cancer more than auto fatalities
Diffusion in space and time	The greater the separation of incidence in both space and time, the more acceptable the risk	A hazard that kills 100 geographically dispersed people throughout the year is generally less feared than one that kills 100 people in one town in one day

affect a sense of outrage (Slovic *et al.*, 1982; Viscusi and Magat, 1987). These factors include the voluntariness of the risk, the degree to which the consumer has control, the fairness or distribution of risk across groups of individuals, the process by which the consumer is involved in the risk management and control activities, the moral or ethical perception of the nature of the risk, familiarity with the risk, memorability of or personal experience with the risk, dread or fear associated with the hazard, and the diffusion in space and time of the incidence of adverse outcomes. These risk attributes are defined and illustrated with examples in Table II. These factors help explain why some consumers are more concerned with the risk from pesticide residues than they are with that from automobiles or radon. A critical finding of psychological research on risk perceptions is that it is multidimensional and goes far beyond the simple statistical characterizations of risk. Consequently, effective risk communication strategies must address these facets and not dismiss them without due consideration.

Research relating consumer perceptions of risk provide valuable insight into understanding consumer response to information about different types of foodborne hazards. This information is useful in evaluating public demand for regulatory risk management strategies, public acceptance of new "risky" food technologies, and public reaction to new information about risk attributes of particular commodities (Covello *et al.*, 1988).

C. Consumer Perceptions Compared with Scientific Expertise

Consumer perceptions of food safety risks, when they differ from real risks, can lead to market distortions and inefficiencies. Remedies for these market imperfections include product labeling, brand identification, consumer education programs (possibly financed through generic promotion), increased government regulation and surveillance, and private certification programs.

Quantitative analysis of safety risks is a difficult and inexact science (Graham *et al.*, 1988). It is difficult because any quantification of disease statistics will be considered cold and unfeeling by someone who considers a singe case of cancer to be unacceptable. It is inexact because controlled experiments with humans are not possible. Moreover, extrapolation of animal data to human risk is not clear cut, and epidemiological data is obscured by multifactor causation. Quantitative analysis can help compare relative risks and put these risks into perspective.

Technical experts generally agree that health risks from pesticide residues are extremely small. Environmental Protection Agency (EPA) experts estimate that, at most, pesticide residues add 6000 cancers per year (EPA, 1987), less than 2% of the over 300,000 cancer fatalities per year from all sources. However, the estimate for cancers due to pesticides is overstated because not all cancers result in death and the number is extrapolated from worst case scenarios in which sensitive rodent species are fed dosages typically thousands of times higher than actual human exposure. Consequently, the actual number of cancers is likely to be smaller because of the conservative assumptions made by regulators. The maximum acceptable risk allowed by food chemical regulators is generally no more than 1 increase in cancer per million lifetimes of exposure. This means that fewer than 4 new cases of cancer each year would develop in maximally exposed individuals (assuming a 70-yr lifetime of exposure). This message has not been communicated well to consumers.

Conservative (worst case) risk assessments often are misrepresented to consumers as likely outcomes. However, the most likely or probable outcomes are often much smaller. For example, the EPA (1987) estimated the worst case risk from Alar® as 45 additional cancers per million individuals exposed over a lifetime. A study by the California Department of Agriculture (1989) estimated the upper limit as 2.6 excess cancers per million population and the most probable estimate as 3.5 excess cancers per trillion population. The most likely estimate was about one million times smaller than the upper limit. Although no one wants to argue for more cancer, the scientific measure of risk from pesticides is many times smaller than that from many other hazards that individuals readily accept. For example, 50,000 people die in motor vehicle accidents each year and 5,000–20,000 people die from indoor radon exposure (EPA, 1987).

The Food and Drug Administration (FDA), the federal agency responsible for enforcing pesticide residue levels in fresh produce, continually tests and monitors produce for residue levels. In 1989, less than 1% of the 5546 domestic fruit and vegetables samples contained residues that exceeded tolerances (FDA, 1990). Tolerances are the maximum allowed residue levels established by the EPA. Some 56% of the fruit samples and 68% of the vegetable samples showed no detectable residues at all, and less than 1% revealed residues of pesticides that have not received approval for the tested product. EPA risk analysis procedures insure that consumption of residues at the seldom reached tolerance level presents only a negligible risk (less than 1 increase in cancer per million lifetimes of exposure).

Many individuals regard health risks associated with pesticide residue levels in our food as insignificant. Recent survey information from the Pesticide Enforcement Branch of the California Department of Agriculture (1989) provides perspective and indicates that consumers may be overly concerned. The marketplace components of this 1989 California survey (9403 samples) revealed no detectable residue in 78% of the samples. Of the 22% of the samples in which residues were detected, 20% bore less than 50% of the allowable tolerance. Illegal residue levels were present in 0.71% of samples (National Food Processors Association, 1990). Despite evidence that consumers are not being exposed to dangerously high dietary pesticide levels, public perception of risk remains a very powerful force. Public concern over food safety is seen as an extremely important motivating factor in elevating integrated pest management (IPM) to new levels of acceptance in vegetable and fruit production.

Experts (Wodicka, 1985), rank foodborne hazards in decreasing order of importance as

- microbial contaminants
- malnutrition
- environmental contaminants
- toxic natural constituents
- pesticides
- food additives

Pesticides are ranked next to last in importance in the six foodborne hazards listed. However, the annual FMI survey (1990) of food shopper attitudes continues to rank pesticides high on the list of food safety concerns (Table I). In 1990, 4 of every 5 respondents in the FMI survey reported pesticides as a serious food safety hazard.

In addition, nearly 2 of 5 reported that irradiation (an alternative to fumigation) and nitrites (a preservative) are serious health concerns. Thus, consumer concerns and perceptions of food safety will continue to be major factors in the marketing of fresh fruits and vegetables in the foreseeable future.

II. Chemical and Pesticide Safety

Integrated pest management (IPM) is the biologically and economically enlightened, management-intensive successor to traditional chemical pest control. Pest management makes use of all available tools, including pesticides, to keep pest populations below damaging levels. Conventional crop protection, especially as practiced in high-value crops such as fresh vegetables and fruits is based on preventive pest control. IPM provides crop protection as needed. Growers using IPM substitute knowledge of pest biology and careful systematic crop and pest monitoring for preventive spraying without suffering loss in yield or quality.

The central concern and emphasis of this chapter is the role of food microbiology in maintaining a safe food supply. Food items injured by pest infestation are more likely to pose a health risk because of contamination by pathogenic microorganisms. Consumers and produce distributors base purchase decisions on visual appeal, insuring a widespread commitment to production systems that minimize pest injury. Market intolerance of even minimal pest injury is in ongoing conflict with public concern about pesticides in food and in the environment (Institute of Food Technologists, 1990). The contribution of IPM to food safety is subjective and inadequately quantified.

Historically, growers of such high-volume crops have not used IPM [National Research Council (NRC), 1989], probably because of its greater management requirement and its increased risk. When evaluated over long periods of time, IPM lowers pesticide use in most crops (Allen *et al.*, 1987). The intuitive sense that less spraying means less pesticide residue is neither certain nor consistent, but this premise is accepted widely.

Practitioners of IPM make prudent use of all available tools, including pesticides, to deal with pest problems. Pest management is long term in its outlook. By keeping pest injury below damaging levels, IPM seeks to maintain a rational compromise between biological and economic concerns. Pest management draws from and weaves specific management strategies from the tools at hand. Host plant resistance, cultural controls, conservation and nurture of naturally occurring biological control agents, importation of natural enemies to control exotic pests, and pesticides are all important tools. Unfortunately, nature seldom affords us access to all these IPM tools in a single commodity. Too frequently, we are forced to base IPM strategies on crop and pest monitoring with use of pesticides as needed (NRC, 1989).

The advantages that lead to grower IPM use include a general reduction in pesticide use and accompanying improved profitability (Allen *et al.*, 1987), improved coping with pest resistance to pesticides (NRC, 1989), reduced worker exposure to pesticides, and a more gentle effect on the environment. Pest management reduces our need for pesticides on many fronts. Two examples include use of pheromone mating disruption for insect control (Rothschild, 1982) and use of postharvest rinses that, although actually used to remove surface blemishing fungal lesions from apples, also have shown promise in washing off pesticide residues (Horton *et al.*, 1991). Growers, no matter how noble their ideals may be, must run their farms profitably. Successful IPM programs will spray less, or at least spray more effectively, by precisely timing pesticide applications. Pest management programs also improve management of pesticide resistance. Pest resistance to pesticides is a product of natural selection. Individuals that are able to withstand or avoid exposure to pesticides survive to reproduce. Repeated exposure to pesticides selects for resistant individuals. Pest resistance is a very serious problem. Georghiou (NRC, 1986) noted 447 insect tick and mite species, 100 plant diseases, and 55 weed species that are resistant to pesticides. Pest management, because it uses less pesticide, places less selective pressure on pests to develop resistance. Pest susceptibility to pesticides is a precious, finite, nonrenewable resource that must be protected (NRC, 1986). If pest management lacked any other attributes, its ability to slow or avoid the development of pest resistance is alone sufficient reason to use IPM.

Food safety concerns and the accompanying desire to minimize pesticides on our vegetables and fruits have increased IPM use in these crops. Under field conditions, pesticides degrade and dissipate. IPM replaces preventive pesticide use with spraying as needed. Often, IPM will reduce the amount of pesticide used in a given season, thus reducing the potential pesticide burden of the food crop. However, pesticides applied shortly before harvest will figure most prominently in the pesticide load of any crop. Disease or insect outbreaks at harvest may be devastating (Smith and Barfield, 1982). As harvest approaches, IPM or conventional growers are constrained primarily by the legal restrictions limiting the time period between pesticide application and harvest. Growers and IPM advisors are not toxicologists; they accept government assurances that respecting the preharvest application interval of each pesticide will preclude unsafe pesticide residue levels. Consideration of how a needed spray might influence residual pesticide levels seldom is exercised unless the crop is bound for a specific food safety conscientious market.

Given the variations of IPM pesticide use as needed, one can expect that IPM-grown produce generally will have lower pesticide residues than produce grown under more conventional practices. Whether this lowering of pesticide residues in IPM-produced vegetables and fruits creates safer food is debatable. Overall, the lowest realistically attainable amount of pesticide in food and in the environment is best. Pest management oriented production systems seek these goals and maintain crop quality and yields.

III. Microbiological Safety

A. Organisms of Concern

Microbiological safety is an extremely important component of food quality. Consumers as well as regulatory agencies insist that foods be safe to eat, regardless of any other quality indices. Consequently, more and more pressure is being placed on the food industry to insure that no harmful microorganisms, or as few as possible, be present on their products.

The three classes of microorganisms that affect the safety of fresh fruits and vegetables are viruses, bacteria, and parasites. Each group presents specific problems and circumstances that affect safety. However, the solutions to these problems often have much in common, irrespective of the organisms. This section will focus on those problems and solutions and will detail characteristics of the organisms of most concern in fresh produce. In addition, we will address those organisms that are not presently of concern but could increase in importance in the future.

1. Viruses

Until recent years, the role of viruses in foodborne illness remained somewhat of a mystery. Microbiologists suspected that viruses were responsible for illness. However, it was difficult to demonstrate the presence of an incriminated virus in a food suspected of causing illness. The reason for this difficulty was that methodology for recovering viruses from foods was inefficient. In recent years, newer techniques such as gene probes have provided microbiologists with a sensitive means with which to detect viruses in foods.

Many different types of viruses can infect humans. However, those gaining access through the intestinal tract are the ones of most concern in the case of fresh produce. Those viruses most likely to contaminate foods are listed in Table III. Viruses commonly cited as occurring in fresh fruits and vegetables include polioviruses, coxsackie viruses, echoviruses (Larkin, 1981), and hepatitis A virus (Cliver et al., 1984).

Viruses potentially present in fresh produce cause a wide range of symptoms ranging from minor bouts of gastroenteritis to death. For example, the so-called "Norwalk" virus is a common cause of viral diarrhea (Cliver et al., 1984). Hepatitis A causes a serious and potentially life-threatening illness involving liver injury. This virus is primarily a contaminant of raw shellfish, but outbreaks of hepatitis also have been traced to salads and fruits (Cliver et al., 1984). Polioviruses similarly cause a serious illness that can lead to paralysis and death. The number of virus particles necessary to cause illness varies considerably, but can be as few as 1 to 10 (Larkin, 1981).

Table III
Human Viruses Likely to Contaminate Foods[a]

Picornaviruses	Reoviruses
Polioviruses	Reovirus
Coxsackie virus A	Rotavirus
Coxsackie virus B	Adenoviruses
Echovirus	Human adenoviruses
Enterovirus	Papovaviruses
Hepatitis A	Human BK and JC viruses
Parvoviruses	
Human gastrointestinal viruses	

[a]Adapted from Larkin (1981).

Most viruses of concern are fecal or urinary contaminants (Larkin, 1981). As such, the viruses become a threat to human health when raw sewage and polluted waters are allowed to contact fresh produce. This situation is most likely to occur when human or animal waste is disposed of on land or when polluted waters are used for irrigation. It is also possible for infected produce workers to contaminate products during harvesting or other handling procedures.

Many viruses can survive on fresh produce long enough to infect consumers. Badawy *et al.* (1985) inoculated fresh lettuce, radishes, and carrots with rotavirus and then determined survival of the virus at 5°C storage. This virus, which causes life-threatening diarrhea, survived on the vegetables from 5 to 25 days. However, the virus survived longest on lettuce and least on carrots. Likewise, Bagdasaryan (1964) found that poliovirus, echovirus, and coxsackie viruses could survive in soils for up to 170 days and on vegetables for up to 15 days.

Several practices have been recommended for treating fresh fruits and vegetables suspected of harboring viruses. Larkin (1981) recommends blanching contaminated items for 2 min in 80°C water so all surfaces of the vegetables reach at least 70°C. Chlorine sanitizers are active against some viruses (Jay, 1986). However, improperly maintained chlorine dips or sprays (see Chapter 6) can serve to spread undesirable microorganisms. Thus, the best course of action for dealing with viruses is minimizing contamination by practicing proper sanitation and hygiene during production and handling of fruits and vegetables.

2. Bacteria

Bacteria currently are recognized as the single greatest cause of microbial foodborne illness. Foods most likely to harbor pathogenic bacteria are those of animal origin. Still, fruits and vegetables are responsible for 2–7% of confirmed cases of foodborne illness (Bryan, 1988b; Bean and Griffin, 1990). Although any foodborne pathogen conceivably could contaminate fresh produce and cause illness, only a few (Table IV) are responsible for the greatest threat. Only these bacteria will be discussed in this section.

The gram-negative bacteria are the most common sources of bacterial foodborne

Table IV
Pathogenic Bacteria and Parasites of Potential Concern in Fruits and Vegetables

Gram negative
Salmonella species
Shigella species
Escherichia coli
Pleisiomonas shigelloides
Aeromonas hydrophilia
Gram positive
Clostridium botulinum
Listeria monocytogenes
Parasites
Entamoeba histolytica
Giardia lamblia

illness associated with fresh fruits and vegetables. In general, these bacteria are enteric pathogens and cause various gastrointestinal illnesses.

Shigella is one of the most common gram-negative pathogens associated with fruits and vegetables. Shigellae are small rod-shaped bacteria closely related to the genus *Escherichia* (discussed later in this section). There are four major species of *Shigella*. *Shigella dysentariae* was the primary disease-causing species in the past and is still important in developing countries. However, *S. flexneri* and *S. sonnei* have been responsible for most of the outbreaks of shigellosis in recent years.

Shigellae are not particularly hardy and are killed easily by normal cooking. Freezing and thawing can also reduce the number of bacterial cells in foods (Wachsmuth and Morris, 1989). Moreover, many common food ingredients such as vinegar, salt, and sodium benzoate inhibit the growth of these bacteria. Shigellae also compete poorly with other enteric flora but can survive from weeks to months in foods and up to 120 days in water (Wachsmuth and Morris, 1989). In general, *Shigella* grows best at 37°C and is not considered a psychrotroph.

Shigellosis is primarily an infection that requires live bacterial organisms to cause illness. The specific symptoms of shigellosis can vary considerably, depending on the severity and nature of the illness, and the strain involved. Infections can be quite mild and be asymptomatic or involve only minor diarrhea. More serious symptoms include high fever, chills, dehydration, and sometimes bloody or mucus-containing diarrhea (Wachsmuth and Morris, 1989). The overall mortality is less than 1% (Wachsmuth and Morris, 1989).

Shigellosis is unlike many other foodborne illnesses because very few cells are required to initiate infection. Wachsmuth and Morris (1989) reported that ingesting as few as 10 cells could cause shigellosis in adult humans. The incubation period for shigellosis is about 1–7 days after ingestion. The illness usually persists for about 2 wk (Wachsmuth and Morris, 1989).

Foodborne *Shigella* infections are spread primarily through foods contaminated by human feces. The disease is disproportionately common in poorly educated or low-income individuals and migrant workers. Such individuals often are employed in the fruit and vegetable industry. Although any food conceivably can become contaminated with *Shigella,* those foods that receive no terminal heat treatment are of most concern. Consequently, fresh or minimally processed produce sometimes is implicated in outbreaks of shigellosis. One large outbreak of shigellosis involving 347 people was traced to commercially distributed shredded lettuce (Davis *et al.,* 1988). In this case, an infected food handler contaminated the lettuce during shredding. Once an outbreak such as this occurs, it often spreads further by person-to-person contact from infected to previously uninfected people.

The best way to control shigellosis is through proper hygiene practices, especially thorough hand washing. Workers must be educated about the need for such practices. In addition, workers exhibiting diarrheal illness should be barred from areas in which foods are being handled.

Salmonella is among the most well-known and largest causes of foodborne illness. These organisms are similar to *Shigella* because they are gram-negative rods. This genus is quite diverse and contains over 2000 different species and strains. Although the presence of any *Salmonella* in food is considered unacceptable, some species are thought to be pathogenic only to specific animals. For example,

S. gallinarum and *S. pullorum* affect only poultry. In contrast, other *Salmonella* species that are highly pathogenic to humans may colonize the intestines of poultry but cause no illness in the birds. Species of *Salmonella* most often encountered in foodborne outbreaks of salmonellosis include *S. typhimurium, S. enteritidis, S. infantis,* and *S. heidelberg.*

Salmonellae can grow in a temperature range of 5 to 47°C, but grow best at 35–37°C. In addition, some strains have been reported to grow at temperatures as low as 4°C, although *Salmonella* is not considered a psychrotroph. Freezing injures salmonellae to some degree, but they generally are regarded as fairly resistant to freezing (D'Aoust, 1989). Heat resistance varies with the composition of the food and the strain in question. In general, normal cooking temperatures for fruits and vegetables are sufficient to kill salmonellae. Growth of *Salmonella* also is inhibited by many food ingredients and acidic pH.

Like *Shigella, Salmonella* is an enteric pathogen and live cells are required to cause illness. The normal symptoms of typical cases of salmonellosis include non-bloody diarrhea, abdominal pain, fever, nausea, vomiting, and prostration. These symptoms usually appear 8–72 hr after ingestion of the cells. The symptoms in healthy individuals normally abate after several days. However, illness is more intense and prolonged and can be life-threatening in individuals with underlying health problems. A more serious form of salmonellosis, enteric (typhoid) fever, is caused by *S. typhi* and *S. paratyphi.* Symptoms of the enteric fever form of illness include headache, high fever, and septicemia (blood infection). This form of illness is very serious and is considered life threatening. The number of *Salmonella* cells required to cause illness depends on the strain and the general health of the patient. Usually, more than 10^5 cells are required to initiate infection, although particularly virulent strains require only hundreds of cells or less (D'Aoust, 1989). Salmonellae are primarily a problem in foods of animal origin. However, fresh fruits and vegetables have been linked to salmonellosis. Recently, several outbreaks of salmonellosis were linked epidemiologically to fresh produce, although salmonellae were never recovered from the implicated products. In the first case, a large outbreak of *S. chester* involving over 250 cases was linked to both fresh and cut cantaloupe produced in Latin America (Unrein, 1990a). Investigators hypothesized that the rinds of the cantaloupes became contaminated when contaminated water was used to wash the melons. In the second incidence, at least 148 cases of *S. javiana* gastroenteritis occurring in the midwestern United States were linked to tomatoes (Unrein, 1990b). The tomatoes later were determined to have been produced in the southeastern United States. This case was especially alarming because *Salmonella* normally would not be expected to survive in acidic tomatoes.

Aeromonas hydrophila, another gram-negative rod, is one of the "emerging pathogens" in food microbiology (Buchanan and Palumbo, 1985). This organism causes illness but its role in foodborne illness is not yet defined. *Aeromonas hydrophila* causes two related forms of gastroenteritis (Stelma, 1989). The first is a so-called "cholera-like" illness characterized by watery diarrhea and mild fever. The second form, a "dysentery-like" illness, is more serious and involves a bloody or mucus-containing diarrhea. The illness caused by *A. hydrophila* is usually self-limiting and mild, however.

Aeromonas possesses several characteristics of concern in fresh produce. First,

Aeromonas is a true psychrotroph and some strains grow at temperatures as low as 1°C (Palumbo, 1987). Second, *Aeromonas* differs from *Salmonella* and *Shigella* because it is not primarily a fecal pathogen. In fact, *Aeromonas* is ubiquitous in nature (Buchanan and Palumbo, 1985) and is especially common in water. *Aeromonas* is fairly tolerant of acidity and will grow at pHs as low as 4.0 (Hazen *et al.*, 1978). Finally, the bacterium also grows well in both ambient and modified atmospheres (Berrang *et al.*, 1989a). Consequently, *Aeromonas* could be expected to grow in most conditions under which fresh produce is stored.

Since this organism is so common in nature, it should not be surprising that it is also found in fresh produce. Callister and Agger (1987) found *A. hydrophila* on virtually 100% of the 12 different produce items surveyed. Moreover, populations were as high as 10^4 cells/gm at the time of purchase. Likewise, Berrang *et al.* (1989a) found *A. hydrophila* to be present naturally on fresh asparagus, broccoli, and cauliflower. They also noted that the organism grew to populations as high as 10^6 cells/gm on all three vegetables during storage, although growth differed for the various vegetables.

Several other gram-negative bacteria are also of potential concern in fresh produce. *Escherichia coli* is a common human enteric bacterium that is usually only thought of as an indicator of fecal contamination. However, several strains of *E. coli* are also pathogenic to some degree. In addition, some of these pathogenic strains are psychrotrophic (Palumbo, 1987).

The primary illness caused by pathogenic *E. coli* is the so-called "traveler's disease" or "Montezuma's revenge." This illness is similar to shigellosis or salmonellosis, that is, symptoms normally involve gastroenteritis. One strain, O157:H7, causes a more serious illness known as hemorrhagic colitis. Symptoms of this form of illness include profuse bloody diarrhea and may progress to include serious kidney damage and death (Doyle and Padhye, 1989). *Escherichia coli* O157:H7 primarily has been a problem in improperly cooked meats, but could also occur in vegetables.

Plesiomonas shigelloides is another uncommon pathogen that could potentially contaminate fresh produce. This organism is a close relative of *A. hydrophila* and shares many characteristics with the latter, including psychrotrophic growth (Koburger, 1989). *Plesiomonas shigelloides* also produces a diarrheal illness similar to that caused by *A. hydrophila*. Gastroenteritis caused by *P. shigelloides* primarily has been associated with fish and crustaceans (Koburger, 1989). Nevertheless, its similarity to *A. hydrophila* makes it a bacterium of which individuals involved with fresh produce also should be aware.

Listeria monocytogenes is presently one of the bacteria of most concern in fresh produce. Until the last decade or so, few individuals in the food industry were familiar with this organism. Then, a series of listeriosis outbreaks caused by contaminated foods brought *L. monocytogenes* to the forefront as an important foodborne pathogen. Indeed, an outbreak traced to a vegetable product (Schlech *et al.*, 1983) offered the first evidence that *L. monocytogenes* was a foodborne pathogen.

Listeria monocytogenes differs in many ways from the bacteria discussed thus far. First, it is a gram-positive bacterium. In addition, it is a true psychrotroph and is reported to grow at temperatures as low as 0°C (Khan *et al.*, 1973). *Listeria monocytogenes* also differs from most other foodborne pathogens because it is an environmental contaminant whose ecological niche is soil and decaying vegetation (Brackett, 1987a). *Listeria monocytogenes* also is found commonly in drains, condensate, and other soiled areas of food processing facilities. Moreover, both humans and animals are often carriers of this organism (Lovett, 1989). Consequently, *L. monocytogenes* has ample opportunity to contaminate both raw and processed foods.

The main reason *L. monocytogenes* has become such a concern in foods is related to the illness it causes. In healthy people, *L. monocytogenes* causes a mild flu-like illness that is sometimes even unrecognizable. However, listeriosis can be a far more serious illness in those individuals whose health is, for some reason, compromised. These individuals include pregnant women, the elderly, and those suffering from chronic diseases such as diabetes, cancer, or AIDS. In such individuals, listeriosis can present as meningitis and encephalitis. The mortality rate can be has high as 30% (Ciesielski *et al.*, 1988).

Plant products are an important component in the cycle of infection of *L. monocytogenes* (Fig. 1). The organism, being an environmental contaminant, can survive in soils for many months and often is found on both agronomic and vegetable crops (Beuchat *et al.*, 1990). Thus, it should be no surprise that this bacterium has been found on and illness has been associated with fresh fruits and vegetables.

Blenden and Szatalowicz (1967) were among the first to suggest that lettuce and other fresh vegetables could be responsible for cases of listeriosis. Only a decade or so later, an outbreak occurred in which salad vegetables actually were implicated epidemiologically (Ho *et al.*, 1986). The specific vegetables implicated included raw celery, tomatoes, and lettuce. However, *L. monocytogenes* never actually was isolated from the implicated products. A now well-publicized outbreak of listeriosis also was linked epidemiologically to a fresh vegetable, in this case, cabbage (Schlech *et al.*, 1983). In this case, however, identical serotypes of *L. monocytogenes* were isolated from both the cole slaw made from the incriminated cabbage and the patients. A review of the cabbage producer's farming practices revealed that the

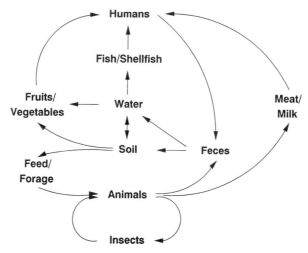

Figure 1 Hypothesized cycle of infection for *Listeria monocytogenes*. (Reprinted with permission from Brackett, 1988).

cabbage had been fertilized with sheep manure from sheep that also had exhibited symptoms of listeriosis. More recently, an outbreak of listeriosis was traced to refrigerated salted mushrooms (Junttila and Brander, 1989).

The actual extent of contamination of fruits and vegetables is unknown. However, Heisick *et al.* (1989) found the bacterium in 21% of potatoes, 14% of radishes, and up to 2% of cucumbers and cabbage. Other investigators (Beuchat *et al.*, 1990) have estimated that as much as 30–60% of raw salad vegetables in the United States may be contaminated with *L. monocytogenes*.

Once present, *L. monocytogenes* not only survives but grows well in vegetables. For example, *L. monocytogenes* strain LCDC 81-861 [the outbreak strain reported by Schlech *et al.* (1983)] was able to grow from about 10^4 to 10^8 cells/gm in raw shredded cabbage stored at 5°C for 25 days (Conner *et al.*, 1986). Similarly, Berrang *et al.* (1989b) found that *L. monocytogenes* could grow up to over 10^6 cells/gm in asparagus, broccoli, and cauliflower (Table V) and in shredded and packaged lettuce (Beuchat and Brackett, 1990b). Storing the vegetables in modified atmosphere did inhibit the growth of *L. monocytogenes*.

Listeria monocytogenes will not grow on all vegetables to the same degree. For instance, Beuchat and Brackett (1990a) found that raw carrots were bactericidal to *L. monocytogenes*. They also determined that tomatoes were too acidic to allow growth of the organism but did allow original populations to be maintained for at least 2 wk (Beuchat and Brackett, 1991).

Most handling procedures in minimally processed produce have little effect on *L. monocytogenes*. A chlorine dip (200 μg/ml), often used to sanitize vegetables, only reduced the populations of *L. monocytogenes* on brussels sprouts by about 2 $\log_{10}$ CFU/gm whereas dipping in water alone reduced populations by about 1 $\log_{10}$ CFU/gm (Brackett, 1987a). Similar concentrations of chlorine were ineffective in inhibiting growth of *L. monocytogenes* on lettuce (Beuchat and Brackett, 1991).

Listeria monocytogenes is also fairly resistant to NaCl and acid. Conner *et al.* (1986) found that the bacterium remained viable for up to 70 days in cabbage juice containing up to 4.5% NaCl. *Listeria monocytogenes* grew well in cabbage at pH 5.0 but death occurred as pH values approached 4.6. However, George *et al.* (1988) found that *L. monocytogenes* grew in trypticase soy broth at pH 4.39. Thus, the

Table V

Growth of *Listeria monocytogenes* LCDC 81-861 on Raw Vegetables

Vegetable	Storage temperature (°C)	$\log_{10}$ CFU/gm (day 0)	Number of days edible	$\log_{10}$ CFU/gm (on last day edible)
Broccoli	4	3.94	14	3.27
	15	5.07	6	7.78
Cauliflower	4	2.69	14	2.11
	15	3.44	4	6.57
Asparagus	4	4.30	14	5.22
	15	3.94	4	7.06

Source: Beuchat *et al.* (1990).

sensitivity of the organism to normally adverse conditions is likely to be related to other environmental or nutritional factors and should not be assumed to inhibit growth and survival. Indeed, the pH of the tomatoes on which Beuchat and Brackett (1991) found *L. monocytogenes* to survive was 4.0.

Like *Salmonella,* the term botulism is one with which the layperson is likely to be familiar. Botulism is caused by the bacterium *Clostridium botulinum. Clostridium botulinum* again differs from most of the previous organisms discussed because it is gram positive. In addition, *C. botulinum* produces heat-resistant endospores, is anaerobic, and usually only grows at pH 4.6 or greater.

Botulism is a serious and potentially life-threatening illness caused when a person consumes botulinum toxin. This toxin, among the most potent nerve toxins known, is produced when *C. botulinum* grows in foods. The initial symptoms of botulism can include gastrointestinal disturbances, including diarrhea or constipation. Later, the illness progresses to include severe paralysis and sometimes death.

Botulism is primarily a problem in canned low-acid foods. However, botulism conceivably can be a problem in fresh and minimally processed fruits and vegetables as well. For example, molds can sometimes make conditions favorable for the growth of *C. botulinum,* even in acidic products. Mundt and Norman (1982) found that molds growing on tomatoes could raise the pH as high as 8.1. Later, Draughon *et al.* (1988) also demonstrated that *C. botulinum* would grow after the pH was raised in this manner.

Hintlian and Hotchkiss (1986) expressed concern that modified atmosphere packaging also might increase the chances for *C. botulinum* to grow. Of particular concern in this respect are the barrier packaging techniques being proposed for respiring products such as fresh produce. Sugiyama and Yang (1975) demonstrated with mushrooms that this concern is not without merit. They were able to show that *C. botulinum* was able to grow and produce toxin when fresh mushrooms were overwrapped in a barrier film. These findings, combined with the potential for some strains of *C. botulinum* to grow at refrigeration temperatures (Hauschild, 1989), make this organism a potential threat in fresh as well as canned produce.

3. Fungi

Growth of fungi in fruits and vegetables results primarily in reduced organoleptic quality. However, some fungi also can affect safety by producing toxic metabolites known as mycotoxins. Mycotoxins are produced primarily by fungi in the genera *Aspergillus, Penicillium, Alternaria,* and *Fusarium.* These compounds are often carcinogenic, mutagenic, teratogenic, or otherwise highly toxic. The severity of the toxicity depends on the type of mycotoxin in question, the dose, and the animal consuming the toxin (Davis and Diener, 1987).

Mycotoxin contamination is primarily a problem in grains and nuts (Davis and Diener, 1987), although they can contaminate fresh fruits also. Vegetables rarely are affected because bacterial growth usually spoils the product before fungi have a chance to grow and produce mycotoxin (Pitt and Hocking, 1985). Although there are many different types of mycotoxins, only patulin is significant in fruits. This mycotoxin is both acutely and chronically toxic and may be weakly carcinogenic. Patulin is primarily produced by *P. expansum,* although *Byssochlamys nivea,*

Aspergilles clavatus, A. giganteus, and *A. terreus* also produce the mycotoxin. Fruits most often contaminated by patulin include apples, pears, and stone fruits. However, patulin-producing fungi also affect cherries, grapes, and quinces. Patulin typically is found in an around decayed areas of the affected fruit. Therefore, one should avoid using decayed or wounded fruits as food, even in juices.

4. Parasites

Parasites largely have been considered to be a problem of developing and tropical countries. However, parasites do exist in industrialized nations and cause illness (Jackson, 1983). Some of the more well-known parasitic illnesses, such as trichinosis, result from consuming undercooked meats or fish, but several parasitic illnesses can come from consuming contaminated fresh produce also.

Giardia lamblia is a flagellated protozoan that is a growing problem in North America and Europe. Many of the cases of giardiasis have occurred in institutions such as nurseries and daycare facilities. However, this organism can contaminate surface waters and wells also and can grow on fresh fruits and vegetables.

Giardia lamblia causes a diarrheal illness known as giardiasis. Symptoms of the illness can range from asymptomatic to severe gastrointestinal distress including cramps, vomiting, diarrhea, and occasionally, constipation (Jackson, 1990). It is estimated that 1.5–20% of Americans have been infected with the parasite and may be asymptomatic carriers (Ayres *et al.,* 1980). In addition, animals such as dogs, hamsters, and beavers often serve as reservoir hosts for the human giardia (Jackson, 1990).

Some cases of giardiasis result from direct transmission of the organism between infected humans. However, food handlers also have been responsible for outbreaks (Jackson, 1990), and other cases result from waterborne parasites (Barnard and Jackson, 1984). Raw fruits and vegetables, including lettuce and strawberries, have been implicated in some cases of giardiasis (Barnard and Jackson, 1984).

Entamoeba histolytica, an ameba, is among the most well-known and documented parasites that may cause foodborne illness (Jackson, 1990). The illness caused by this organism, amebic dysentery or amebiasis, often is characterized by abdominal cramps, fever, vomiting, and other symptoms similar to those of shigellosis (Jackson, 1990). In contrast, the infection may be quite mild and asymptomatic. As in the case of giardiasis, some infections result in the patients becoming asymptomatic carriers.

This parasite generally is associated with fecal contamination. Consequently, it is most likely to become a problem in fruits and vegetables that have been irrigated with polluted water. In addition, infected human handlers, animals, and even insects can cause direct contamination of produce. The cysts of *E. histolytica* are destroyed readily by drying, sunlight, and chlorination (Ayres *et al.,* 1980). However, proper personal hygiene and sanitation are of critical importance in preventing amebiasis.

Ascaris lumbricoides is a large nematode of humans that also has the potential of being spread by fruits and vegetables. This parasite differs from the previous two because it does not cause a diarrheal disease. Rather, *A. lumbricoides* causes an infection of vital organs, nerve tissue, and other body tissue (Ayres *et al.,* 1980). The organism usually develops in the intestinal tract, crosses into the circulatory system, and then migrates to other parts of the body. Symptoms of infections are often so mild that they go unnoticed. More serious infections are characterized by fever, labored breathing, and cough.

Ascaris lumbricoides is a very hardy organism. It is resistant to drying and to many chemical disinfectants. In addition, the worm may survive for years in soils and consequently contaminate vegetables grown in that soil (Jackson, 1990). Most techniques used in minimal processing of fruits and vegetables are ineffective in destroying this parasite.

The relative hardiness of *A. lumbricoides* has been one reason for some concern that it is a potential problem in fruits and vegetables. Because *A. lumbricoides* may survive treatment procedures, Jackson (1977) warned that the use of municipal wastewater or sewage sludge could increase the risk of infection by this organism.

B.   Factors Leading to Safety Problems

A number of practices employed in the production, processing, and transportation of fresh produce can influence microbial safety. Virtually every step from production through consumption may affect the microbial ecology of a food. Such changes also influence the growth and survival of pathogenic microorganisms. Although each of the major influences will be discussed separately, it should be remembered that each practice influences the impact of the others.

Many people, including some in the food industry, think that insuring food safety begins during processing and distribution. This belief is not only erroneous but can be potentially disastrous in the case of the fresh produce industry. Many actions taken during production will have a direct influence on the safety of fresh produce.

The type of fertilizer used is among the most important factors influencing the safety of fresh fruits and vegetables. Using untreated animal, especially human, wastes for fertilizer is extremely risky; such fertilizers should never be used for products destined to be eaten fresh. Such wastes always must be assumed to be contaminated with pathogenic microorganisms. Thus, produce fertilized with such waste also should be expected to be contaminated with these organisms (Goepfert, 1980). Goepfert (1980) pointed out that as much as 20% of recurrent infections of enteric illness in the Orient were due to use of "night soil" (human waste) as fertilizer. Similarly, Bryan's (1977) review revealed that contamination of foods, including fruits and vegetables, with raw or partially treated sewage has been responsible for many outbreaks of illness.

If present, many pathogenic organisms can persist on fresh fruits and vegetables long enough to infect consumers. Konowalchuk and Speirs (1975) discovered that several enteric viruses could survive on fresh vegetables long enough to infect consumers. Similarly, Berrang *et al.* showed that *A. hydrophila* (1989a) and *L. monocytogenes* (1989b) will grow and survive on fresh vegetables, even beyond the normal shelf life. Other enteric organisms also will survive in fresh produce (Nichols *et al.*, 1971). Finally, treatments such as washing with chlorinated water cannot be relied on to eliminate pathogens from fruits or vegetables. Brackett (1987a) found that at least one pathogenic microorganism present in vegetables could survive washing with chlorine solutions.

The use of treated sewage, although less risky than the use of untreated sewage, still carries risk of contaminating products with pathogenic microorganisms. For example, *L. monocytogenes* survives sewage treatment (Al-Ghazali and Al-Azawi, 1990) and frequently is found in sewage effluent (Watkins and Sleath, 1981). Moreover, Watkins and Sleath (1981) found that populations of this bacterium remained unchanged in soil to which contaminated sewage sludge was applied. Jackson (1977) similarly found that some parasites can survive sewage treatment and could contaminate vegetables fertilized with sludge.

Contaminated irrigation waters are a major source of pathogens and can compromise microbiological safety (Jones and Watkins, 1985). Run-off from agricultural lands or sewage effluents can contaminate surface waters with pathogenic organisms. Irrigating fruits or vegetables with this contaminated water can contaminate these products as well and has been responsible for illness in the past (Bryan, 1977).

Workers are another source of fecal contamination and pathogenic microorganisms. Proper in-field sanitary facilities often are overlooked or considered unimportant. However, workers who do not have access to proper facilities will resort to using fields or ditches for excretory functions. Such practices not only contaminate growing areas or water but also other workers. All workers, particularly those handling fresh produce, should have relatively close access to portable toilets and hand-washing facilities. Moreover, workers must be told the importance of using such facilities and must be expected to do so.

1. Processing and packing

Fruits and vegetables usually undergo some handling or minimal processing after they are harvested. Most steps provide at least some opportunity for contamination with pathogenic microorganisms. Preventing or minimizing contamination with or growth of pathogens is especially important in fresh products because such foods receive no terminal heat treatment before consumption.

As stressed earlier, workers must be instructed and expected to practice proper hygiene. Another often ignored aspect of food safety in the produce industry is the importance of properly maintained equipment. Improperly maintained equipment not only can damage fruits and vegetables and reduce quality, but can encourage undesirable microorganisms to grow. Cuts or other lesions in produce allow juices to leak and serve as a source of nutrients for bacteria and molds. Often, biofilms containing pathogenic or spoilage microorganisms form on contaminated equipment. For example, drains and equipment are among the most common sources of *L. monocytogenes* contamination in food processing facilities (Ellis, 1989; Nelson, 1990).

Various procedures involved in minimal processing also can influence the safety of fruits and vegetables. Traditional and routine handling procedures such as cutting, slicing, or peeling can serve to contaminate products if strict sanitary precautions are not followed. In addition, many of the new processing techniques can affect the growth and survival of pathogenic microorganisms.

Modified atmosphere packaging or storage can influence microbial safety of fruits and vegetables in several ways. Changes in atmospheric gases can select for different types of microorganisms than would be present in conventionally stored products (Brackett, 1987b). Such changes sometimes can encourage the growth of pathogenic organisms (Sugiyama and Yang, 1975; Hintlian and Hotchkiss, 1986; Brackett, 1987b).

Modified atmosphere storage is used primarily to extend shelf life of the product. Although extending shelf life may be advantageous from a marketing point of view, it is not without risks. The perception of extended shelf life usually results from evaluation of sensory qualities alone. However, sensory qualities may or may not reflect the microbiological quality of a product. For example, Berrang *et al.* (1989a,b) found that modified atmosphere extended the useful life of several vegetables but allowed continued growth of pathogenic microorganisms. Thus, consumers unknowingly may have eaten higher populations of pathogens then they would have if conventionally stored products were purchased. The point is that food processors should never base quality assessments on sensory evaluation alone. Potential microbial safety problems should always be considered and tested for if necessary.

Packaging can influence safety, even without modifying storage atmospheres. Because packaging materials usually contact the product directly, opportunity for contamination of the food by the packaging materials exists. Thus, packaging must be handled and stored as if it were an ingredient in the food. Thus, packaging materials must be microbiologically safe and must not be prone to contaminating foods with unapproved chemicals such as plasticizers. Finally, fresh or minimally processed fruits and vegetables are being used in so-called "new generation" consumer foods. Most notable in this regard are salad-type foods containing both fresh fruits and vegetables and cooked meats or seafoods. Such products offer unique challenges to food safety and have been discussed in more detail elsewhere (National Food Processors Association, 1988).

Cooked meats and seafood are highly nutritious, so most foodborne pathogens grow extremely well on these foods. Consequently, cooked meats and seafood are among the most hazardous types of foods if not handled and stored properly. Mixing with raw food products, such as vegetables or fruits, greatly increases the opportunity for the meats or seafood to become contaminated with pathogenic microorganisms. Thus, extra care in maintaining both a high degree of sanitation and temperature control are essential. However, even these precautions may not be sufficient to prevent growth of psychrotrophic pathogens (Palumbo, 1987). In some cases, it may be necessary to control tightly the length of time the product is allowed to be marketed.

2. Transportation and distribution

The same precautions and requirements for insuring product safety during production and processing apply to distribution and marketing. Both trucks and storage facilities are critical points in insuring safety. Truckers and warehouse managers sometimes overlook or ignore proper sanitation and temperature control. Often these omissions result from ignorance, but sometimes economic concerns dictate behavior. For example, the practice of "back-hauling" hazardous nonfood items in the same trucks used for hauling foods resulted from both ignorance and economy.

Produce retailers also play a major part in insuring food safety. As mentioned earlier, many of the "new generation" food products rely either solely or primarily on refrigeration to insure safety. In most cases, the retailer is responsible for maintaining temperature control and is therefore the final safeguard. Again, ignorance or carelessness by the grocer in maintaining refrigeration can nullify all the precautions the processor might have taken. For this reason, some food companies have resorted to placing their products in their own kiosks in stores. Other companies choose to use time–temperature indicators to insure that their foods are being marketed safely.

Finally, the consumer also plays a role in maintaining or affecting food safety. In fact, home storage may be the most hazardous stage in the food production and delivery system. It is difficult, if not impossible, to prepare foods so a careless or ignorant consumer cannot compromise safety. Chilled and extended shelf life foods are especially vulnerable for several reasons. First, these foods are often highly dependent on the maintenance of temperature control and strict sanitation. These steps often are ignored or unappreciated by consumers. Second, consumers often rely solely on sight or smell as indications of safety, Thus, they may ignore "use by" dates and assume that the product is safe as long as it looks or smells good. Consequently, they may eat a product that is beyond the intended shelf life and risk contracting a foodborne illness. Unfortunately, that same consumer may place blame for his or her illness unjustly on the producer or processor. Food producers and processors should consider such a scenario when marketing foods and should attempt to take appropriate steps to minimize the chances of such an event occurring. The use of time–temperature indicators, as mentioned earlier, may be one way to warn the consumer that a product should not be eaten.

C.   Avoiding Problems

Avoiding food safety problems usually is easier than correcting them. Moreover, a systems approach that takes into account all aspects of fruit and vegetable production will offer the greatest chances for success. Each step of the food production, processing, and delivery system affects food safety and affects or is dependent on other steps. Thus, only with consideration of the whole system can potential food safety problems be predicted and remedied adequately. However, there are an enormous number of aspects to consider and not all aspects of the system are equally important in insuring safety. How does one choose which aspects are most important?

One of the most successful techniques of incorporating a systems approach to insure food safety is the Hazard Analysis Critical Control Point (HACCP) system. A detailed discussion of HACCP is beyond the scope of this chapter. However, the reader is directed to one of several excellent reviews on the subject (Bryan, 1988a; Silliker et al., 1988). In summary, HACCP is a systematic means to identify and control potential food safety hazards in food production and delivery systems. In order to be performed effectively, HACCP requires a detailed and often documented knowledge of the system under consideration. Usually, HACCP involves six discrete procedures (Bryan, 1988a).

- Determine hazards and assess their severity and risks.
- Identify critical control points, that is, those steps in the food system

at which a control measure can be exercised that will eliminate, prevent, or minimize a hazard.

- Institute control measures and establish criteria to insure control, for example, determine that temperatures should be no higher than 10°C.
- Monitor critical control points to be sure that control measures are being adhered to.
- Take action whenever monitoring results indicate criteria are not being met, for example, discard product and repair refrigeration equipment if the temperature exceeds 10°C.
- Verify that the HACCP system is functioning as planned.

Because specific products and processing techniques differ, the specific hazards and critical control points also differ depending on the product. Thus, a separate HACCP system should be designed for each product.

IV. Conclusions

Food safety is an important and essential component of quality. As such, safety also must be a focus of the systems approach to quality maintenance. The present scientific knowledge indicates that pesticides are more a perceived than a real threat to human health. In contrast, microbial safety problems are both a real and a significant threat to consumers. The specific safety issues that will need to be addressed in future years will depend not only on scientific knowledge but on consumer and political pressures. Regardless of the issue, however, the systems approach can serve as a basic foundation to address the concerns more quickly.

Bibliography

Al-Ghazali, M. R., and Al-Azawi, S. K. (1990). *Listeria monocytogenes* contamination of crops grown on soil treated with sewage sludge cake. *J. Appl. Bacteriol.* **69**, 642–647.

Allen, W. A., Rajotte, E. G., Kazmierczak, R. F., Jr., Lambur, M. T., and Norton, G. W. (1987). "The National Evaluation of Extension's Integrated Pest Management (IPM) Programs." VCES Publication 491-010. Virginia Cooperative Extension Service, Blacksburg.

Ayres, J. C., Mundt, J. O., and Sandine, W. E. (1980). Nonmicrobial foodborne illness. *In* "Microbiology of Foods," pp. 531–573. Freeman, San Francisco.

Badawy, A. S., Gerba, C. P., and Kelley, L. M. (1985). Survival of rotavirus SA-11 on vegetables. *Food Microbiol.* **2**, 199–205.

Bagdasaryan, G. A. (1964). Survival of viruses of the enterovirus group (poliomyelitis, echo, coxsackie) in soil and on vegetables. *J. Hyg. Epid. Microbiol. Immun.* **8**, 497–505.

Barnard, R. J., and Jackson, G. J. (1984). The transfer of human infections by foods. *In* "Giardia and Giardiasis" (S. L. Erlandsen and E. A. Mayer, eds.), pp. 365–378. Plenum Press, New York.

Bean, N. H., and Griffin, P. M. (1990). Foodborne disease outbreaks in the United States, 1973–1987: Pathogens, vehicles, and trends. *J. Food Prot.* **53**, 804–817.

Berrang, M. E., Brackett, R. E., and Beuchat, L. R. (1989a). Growth of *Aeromonas hydrophila* on fresh vegetables stored under a controlled atmosphere. *Appl. Environ. Microbiol.* **55**, 2176–2171.

Berrang, M. E., Brackett, R. E., and Beuchat, L. R. (1989b). Growth of *Listeria monocytogenes* on fresh vegetables stored under a controlled atmosphere. *J. Food Prot.* **52**, 702–705.

Beuchat, L. R., and Brackett, R. E. (1990a). Inhibitory effects of raw carrots on *Listeria monocytogenes*. *Appl. Environ. Microbiol.* **56,** 1734–1742.

Beuchat, L. R., and Brackett, R. E. (1990b). Growth of *Listeria monocytogenes* on lettuce as influenced by shredding, chlorine treatment, modified atmosphere packaging, and temperature. *J. Food Sci.* **55,** 755–758, 870.

Beuchat, L. R., and Brackett, R. E. (1991). Behavior of *Listeria monocytogenes* inoculated into raw tomatoes and processed tomato products. *Appl. Environ. Microbiol.* **57,** 1367–1371.

Beuchat, L. R., Berrang, M. E., and Brackett, R. E. (1990). Presence and public health implications of *Listeria monocytogenes* on vegetables. *In* "Foodborne Listeriosis" (A. J. Miller, J. L. Smith, and G. A. Somkuti, eds.), pp. 175–181. Elsevier, New York.

Blenden, D. C., and Szatalowicz, F. T. (1967). Ecological aspects of listeriosis. *J. Am. Vet. Med. Assn.* **151,** 1761–1766.

Brackett, R. E. (1987a). Antimicrobial effect of chlorine on *Listeria monocytogenes*. *J. Food Prot.* **50,** 999–1003.

Brackett, R. E. (1987b). Microbiological consequences of minimally processed fruits and vegetables. *J. Food Qual.* **10,** 195–206.

Brackett, R. E. (1988). Presence and persistence of *Listeria monocytogenes* in food and water. *Food Technol.* **42(4),** 162–178.

Bryan, F. L. (1977). Diseases transmitted by foods contaminated by wastewater. *J. Food Prot.* **40,** 45–56.

Bryan, F. L. (1988a). Hazard analysis critical control point: What the system is and what it is not. *J. Environ. Health* **50,** 400–401.

Bryan, F. L. (1988b). Risks associated with vehicles of foodborne pathogens and toxins. *J. Food Prot.* **51,** 498–508.

Buchanan, R. L., and Palumbo, S. A. (1985). *Aeromonas hydrophila* and *Aeromonas sobria* as potential food poisoning species: A review. *J. Food Safety* **7,** 15–29.

California Department of Agriculture (1989). "Analysis of Natural Resources Defense Council Report Intolerable Risks: Pesticides in our Children's Food Supply." California Department of Agriculture, Sacramento.

Callister, S. M., and Agger, W. A. (1987). Enumeration and characterization of *Aeromonas hydrophila* and *Aeromonas caviae* isolated from grocery store produce. *Appl. Environ. Microbiol.* **53,** 249–253.

Ciesielski, C. A., Hightower, A. W., Parsons, S. K., and Broome, C. V. (1988). Listeriosis in the United States: 1980–1982. *Arch. Intern. Med.* **148,** 1416–1419.

Cliver, D. O., Ellender, R. D., and Sobsey, M. D. (1984). Foodborne viruses. *In* "Compendium of Methods for the Microbiological Examination of Foods," 2d Ed. (Marvin Speck, ed). pp. 508–541. American Public Health Association, Washington, D.C.

Conner, D. E., Brackett, R. E., and Beuchat, L. R. (1986). Effect of temperature, sodium chloride, and pH on growth of *Listeria monocytogenes* in cabbage juice. *Appl. Environ. Microbiol.* **5,** 59–63.

Covello, V. T., Sandman, P. M., and Slovic, P. (1988). "Risk Communication, Risk Statistics, and Risk Comparisons: A Manual for Plant Managers." Chemical Manufacturers Association, Washington, D.C.

D'Aoust, J.-V. (1989). *Salmonella*. *In* "Foodborne Bacterial Pathogens" (M. P. Doyle, ed.), pp. 327–445. Marcel Dekker, New York.

Davis, H., Taylor, J. P. Perdue, J. N., Stelma, G. N., Jr., Humphreys, J. M., Jr., Rowntree, R. III, and Greene, K. D. (1988). A shigellosis outbreak traced to commercially distributed shredded lettuce. *Am. J. Epidemiol.* **128,** 1312–1321.

Davis, N. D., and Diener, U. L. (1987). Mycotoxins. *In* "Food and Beverage Mycology" (L. R. Beuchat, ed.), pp. 517–570. Van Nostrand Reinhold, New York.

Doyle, M. P., and Padhye, V. V. (1989). *Escherichia coli*. *In* "Foodborne Bacterial Pathogens" (M. P. Doyle, ed.), pp. 235–281. Marcel Dekker, New York.

Draughon, F. A., Chen, S., and Mundt, J. O. (1988). Metabiotic association of *Fusarium, Alternaria,* and *Rhizoctonia* with *Clostridium botulinum* in fresh tomatoes. *J. Food Sci.* **53,** 120–123.

Ellis, R. F. (1989). Control of *Listeria* in processed food plants. *Food Proc.* **50(10),** 201–203.

Environmental Protection Agency (1987). "Unfinished Business: A Comparative Assessment of Environmental Problems." U.S. Govt. Printing Office Washington, D.C.

Fitch, J. (1989). Estimates of the loss from Alar crisis vary: Impacts now clearer. *Good Fruit Grower*. Nov. 15, 1989. **40(19),** 65–70.

Food and Drug Administration (1990). "Residues in Food 1989." U.S. Govt. Printing Office, Washington, D.C.

Food Marketing Institute (1990). "Trends: Consumer Attitudes and the Supermarket 1990." Food Marketing Institute, Washington, D.C.

George, S. M., Lund, B. M., and Brocklehurst, T. F. (1988). The effect of pH and temperature on initiation of growth of *Listeria monocytogenes*. *Lett. Appl. Microbiol*. **6,** 153–156.

Goepfert, J. M. (1980). Vegetables, fruits, nuts, and their products. *In* "Microbial Ecology of Foods" (J. H. Silliker, R. P. Elliott, A. C. Baird-Parker, F. L. Bryan, J. H. B. Christian, D. S. Clark, J. C. Olson, Jr., and T. A. Roberts, eds.), Vol. II, pp. 606–642. Academic Press, New York.

Graham, J. D., Green, L. C., and Roberts, M. J. (1988). "In Search of Safety—Chemical and Cancer Risks." Harvard University Press, Cambridge.

Hammitt, J. K. (1986). "Estimating Consumer Willingness to Pay to Reduce Food-Borne Risk." U.S. Environmental Protection Agency R-3447-EPA. U.S. Govt. Printing Office, Washington, D.C.

Hauschild, A. H. W. (1989). *Clostridium botulinum*. *In* "Foodborne Bacterial Pathogens" (M. P. Doyle, ed.), pp. 111–189. Marcel Dekker, New York.

Hazen, T. C., Fliermans, C. B., Hirsch, R. P., and Esch, G. W. (1978). Prevalence and distribution of *Aeromonas hydrophila* in the United States. *Appl. Environ. Microbiol*. **36,** 731–738.

Heisick, J. E., Wagner, D. E., Nierman, M. L., and Peeler, J. T. (1989). *Listeria* spp. found on fresh market produce. *Appl. Environ. Microbiol*. **55,** 1925–1927.

Hintlian, C. B., and Hotchkiss, J. H. (1986). The safety of modified atmosphere packaging: A review. *Food Technol*. **40(12),** 70–76.

Ho, J. L., Shands, K. N., Friedland, G., Eckind, P., and Fraser, D. W. (1986). An outbreak of type 4b *Listeria monocytogenes* infection involving patients from eight Boston hospitals. *Arch. Intern. Med*. **146,** 520–524.

Horton, D. L., Pfeiffer, D. G., and Hendrix, F. F., Jr. (1991). Southeastern apple integrated pest management. *In* "Sustainable Agricultural Research and Education in the Field: A Proceedings," pp. 165–182. National Academy Press, Washington, D.C.

Institute of Food Technologists (1990). Quality of fruits and vegetables. A scientific status summary. *Food Technol*. **44(6),** 99–106.

Jackson, G. J. (1977). Recycling of refuse into the food chain: The parasite problem. *In* "Proceedings of the Conference on Risk Assessment and Health Effects of Land Application of Municipal Wastewater and Sludges" (B. P. Sagik and C. A. Sorber, eds.), pp. 116–131. Center for Applied Research and Technology, San Antonio, Texas.

Jackson, G. J. (1983). Examining food and drink for parasitic, saprophytic, and free-living protozoa and helminths. *In* "CRC Handbook of Foodborne Diseases of Biological Origin" (R. Miloslav, ed.), pp. 247–255. CRC Press, Boca Raton, Florida.

Jackson, G. J. (1990). Parasitic protozoa and worms relevant to the U.S. *Food Technol*. **44(5),** 106–112.

Jay, J. M. (1986). Other proven and suspected food-borne pathogens. *In* "Modern Food Microbiology," pp. 541–575. Van Nostrand Reinhold, New York.

Jones, F., and Watkins, J. (1985). The water cycle as a source of pathogens. *J. Appl. Bacteriol. Symp. Supp*. **59,** 27S–36S.

Junttila, J., and Brander, M. (1989). *Listeria monocytogenes* septicemia associated with consumption of salted mushrooms. *Scand. J. Infect. Dis*. **21,** 339–342.

Kaufman, P., and Newton, D. (1990). Retailers explore food safety assurance options. *National Food Rev*. **13,** 11–15.

Khan, M. A., Palmas, C. V., Seaman, A., and Woodbine, M. (1973). Survival versus growth of a facultative psychrotroph. Meat and products of meats. *Zentralbl. Bakteriol. Hyg., I. Abt. Orig. B* **157,** 277–282.

Koburger, J. A. (1989). *Plesiomonas shigelloides*. *In* "Foodborne Bacterial Pathogens" (M. P. Doyle, ed.), pp. 311–325. Marcel Dekker, New York.

Konowalchuk, J., and Speirs, J. I. (1975). Survival of enteric viruses on fresh vegetables. *J. Milk Food Technol.* **38**, 469–472.

Larkin, E. P. (1981). Food contaminants-viruses. *J. Food Prot.* **44**, 320–325.

Lovett, J. (1989). *Listeria monocytogenes. In* "Foodborne Bacterial Pathogens" (M. P. Doyle, ed.), pp. 283–310. Marcel Dekker, New York.

Mundt, J. O., and Norman, J. M. (1982). Metabiosis and pH of moldy fresh tomatoes. *J. Food Prot.* **45**, 829–832.

National Academy of Sciences (1989). "Diet and Health: Implications for Reducing Chronic Disease." National Academy Press, Washington, D.C.

National Food Processors Association (1988). Safety considerations for new generation refrigerated foods. *Dairy Food Sanit.* **8**, 5–7.

National Food Processors Association (1990). "National Food Processors Information Letter." September 29, 1990.

National Research Council (1986). "Pesticide Resistance: Strategies and Tactics for Management." National Academy Press, Washington, D.C.

National Research Council (1989). "Alternative Agriculture." National Academy Press, Washington, D.C.

Nelson, J. H. (1990). Where are *Listeria* likely to be found in dairy plants? *Dairy Food Environ. Sanit.* **10**, 344–345.

Nichols, A. A., Davies, P. A., King, K. P., Winter, E. J., and Blackwall, F. L. C. (1971). Contamination of lettuce irrigated with sewage effluent. *J. Hort. Sci.* **46**, 425–433.

Ott, S. (1989). "Pesticide Residues: Consumer Concerns and Direct Marketing Opportunities." Georgia Agricultural Experiment Station Report No. 574, University of Georgia, Griffin.

The Packer Focus (1991). "Fresh Trends '91: A Profile of Fresh Produce Consumers." Vance Publishing, Overland Park, Kansas.

Palumbo, S. A. (1987). Can refrigeration keep our foods safe? *Dairy Food Sanit.* **7**, 56–60.

Pitt, J. I., and Hocking, A. D. (1985). The ecology of fungal food spoilage. *In* "Fungi and Food Spoilage," pp. 5–18. Academic Press, New York.

Rothschild, G. H. L. (1982). Suppression of mating in codling moths with synthetic sex pheromone and other compounds. *In* "Insect Suppression with Controlled Release Pheromone Systems" (A. F. Kydonieus and M. Beroza, eds,) Vol. II, P. 117–134. CRC Press, Boca Raton, Florida.

Schlech, W. F., Lavigne, P. M., Bortolussi, R. A., Allen, A. C., Haldane, E. V., Wort, A. J., Hightower, A. W., Johnson, S. E., King, S. H., Nicholls, E. S., and Broome, C. V. (1983). Epidemic listeriosis-evidence for transmission by food. *New Engl. J. Med.* **308**, 203–206.

Silliker, J. H., Baird-Parker, A. C., Bryan, F. L., Christian, J. H. B., Roberts, T. A., and Tompkin, R. B. (eds.) (1988). "HACCP in Microbiological Safety and Quality." Blackwell Scientific, Boston.

Slovic, P., Fischhoff, B., and Lichtenstein, S. (1982). Facts versus fears: Understanding perceived risk. *In* "Judgment under Uncertainty: Heuristics and Biases" (D. Kahneman, P. Slovic, and Amos Tversky, eds.), pp. 463–489. Cambridge University Press, Cambridge.

Smith, J. W., and Barfield, C. S. (1982). Management of preharvest insects. *In* "Peanut Science and Technology" (H. E. Pattee and C. T. Young, eds.), pp. 250–325. American Peanut Research and Education Society, Yoakum, Texas.

Stelma, G. N., Jr. (1989). *Aeromonas hydrophila. In* "Foodborne Bacterial Pathogens" (M. P. Doyle, ed.), pp. 1–19. Marcel Dekker, New York.

Sugiyama, H., and Yang, K. H. (1975). Growth potential of *Clostridium botulinum* in fresh mushrooms packaged in semi-permeable plastic film. *Appl. Microbiol.* **30**, 964–969.

Unrein, J. (1990a). Melons tested for outbreak link. *The Packer* **97(13)**, 3A.

Unrein, J. (1990b). Tainted tomato source to be named. *The Packer* **97(34)**, 4A.

Viscusi, W. K., and Magat, W. (1987). "Learning about Risk." Harvard University Press, Cambridge.

Wachsmuth, K., and Morris, G. K. (1989). *Shigella. In* "Foodborne Bacterial Pathogens" (M. P. Doyle, ed.), pp. 447–462. Marcel Dekker, New York.

Watkins, J., and Sleath, K. P. (1981). Isolation and enumeration of *Listeria monocytogenes* from sewage, sewage sludge, and river water. *J. Appl. Bacteriol.* **50**, 1–9.

Wodicka, V. (1985). Prioritizing risks associated with the food supply. *In* "The American Food Supply: Are We at Risk?" ICET Symposium II, Cornell University, Ithaca, New York.

CHAPTER 16

INTERDISCIPLINARY SOLUTIONS TO CHALLENGES IN POSTHARVEST HANDLING

Robert L. Shewfelt and Stanley E. Prussia

In this book we present a new paradigm for postharvest research. We suggest that future progress in postharvest handling will stem from the greater interaction of

- crop production and postharvest handling practices
- postharvest physiology and postharvest technology
- postharvest handling and retail marketing
- technical operations and management practices
- university research and commercial operations

We propose a systems approach to provide the conceptual framework for the integration needed. This book describes the implementation of a systems approach for postharvest handling. This chapter places the book in the context of current thought and future opportunities for improving handling systems.

I. Current Disciplinary Perspectives

When we started research on postharvest handling over a decade ago, we learned that interdisciplinary research was constrained by disciplinary perspectives. Certain differences were not difficult to recognize, confront, and resolve. The more difficult issues were the more subtle ones, such as the use of clear, unambiguous terms that have distinctly different meanings in different disciplines or unwritten assumptions within a discipline that are never expressed but always present. In the intervening years, some of these barriers have come down while others have remained firmly entrenched. Concise statements of relevant disciplinary perspectives follow.

A. Crop Production

The overriding concern of the system is to produce a crop of high-yield, marketable product. Each crop is unique. Cultural practices must be tailored to the specific cultivar, the local soil and growing conditions, and the capabilities of the grower. Crop production becomes more complex with a proliferation of available cultivars, greater pest pressures and governmental regulations, fewer agricultural chemicals to control pests, more emphasis on appearance of harvested crops, more competition from domestic and foreign sources, and lower prices for many crops. The grower takes the greatest risks in the handling system and must improve efficiency continually simply to maintain current economic status.

B. Engineering

Products, equipment, systems, and even social situations are viewed from the perspective of "How can it be improved?" Action is directed at making changes thought to lead to improvements. The purpose of collecting data and completing analyses is to develop and implement a new approach that solves a problem. Developing a device, equipment, process, or system that works smoothly, efficiently, and effectively is valued highly. Examples include the development of refrigeration equipment for cooling fresh fruits and vegetables and the application of systems engineering to optimizing hydrocooler size relative to the arrival rate of transport vehicles (Thai and Wilson, 1988).

C. Food Science

Fruits and vegetables are raw materials that serve as components of processed food products. Food scientists develop quality specifications for raw materials that will insure optimal quality in the finished product and test the raw materials to insure that they are in compliance. Specifications vary depending on the end use of the item. New technologies such as modified atmosphere packaging and low-dose irradiation produce minimally processed products. These processes are difficult to control by conventional food quality assurance methodology.

D. Economics

Value (quality/price) is an integrating factor for all fruit and vegetable products. Markets are consumer driven, that is, products that provide the greatest value and meet consumer needs eventually will succeed. The market system, with minimum constraints, will reward efficient growers, handlers, and distributors and will weed out inefficient firms.

E. Interdisciplinary Research

Successful interdisciplinary research requires careful consideration of other disciplinary perspectives, serious questioning of cherished assumptions within a discipline, and integration of these perspectives at problem definition and experimental design levels. A systems approach provides the necessary tools to cross disciplinary lines in postharvest research.

II.   Postharvest Challenges

In Chapter 2, we framed postharvest challenges in the context of different steps in the handling system, identifying specific issues amenable to systems solutions. Subsequent chapters provided a current assessment of systems thinking in each of these areas. In this chapter, we frame challenges facing postharvest handling at a different level.

A. Understanding the Consumer

Surveys (Resurreccion, 1986; Ott, 1989; Zind, 1990; Bruhn *et al.*, 1991) help improve our understanding of the consumer. Unfortunately, the desires of consumers frequently appear to be confusing and contradictory. Although consumers are concerned about the safety of fruits and vegetables, appearance and price appear to be more important factors in purchase decisions than safety. Despite safety concerns, fresh produce consumption is increasing, but "organic" products are not making significant gains in market share in the United States. A typical response by a scientist or distributor is that the consumer is ignorant; the scientist advocates consumer education, and the distributor searches for a marketing angle. A challenge that will be met by a successful entrepreneur of the future is to explain the apparent inconsistencies in consumer desires and provide a product that meets or exceeds expectations at an acceptable price.

B. Quantifying Preharvest Effects on Postharvest Quality

Biological variability represents the greatest source of variation in quality in the postharvest system. Thus, to maintain consistent quality during handling, we must return to the field or orchard. It is well established that crop production practices affect quality of fruits and vegetables at harvest. Strong evidence suggests that these practices also affect storage stability of fruits and vegetables after harvest. Unfortunately, preharvest causes of postharvest losses in quality are multiple and not amenable to simple controlled experiments. An example of the data requirements for the integration of a simple crop production experiment with a simple postharvest experiment is shown in Fig. 1. Even when such studies are designed and controlled carefully, the results are difficult to analyze and interpret, as noted by Brennan (1986) and Omoloh (1986). The challenge is to develop a new methodology that integrates field-plot with postharvest storage studies and produces meaningful results.

C. Providing Economic Incentives for Improved Quality and Value

Fresh fruits and vegetables are "owned" by many firms in the handling system (Chapter 1; How, 1991). Since each firm attempts to maximize its profit, system efficiency and optimal product quality are not always achieved. Firms are reluctant

3	Compounds (nutrients, pesticides, etc.)
×5	Rates
×4	Replications
60	Field plots
×3	Postharvest regimes (handling techniques, storage conditions, etc.)
180	Preharvest/postharvest treatment combinations (production method, handling step, storage time, etc.)
×5	Samples/treatment combinations
900	Samples per period
×5	Evaluation periods (storage times, handling steps, etc.)
4,500	Samples
×5	Quality attributes (color, firmness, Brix, acidity, etc.)
22,500	Measurements

Figure 1 Hypothetical example of an experimental design to test the interaction of preharvest and postharvest factors on quality and shelf life of fresh fruits and vegetables.

to make investments, even when system efficiency of final quality can be improved, unless they are assured of gaining a return on their investment. Even within the same firm, sufficient incentives do not always exist for competing departments to implement desired changes. In such a highly segmented marketing system, the market signal is not always clear. Thus, the challenge is to develop incentive systems that reward efficiency and improved quality.

D. Exploiting Basic Knowledge to Solve Applied Problems

More than 40 years of research has provided a wealth of information about the basic physiology of harvested crops (Weichmann, 1987; Kays, 1991) and about practical management of these crops during handling (Kader, 1985; Wills *et al.*, 1989). Despite these advances, there appears to be very little mutual interchange between postharvest physiology and technology. Most of the technological advances have resulted from empirical observations that become subjects of intense physiological investigation (e.g., controlled atmosphere storage, ripening of fruits with exogenous ethylene, quality degradation resulting from chilling injury). A great wealth of knowledge is accumulating on the molecular biology and cellular physiology of fruits and vegetables. The challenge is to find ways to improve consumer quality of fruits and vegetables using this knowledge.

E. Implementing Changes within the System

Resistance to change is only one reason that managers in postharvest businesses do not modify their operations readily. Typically, managers from retail store produce clerks to farmers are not motivated to change because they believe the problem is not caused by them but by someone in another business. For example, a truck driver may attribute the rejection of his load of products to poor cooling at the packinghouse, whereas the shipper is convinced that the refrigeration unit on the trailer must have had a problem. An understanding of latent damage helps to resolve such differences.

Most people define a problem as a deviation from the norm rather than a difference between what exists and what might be. A retail store manager whose produce department historically experiences 20% shrinkage would take immediate action if shrinkage jumped to 25%. It is unlikely that the same store manager would consider 20% shrinkage a problem simply because someone claimed it was possible to reduce shrinkage to 15%. However, a rapid shift in attitude would result if a competitor made changes that resulted in significantly lower shrinkage. Considerable effort would then be made to find solutions to what previously was not recognized as a problem (i.e., discarded one-fifth of their products).

Economic considerations often inhibit changes that would lead to improvements. Postharvest losses on a national scale are incredible, yet motivation to improve is limited, possibly because of the distribution of losses over many businesses, wide geographical areas, and diverse products. A reason given for delays in making the changes necessary to reap benefits from Project MUM (modularization, unitization, and metrication of containers and pallets) is that the businesses making an investment have no direct way to recover their costs. Likewise, managers are hesitant to invest in changes that have indirect or delayed returns, for example, reducing latent damage (i.e., implementing quality management programs, investing in automatic chlorination equipment, or simply building shaded parking for trucks).

Another deterrent to making change is that new products, equipment, procedures, and other innovations usually are developed as individual components rather than as part of the system in which they must function. Untold numbers of patents and technical articles describe "solutions" that were never implemented because their benefit did not outweigh the costs of changing the rest of the system. Similarly, solutions often are developed without involving final users in the definition of the problem. Adopting new approaches is favorable when totally new systems are installed, as was done for fruit exporting facilities in Chile.

III. Interdisciplinary Solutions

The challenges to improving postharvest handling operations and businesses can be overcome through interdisciplinary cooperation. The capability of teams is enhanced when the members represent a wide range of academic disciplines, business departments, businesses, consumers, and other groups interested in the products grown and marketed. Possible barriers to cooperation must be understood and addressed to achieve favorable results.

A. Interdisciplinary Teams

Teams of people working together have the potential to produce more and better results than the combined output of the individuals. Teams are strengthened when members are selected from diverse backgrounds. The postharvest research conducted by the authors of the chapters of this book has benefited from cooperation with the wide range of specialists contributing to this effort. Motivation and creativity were enhanced through "cross-pollination" of ideas and the enthusiasm that results from shared goals.

Our experiences have revealed the need for teams with even wider diversity, including experts in business administration, anthropology, psychology, and other social sciences. Research directed at solving applied problems would benefit from having team members from cooperative extension departments and the businesses served, such as packinghouses, trucking companies, warehouses, and retail stores. Similarly, teams formed within a business should include members from several departments, and possibly vendors and customers.

An important reason to seek diversity in team members is that most problems are far too complex to be solved by an individual, a single discipline, one department, or even one company. There is even reason for companies and countries to work together as an industry to solve common problems.

Although the value of interdisciplinary approaches to complex problems is widely acclaimed, true interdisciplinary cooperation is rare in universities or industry. Despite the obvious advantages of interdisciplinary teams that interact at the problem definition and project design stage, there is insufficient incentive to individual scientists or handlers to engender such cooperation in most cases. Interaction at the interface of two disciplines is becoming more common, as are large multidisciplinary task forces that serve as information exchange groups. Barriers exist in disciplinary perspectives, funding mechanisms, project management, and institutional credit to encourage these efforts.

B. Barriers to Cooperation

1. Research teams

Researchers wishing to cooperate with others outside of the discipline in which they were educated can be discouraged by the unfamiliar intellectual frameworks used by other disciplines. The framework used by each discipline is called a scientific paradigm, that is, "the meaningful body of knowledge currently used by a particular group of scientists to explain their observations" (Wilson and Morren, 1990). Cooperation requires extra efforts to understand how others give meaning to information, are motivated, define and solve problems, and seek to take action. Likewise, research manuscripts that cross discipline boundaries often are rejected for publication by reviewers who evaluate them using only their own scientific paradigms.

2. Industrial teams

Teams organized for industrial settings also face many barriers. Competition among departments, firms, commodities, industries, and even countries often prevents the organization of teams. Managers often view a problem only from the perspective of their unit or group, rather than from a broader perspective including other functions or firms. Thus, problems are identified and solved by members of the unit rather than by teams that include other units. Such concentration of efforts within a unit is expected when profits and other measures of success place units in competition with each other.

When operations are considered normal, industrial units sense little need to launch a team effort to look for improvements (problems to solve). Similarly, it is natural to believe that problems in a unit are caused by outside forces, such as poor quality products shipped to a retail store. There are real questions about how the extra efforts for team cooperation can provide tangible results. Consequently, there is little enthusiasm to form company, industry, or national teams to evaluate complete systems with the goal of better understanding of how to make continuing improvements (Groocock, 1986).

3. General principles creating barriers

In general, important barriers to the forming and functioning of teams are differences in world views and learning styles. Checkland (1981), Checkland and Scholes (1990), Wilson and Morren (1990), and others use the German word "Weltanschauung" to describe a particular mental framework people use to view, filter out, and give meaning to their experience. A person's or group's weltanschauuungen (the plural), or world views, are based on factors such as values, beliefs, morals, tastes, emotions, feelings, experiences, patterns of reasoning, and stores of knowledge.

Our world views influence the way we give meaning to situations and how we act as individuals or groups. Scientific and management paradigms provide world views that determine how people typically conduct research and take action. It is interesting to note that most breakthroughs result from the adoption of a new world view. Interdisciplinary teams face the difficulties that result when each member reduces the situation of interest to separate components in a way unique to his or her discipline. As a result, not enough of the complex situation is understood during the problem identification stage.

Both the solution of industrial problems and the completion of research depend on the ability of participants to learn about the situation and events surrounding them. Divergent, assimilative, convergent, and accommodative are the four styles of learning identified by Kolb (1986). All four styles must be applied to solve most problems successfully. Individuals suffer when they allow one style to predominate their approach to problems. Teams are strengthened by having members with different preferred learning styles. However, it is likely that team progress will be jeopardized by the conflicts that are normal among people with different preferred learning styles.

Granting agencies tend to support projects that are focused narrowly with clearly defined, achievable objectives. The challenges to postharvest research described in the previous section are, in contrast, broad based with poorly defined objectives. Although narrowly focused studies can be developed for each of these areas, an interdisciplinary systems approach to problem definition and experimental design will probably hurt rather than enhance chances for funding. Likewise, the fresh produce industry is unlikely to support interdisciplinary research at universities unless the research results in observable economic advantages to the supporter of the research, since the primary beneficiaries of postharvest systems research are usually growers and consumers. At agricultural colleges, dwindling state and federal support of research tends to be redirected as seed money for projects that can attract external funding rather than being used to encourage interdisciplinary efforts.

In addition, universities disburse funds through academic units. Accounting procedures to disburse funds or assign laboratory personnel to more than one department are difficult, if not impossible, to follow. Usually, informal arrangements can be made to surmount bureaucratic difficulties, but such arrangements can lead to misunderstandings and confrontations, which increase with personnel turnover. Establishment of new administrative units such as Centers, Institutes, or Laboratories can alleviate some of these problems. However, the establishment of extra-departmental units usually is achieved at a cost of funds or autonomy of existing units, can result in additional administrative costs and burdens which usually diminish research capacity, and may result in an additional bureaucratic structure that is difficult to disband after the mission has been achieved.

Finally, the status of faculty members engaged in interdisciplinary research is usually not as great as that of those who make contributions within a discipline. Innovation, creativity, and credit for a contribution to interdisciplinary research is difficult to apportion. Too often, a disproportionate amount of credit is given to the leader of the group while others who contribute intellectually are not recognized for their contributions.

IV. Future Opportunities

Despite the barriers to interdisciplinary cooperation, we believe that future advances in postharvest handling will come from interdisciplinary efforts. New cultivars will not be successful unless they meet consumer needs. New handling techniques will not be adopted unless they can be shown to be economically feasible. Although there still may be a role for the entrepreneur who is willing to risk everything on the hope that an idea will be successful, future successes are more likely to come through individuals who can manage information from disparate sources and integrate it into providing a product that fits a marketing niche.

Many recent advances are breaking down the barriers that commonly prevent cooperative team efforts. Several newly developed methods provide the opportunity to make major improvements in the way postharvest research is conducted and how businesses are managed. Postharvest researchers and businesses have the potential to benefit from improved understanding of learning methods, innovative management approaches, and new methodologies for structuring investigations into problem situations and debating on the best actions to take.

A. Understanding Learning

The ability to discuss learning styles makes it possible for teams to understand their strengths and overcome the weaknesses presented in the previous section. When forming teams, attention should be given to making sure all four learning styles are represented. It is also important to include at least one member with a broad holistic perspective who can integrate the diverse paradigms of the other members. Wilson and Morren (1990) present a complete treatment of the application of learning styles to team activities.

The Massachusetts Institute of Technology developed a Center for Systems Thinking and Organizational Learning in their Sloan School of Management. "The learning center is based on the premise that the essence of the needed transformation is a shift from seeing parts to seeing wholes, from reacting to the past to creating the future, and from controlling to learning" (Senge, 1990). An interesting part of the center is the introduction of "learning laboratories," in which researchers and industry representatives actively participate in simulated and real management situations. A result of the learning labs and the center is the education of leaders who help their organizations become companies that are good at learning. *The Systems Thinker,* published by Pegasus Communications (Cambridge, Massachusetts), communicates current developments useful to managers.

B. Adopting Industrial Management Approaches

Total Quality Management (TQM) is a management tool that has helped a wide range of industrial businesses. TQM takes a system approach and includes technical issues as well as people issues in an organization. Adoption of TQM approaches to postharvest businesses would provide a framework for addressing the needs presented in this book. Preharvest effects on postharvest quality would be addressed through vendor certification. Information flows from consumer to grower would be designed and monitored. Compliance with procedures for insuring food safety would be monitored and corrected when necessary. Processes would be controlled through statistical process control techniques.

A TQM approach by businesses in the postharvest chain would provide the management structure necessary for delivering consistantly high quality products with a documented history. To be successful, however, implementers must realize the unique aspects of fresh products, incorporating the perspectives of horticulturists, entomologists, pathologists, postharvest physiologists, food microbiologists, food and agricultural engineers, and economists in the design. Food stores are likely to follow other industries in adopting quality management standards such as ISO 9000 issued by the International Standards Organization.

C. New Methodologies for Structuring Improvements

Soft systems methodologies (SSM) provide a proven approach for structuring the investigation and improvement of poorly defined, messy problem situations involving people with different views of the problem. SSMs are especially valuable for clarifying diverse assumptions and their implications. Involvement of the people in the organization in the process improves the result and encourages acceptance of change when decisions are implemented. "The Methodology is system-thinking-based, but the trick is to recast the idea of 'a systems approach' into the form which underlies SSM" (Checkland and Scholes, 1990, p. 277).

Wilson and Morren (1990) and Checkland and Scholes (1990) caution readers that most "systems" are actually a mental image that a person constructs to explain personal observations. Another person viewing the same situation usually has a different concept of what the "system" contains or does. Some interacting components that commonly are called "systems" actually do not exist. For example, the "United States economic system" or "United States educational system" does not meet the basic system requirements of identifiable ownership and self-regulation.

Consequently, an argument can be made that "postharvest systems" do not exist for fruits and vegetables, except when a company has complete vertical integration of operations from growing to retail sales. The movement of fresh produce to packinghouse to warehouse to store really represents the interaction of several business organizations, each of which could be viewed as a separate system. As a combination, they lack the self-regulation and common ownership necessary to be considered a self-contained system.

On that basis, it is inappropriate to try to model a complete postharvest system in the general sense, as for a country or for a specified crop that is owned by more than one company. Future research should benefit by using SSM to learn about the interactions between identifiable systems, for example, each produce deal made between buyer and seller. An example would be to view each deal as a temporary system established only for the shipments ordered.

Similar models could be developed for the growing and packing businesses, the transportation businesses, and retailer and purchaser transactions. The interactions of all the separate systems could be visualized by combining them into one conceptual model (as described in Chapter 3). However, the important point is that no system boundary could be drawn around the combined models because it would not meet the requirements of being a system.

Research efforts will provide the basis for improvements in the handling system. Practical solutions to the complex problems facing the industry today require integrated team efforts. The systems approach provides the tools needed to integrate knowledge from different disciplinary perspectives.

Bibliography

Brennan, P. S. (1986). "Evaluation of Preharvest and Postharvest Treatments on Broccoli Quality." Master's Thesis. University of Georgia, Athens, Georgia.

Bruhn, C. M., Feldman, N., Garlitz, C., Harwood, J., Ivans, E., Marshall, M., Riley, A., Thurber, D., and Williamson E. (1991). Consumer perceptions of quality: Apricots, cantaloupes, peaches, pears, strawberries, and tomatoes. *J. Food Qual.* **14,** 187–195.

Checkland, P. B. (1981). "Systems Thinking, Systems Practice." John Wiley & Sons, Chichester.

Checkland, P. B., and Scholes, J. (1990). "Soft Systems Methodology in Action." John Wiley & Sons, Chichester.

Groocock, J. M. (1986). "The Chain of Quality: Market Dominance through Product Superiority." John Wiley & Sons, New York.

How, R. B. (1991). "Marketing Fresh Fruits and Vegetables." AVI/Van Nostrand Reinhold, New York.

Kader, A. A. (1985). "Postharvest Technology of Horticultural Crops." Agriculture and Natural Resources Publications, University of California, Berkeley.

Kays, S. J. (1991). "Postharvest Physiology of Perishable Plant Products." AVI/Van Nostrand Reinhold, New York.

Kolb, D. A. (1986). "Experiential Learning: Experience as the Source of Learning and Development." Prentice-Hall, Englewood Cliffs, New Jersey.

Omoloh, W. J. (1986). "Quality Changes of Tomatoes within a Simulated Postharvest Handling System." Master's Thesis, University of Georgia, Athens, Georgia.

Ott, S. L. (1989). "Pesticide Residues: Consumer Concerns and Direct Marketing Opportunities." Agricultural Experimental Station Research Report. No. 574. University of Georgia, Athens, Georgia.

Resurreccion, A. V. A. (1986). Consumer use patterns for fresh and processed vegetable products. *J. Cons. Studies Home Econ.* **10,** 317–332.

Senge, P. M. (1990). The leader's new work: Building learning organizations. *Sloan Management Rev.* **32(1),** 7–23.

Thai, C. N., and Wilson, C. P. (1988). Simulation of peach postharvest operations. American Society of Agricultural Engineers Technical Paper No. 88-6058; 17 pp. St. Joseph, Michigan.

Weichmann, J. (1987). "Postharvest Physiology of Vegetables." Marcel Dekker, New York.

Wills, R. B. H., McGlasson, W. D., Graham, D., Lee, T. H., and Hall, E. G. (1989). "Postharvest: An Introduction to the Physiology and Handling of Fruits and Vegetables." AVI/Van Nostrand, New York.

Wilson, K., and Morren, E. B. G., Jr. (1990). "Systems Approaches for Improvement in Agriculture and Resource Management." MacMillan, New York.

Zind, T. (1990). Fresh Trends '90—A profile of fresh produce consumers. *The Packer Focus* **96(54),** 37–68.

INDEX

FOOD SCIENCE AND TECHNOLOGY
A Series of Monographs

Maynard A. Amerine, Rose Marie Pangborn, and Edward B. Roessler, PRINCIPLES OF SENSORY EVALUATION OF FOOD. 1965.

Martin Glicksman, GUM TECHNOLOGY IN THE FOOD INDUSTRY. 1970.

L. A. Goldblatt, AFLATOXIN. 1970.

Maynard A. Joslyn, METHODS IN FOOD ANALYSIS, second edition. 1970.

A. C. Hulme (ed.), THE BIOCHEMISTRY OF FRUITS AND THEIR PRODUCTS. Volume 1–1970. Volume 2–1971.

C. R. STUMBO, THERMOBACTERIOLOGY IN FOOD PROCESSING, second edition. 1973.

Aaron M. Altschul (ed.), NEW PROTEIN FOODS: Volume 1, TECHNOLOGY, PART A–1974. Volume 2, TECHNOLOGY, PART B–1976. Volume 3, ANIMAL, PROTEIN SUPPLIES, PART A–1978. Volume 4, ANIMAL PROTEIN SUPPLIES PART B–1981. Volume 5, SEED STORAGE PROTEINS–1985.

S. A. Goldblith, L. Rey, and W. W. Rothmayr, FREEZE DRYING AND ADVANCED FOOD TECHNOLOGY. 1975.

R. B. Duckworth (ed.), WATER RELATIONS OF FOOD. 1975.

John A. Troller and J. H. B. Christian, WATER ACTIVITY AND FOOD. 1978.

A. E. Bender, FOOD PROCESSING AND NUTRITION. 1978.

D. R. Osborne and P. Voogt, THE ANALYSIS OF NUTRIENTS IN FOODS. 1978.

Marcel Loncin and R. L. Merson, FOOD ENGINEERING: PRINCIPLES AND SELECTED APPLICATIONS. 1979.

J. G. Vaughan (ed.), FOOD MICROSCOPY. 1979.

J. R. A. Pollock (ed.), BREWING SCIENCE, Volume 1–1979. Volume 2–1980.

J. Christopher Bauernfeind (ed.), CAROTENOIDS AS COLORANTS AND VITAMIN A PRECURSORS: TECHNOLOGICAL AND NUTRITIONAL APPLICATIONS. 1981.

Pericles Markakis (ed.), ANTHOCYANINS AS FOOD COLORS. 1982.

George F. Stewart and Maynard A. Amerine, INTRODUCTION TO FOOD SCIENCE AND TECHNOLOGY, second edition. 1982.

Malcolm C. Bourne, FOOD TEXTURE AND VISCOSITY: CONCEPT AND MEASUREMENT. 1982.

Héctor A. Iglesias and Jorge Chirife, HANDBOOK OF FOOD ISOTHERMS: WATER SORPTION PARAMETERS FOR FOOD AND FOOD COMPONENTS. 1982.

Colin Dennis (ed.), POST-HARVEST PATHOLOGY OF FRUITS AND VEGETABLES. 1983.

P. J. Barnes (ed.), LIPIDS IN CEREAL TECHNOLOGY. 1983.

David Pimentel and Carl W. Hall, FOOD AND ENERGY RESOURCES. 1983.

Joe M. Regenstein and Carrie E. Regenstein, FOOD PROTEIN CHEMISTRY: AN INTRODUCTION FOR FOOD SCIENTISTS. 1984.

Maximo C. Gacula, Jr., and Jagbir Singh, STATISTICAL METHODS IN FOOD AND CONSUMER RESEARCH. 1984.

Fergus M. Clydesdale and Kathryn L. Wiemer (eds.), IRON FORTIFICATION OF FOODS. 1985.

Robert V. Decareau, MICROWAVES IN THE FOOD PROCESSING INDUSTRY. 1985.

S. M. Herschdoerfer (ed.), QUALITY CONTROL IN THE FOOD INDUSTRY, second edition. Volume 2–1985. Volume 3–1986. Volume 4–1987.

F. E. Cunningham and N. A. Cox (eds.), MICROBIOLOGY OF POULTRY MEAT PRODUCTS. 1986.

Walter M. Urbain, FOOD IRRADIATION. 1986.

Peter J. Bechtel, MUSCLE AS FOOD. 1986.

H. W.-S. Chan, AUTOXIDATION OF UNSATURATED LIPIDS. 1986.

Chester O. McCorkle, Jr., ECONOMICS OF FOOD PROCESSING IN THE UNITED STATES. 1987.

Jethro Jagtiani, Harvey T. Chan, Jr., and William S. Sakai, TROPICAL FRUIT PROCESSING. 1987.

J. Solms, D. A. Booth, R. M. Dangborn, and O. Raunhardt, FOOD ACCEPTANCE AND NUTRITION. 1987.

R. Macrae, HPLC IN FOOD ANALYSIS, second edition. 1988.

A. M. Pearson and R. B. Young, MUSCLE AND MEAT BIOCHEMISTRY. 1989.

Dean O. Cliver (ed.), FOODBORNE DISEASES. 1990.

Majorie P. Penfield and Ada Marie Campbell, EXPERIMENTAL FOOD SCIENCE, third edition. 1990.

Leroy C. Blankenship, COLORIZATION CONTROL OF HUMAN BACTERIAL ENTEROPATHOGENS IN POULTRY. 1991.

Yeshajahu Pomeranz, FUNCTIONAL PROPERTIES OF FOOD COMPONENTS, second edition. 1991.

Reginald H. Walter, THE CHEMISTRY AND TECHNOLOGY OF PECTIN. 1991.

Tom Brody, NUTRITIONAL BIOCHEMISTRY. 1993.

Herbert Stone and Joel L. Sidel, SENSORY EVALUATION PRACTICES, second edition. 1993.

Robert L. Shewfelt and Stanley E. Prussia, POSTHARVEST HANDLING: A SYSTEMS APPROACH. 1993.

John A. Troller, SANITATION IN FOOD PROCESSING, second edition. 1993.

R. Paul Singh and Dennis R. Heldman, PRINCIPLES OF FOOD ENGINEERING. 1993.

Tilak Nagodawithhana and Gerald Reed, ENZYMES IN FOOD PROCESSING, third edition. 1993.

Dallas G. Hoover and Larry R. Steenson, BACTERIOCINS. 1993.

Takayaki Shibamoto and Leonard Bjeldanes, INTRODUCTION TO FOOD TOXICOLOGY. 1993.

Ronald S. Jackson, WINE SCIENCE: PRINCIPLES AND APPLICATIONS. In preparation.

Robert G. Jensen, Marvin P. Thompson, and Robert Jenness, HANDBOOK OF MILK COMPOSITION. In preparation.